W9-BRK-462

Metallurgy Fundamentals

by
Daniel A. Brandt
Mechanical Engineering Department
Milwaukee School of Engineering

J. C. Warner
Principal Scientist
Warner Consulting, Inc.

Publisher
THE GOODHEART-WILLCOX COMPANY, INC.
Tinley Park, Illinois

 Manufactured in the United States of America.

Library of Congress Catalog Card Number 98-48494
International Standard Book Number 1-56637-543-6

1 2 3 4 5 6 7 8 9 10 99 03 02 01 00 99

Library of Congress Cataloging in Publication Data
Brandt, Daniel A.
Metallurgy fundamentals / Daniel A. Brandt, Jairus C. Warner
p. cm.
Included index.
ISBN 1-56637-543-6
1. Iron—Metallurgy. 2. Steel—Metallurgy I. Warner, Jairus C.
II. Title
TN705.B652 1999
669'.1—dc21
98-48494
CIP

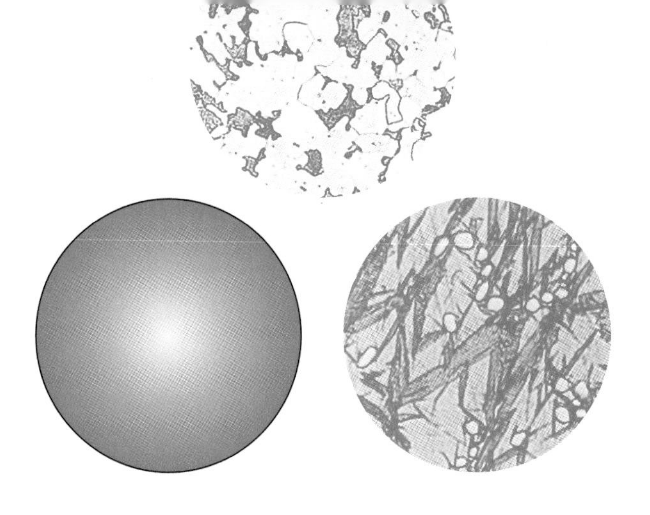

Introduction

Metallurgy Fundamentals provides instruction and information on the basic principles of metallurgy. A knowledge of these principles is invaluable to any person who plans to deal with metals as a future vocation.

Metallurgy Fundamentals emphasizes the practical aspects of metallurgy. It explores the behavior of metals subjected to metallurgical processes. It explains why certain material properties are desired and how these properties are attained.

Metallurgy Fundamentals describes common industrial processes, so that you can confidently discuss the processing of metals with others in the field. These processes are explained in clear, simple terms for easier understanding.

Metallurgy Fundamentals speaks to the reader in down-to-earth language, rather than highly theoretical terms. In many cases, diagrams are used instead of lengthy word descriptions and practical examples are given instead of abstract theories.

A section on nonferrous metallurgy has been added for this edition of *Metallurgy Fundamentals.* This new section discusses the processes used to create nonferrous metals and their alloys. Aluminum, copper, magnesium, zinc, and other nonferrous metals are discussed.

Metallurgy Fundamentals is written for those who want to learn the "basics," for those who want to explore the behavior of metals, and for those who want a broad knowledge of the entire field of metallurgy.

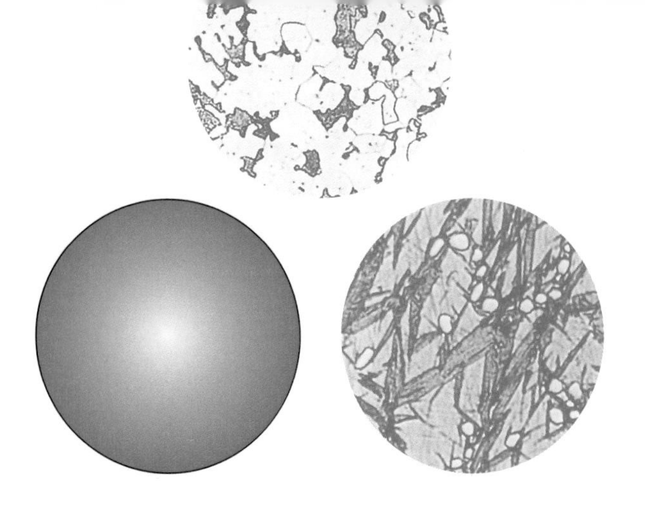

Contents

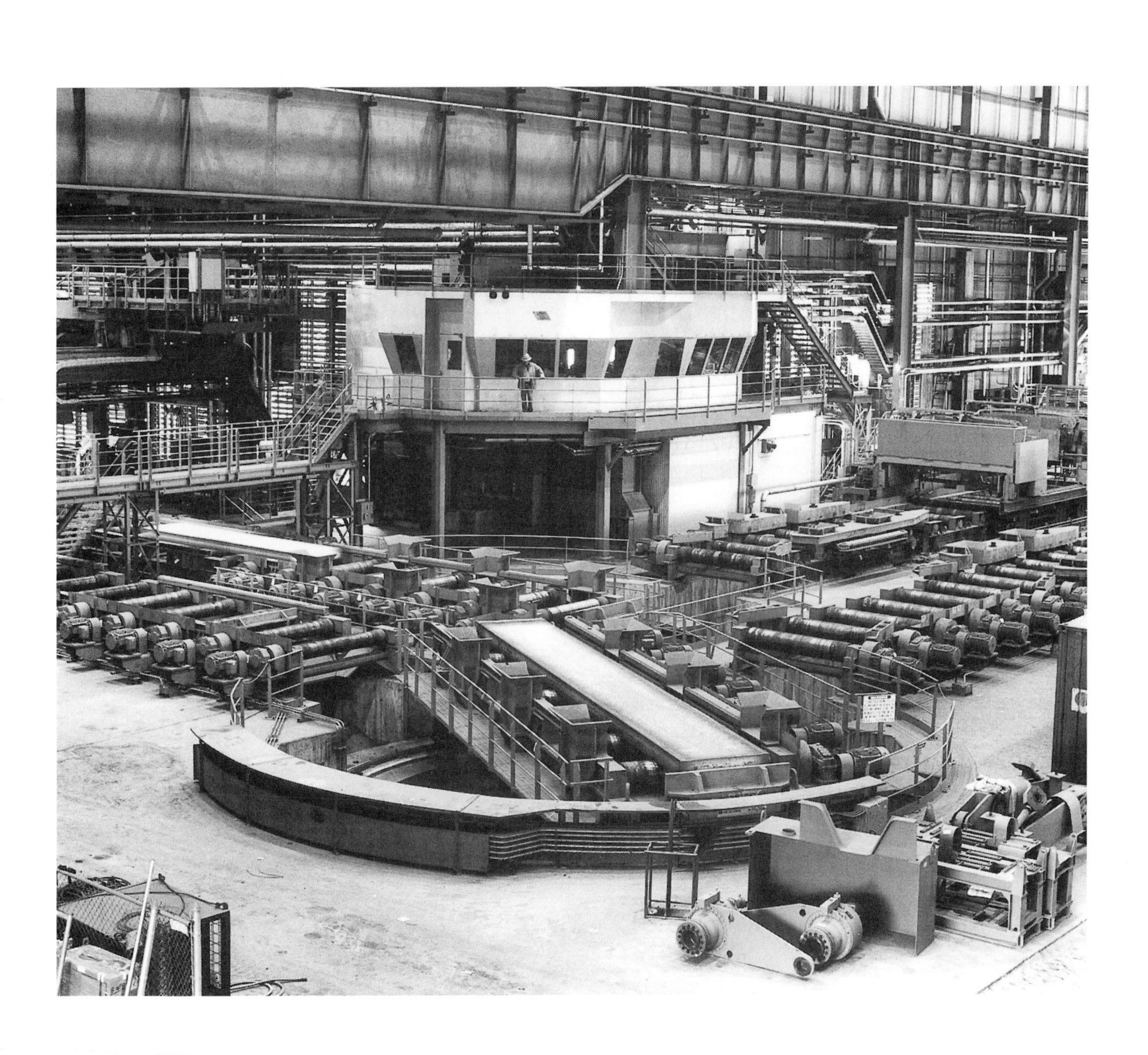

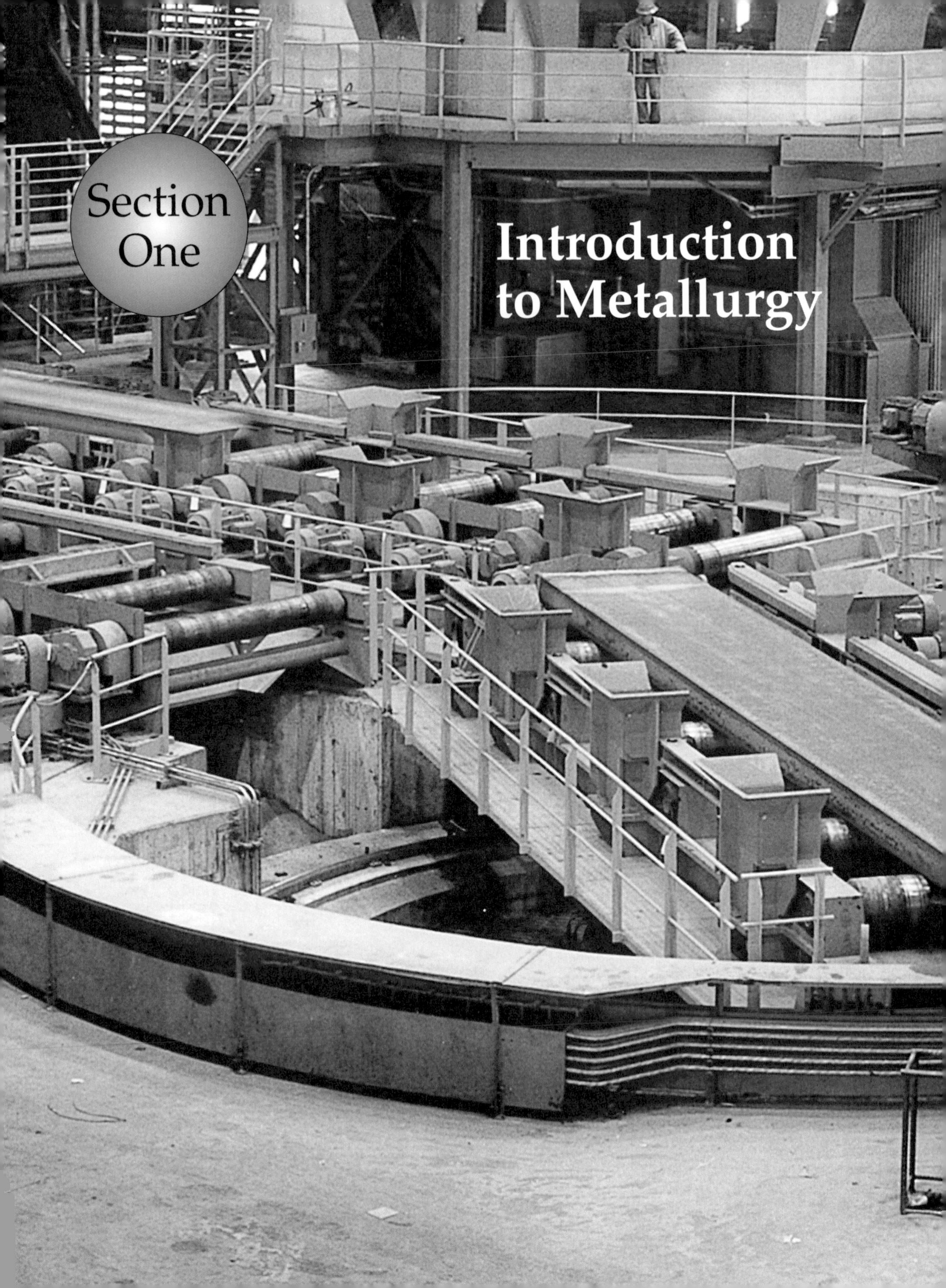

Section One

Introduction to Metallurgy

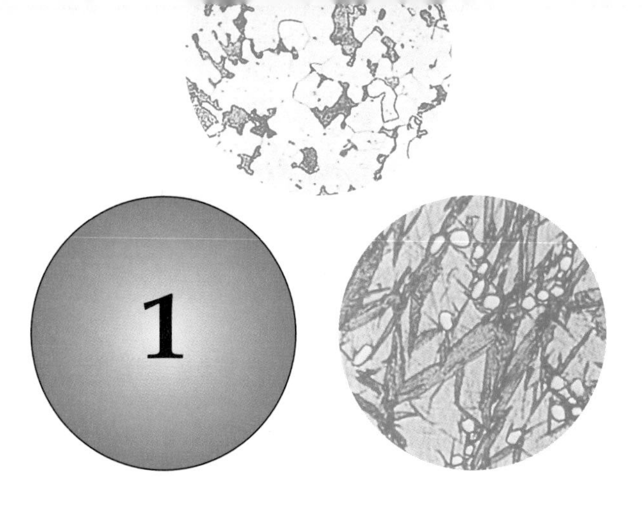

Practical Applications of Metallurgy

After studying this chapter, you will be able to:

- Define metallurgy.
- Explain what a metallurgist does.
- Describe how metallurgy knowledge can be used to solve industrial problems.
- State why the study of metallurgy can be a valuable asset.

Metallurgy and Metallurgists

The dictionary defines *metallurgy* as "the science that explains methods of refining and extracting metals from their ores and preparing them." Today, the subject of metallurgy digs deeper into the heart of metals than that definition describes. Metallurgy is more than examining the refinement and extraction of metals from their ores. Metallurgy is the science that explains the properties, behavior, and internal structure of metals. Metallurgy also teaches us that properties of metals can be changed using various treatments. This allows us to tailor a metal's properties to its specific use.

The study of metallurgy actually explores what makes metals behave the way they do. This exploring is done by *metallurgists,* scientists who probe deeply inside the internal structure of metal. They seek to understand why the metal changes its structure as it is heated and cooled under many different conditions.

Metallurgy involves all metals, Figure 1-1. *Ferrous metallurgy* is the study of metals that use iron as their basic ingredient. *Nonferrous metallurgy* is the study of metals that do *not* use iron as a principal alloying element (such as titanium, aluminum, and copper).

Iron and steel are the main focus of ferrous metallurgy (steel is made primarily from iron). Other types of metals are added to iron to achieve the mechanical properties needed for a particular application. See Figure 1-2.

Predicting the internal behavior of iron and steel during heating, quenching, annealing, tempering, and other heat-treating processes is an exciting challenge. The steel undergoes interesting changes and you—as a metallurgist—can predict the changes that will occur based on the composition of the steel and the treatments to which it is subjected. The examination and knowledge of this predictable behavior of iron and steel is one of the major thrusts of this text.

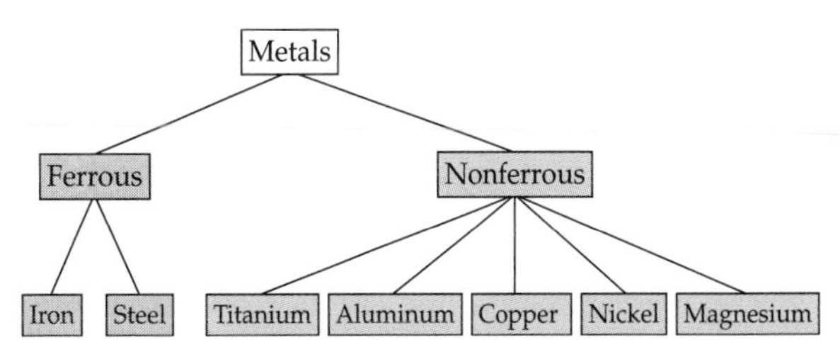

Figure 1-1. *Metals are classified as ferrous and nonferrous. Iron and steel are both ferrous metals.*

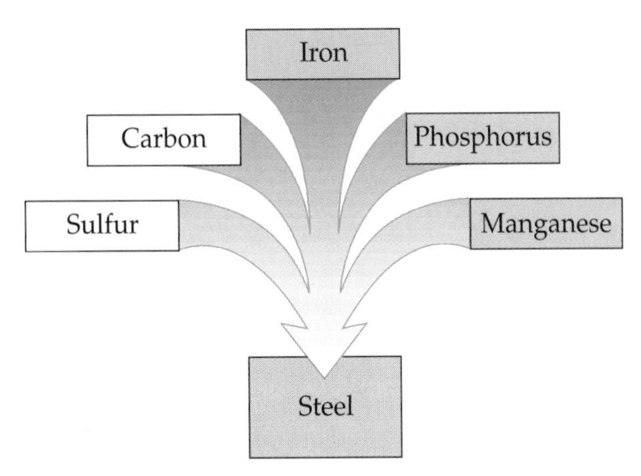

Figure 1-2. *Many alloys are added to iron to produce steel.*

Figure 1-3. *Excessive wear of gear teeth can occur if proper metallurgical processes are not used. (The Falk Corporation, subsidiary of Sundstrand Corporation)*

Practical Examples of Metallurgy in Modern Industry

The way in which metallurgists work is shown in the following examples of specific industrial problems. All metallurgical processes discussed in these examples will be explained in this text.

Practical Example 1

Problem: A gear in a machine that ran continuously was subjected to large forces. As a result, the gear teeth wore rapidly. See Figure 1-3. If a hard, strong material were used to make the gear, it would resist this type of wear. However, most hard and strong materials are also brittle and crack under repeated shock forces.

Solution: To solve this problem, a metallurgical process known as "case hardening" was used. *Case hardening* produces a hard surface on the metal part while the interior core remains relatively soft and ductile (workable, not brittle).

Practical Example 2

Problem: In a particular manufacturing operation, five irregular slots were cut into a large, thin disc. See Figure 1-4. The slots had to be machined to dimensional tolerances closer than ±0.001" (±0.025 mm). After these slots were cut, the disc was installed in a business machine and adjusted until it ran perfectly.

However, problems developed after these machines were shipped to customers. The disc twisted and distorted after being used only two months. The machining of the slots had created internal stresses in the disc. While the disc was in use, these stresses gradually relaxed and caused the disc to twist slightly. This distortion caused friction between the disc and another part, which caused the machine to bind.

Solution: This problem was solved by a metallurgical process known as *process annealing*. Process annealing is a heat-treating process that causes the metal to relax, removing internal stresses. In this "disc" application, process annealing caused the disc to distort *before* leaving the factory. Following this

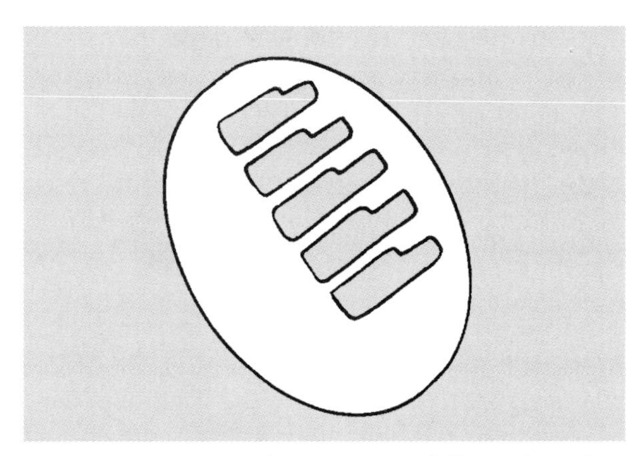

Figure 1-4. *Internal stress caused distortion of this disc until process annealing was used.*

relaxing action, a light machining cut was taken to eliminate the few thousandths of an inch distortion. Then, the machine was shipped, free of internal stresses that could have caused distortion at the customer site.

Practical Example 3

Problem: The cutting tool shown in Figure 1-5 must be very hard. If it is hard, properly ground, and sharpened, it will cut metal cleanly and accurately. However, after a period of use, this cutting tool did not remain sharp. It wore away excessively fast. Then, it did not cut well.

Solution: Again, a knowledge of metallurgy was used to solve the problem. A metallurgical microscopic examination showed that the cutting tool was going through a process known as *tempering* without the machine operator knowing it. See Figure 1-6.

Tempering is a reheating of metal to slightly soften it. Tempering is usually a helpful metallurgical process performed to make the metal more stress-free, distortion-free, and crack-free. However, in the case of this cutter, the unintentional tempering was destroying it.

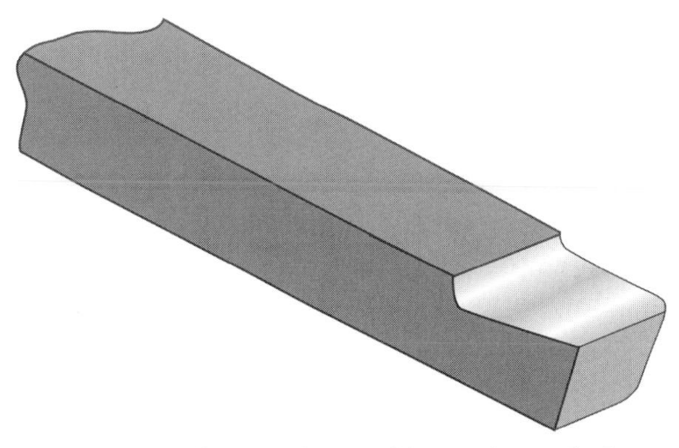

Figure 1-5. *This cutting tool is used on a lathe.*

Practical Example 4

Problem: The cutting blade shown in Figure 1-7 is as sharp as a razor blade. In addition, it must be very hard and strong in order to cut chemically treated paper in a particular industrial application. The problem occurred when the blade did not make clean, smooth cuts.

Solution: It was discovered that a metallurgical process called *water quenching* was used to harden and strengthen the blade. Water quenching, however, also caused distortion that prevented the blade from cutting cleanly in this application. To solve this

Figure 1-6. *When tempering of a cutting tool takes place during a cutting operation, excessive wear of the tool can result.*

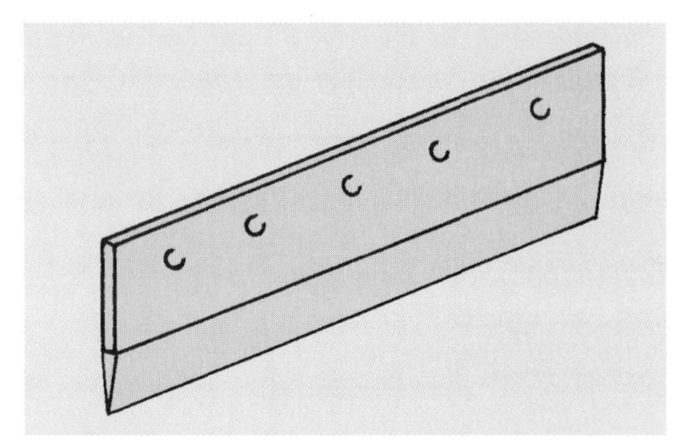

Figure 1-7. *Air quenching produces a hard, sharp cutting blade. The blade will distort (twist slightly out of shape) if water quenching is used.*

problem, a metallurgical process called *air quenching* was substituted for water quenching. In addition, the metallurgist changed the material to a higher alloy tool steel. With these changes, the newly manufactured blades are now making keen and accurate cuts.

Why Study Metallurgy?

Modern industry is dependent on a knowledge of metallurgy. Nearly every kind of manufacturing today is affected by the behavior of metals and alloys. Therefore, anyone who plans a future career in modern industry will find a working knowledge of metallurgical processing to be a valuable asset.

Engineers, technicians, designers, drafters, tool and die makers, and machinists need skills in selecting materials and heat-treating processes. Even production managers and purchasing people can benefit from an understanding of terms such as ductility, hardness, normalizing, and surface hardening. Repair workers, service personnel, and troubleshooters who diagnose causes of equipment failures should be trained to recognize the causes of cracks and excessive wear. They need to know how to examine a material to see whether it has become too hard and brittle.

Iron has been used for more than 5000 years. The tips of spears and other weapons were heat-treated and hardened by metallurgical processes long before the word "metallurgy" existed. Apparently, some important metallurgical methods were stumbled onto accidentally and used long before people knew why they worked.

Today, technology marches forward. Our mass of knowledge has doubled in less than 50 years. More than 90% of the scientists who have ever lived are still alive today. It is a certainty, then, that successful students of metallurgy will find new horizons ahead in their careers.

Test Your Knowledge

Write your answers on a separate sheet of paper. Do not write in this book.

1. What are some things that can be learned from a study of metallurgy?
2. What metal is the main ingredient in steel?
3. Why are other metals added to steel?
4. Metals can be divided into two general categories. Name the two categories.
5. What metallurgical process was used to solve the problem that involved shock forces and wear in Practical Example 1?
6. What type of problem did the metallurgical process known as "process annealing" solve in Practical Example 2?
7. What is *tempering*?
8. Is tempering generally a helpful metallurgical process or does it usually present serious problems?
9. When the metallurgical process known as "water quenching" caused too much distortion, what other metallurgical process was used to solve this problem in Practical Example 4?
10. How would a knowledge of metallurgy benefit an engineer?
11. How would a knowledge of metallurgy benefit a troubleshooter?

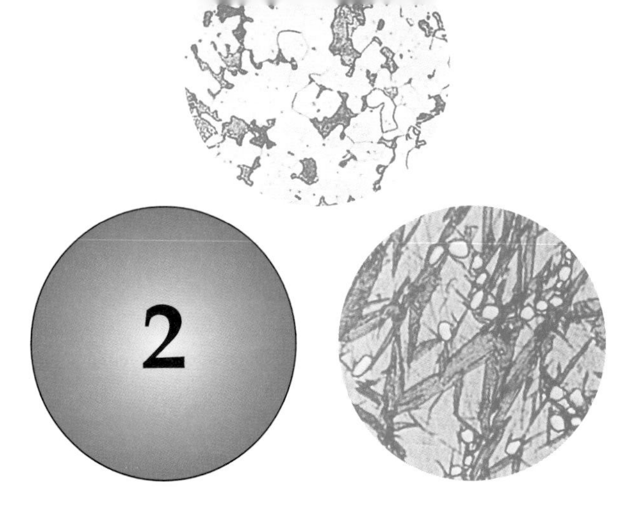

Metallurgical and Chemical Terminology

After studying this chapter, you will be able to:

- State the meaning of basic metallurgy terminology.
- Explain how chemistry is related to metallurgy.
- Define chemical terms such as element, compound, solution, and atom.
- Describe metals and alloys.

There are many chemical and metallurgical terms in the study of metallurgy. Before you get your mind and fingers into the physical operations, you must understand what is going on inside the metal—in fact, inside the metal crystals.

Studying the activity inside the metal can be extremely interesting. For example, iron and carbon are found inside every piece of steel. The carbon is dissolved inside the iron. Understanding the nature of this dissolving action between iron and carbon is basic to the understanding of metallurgy.

Basic Structure of Matter

In order to understand metallurgy, you must begin by understanding the basic chemical components of all matter. Protons, neutrons, and electrons form atoms, which are the smallest particles that retain the properties of an element. Elements combine to form compounds. Elements and compounds form solutions and mixtures, which constitute most of the substances we come in contact with in our daily lives.

Atoms

In ancient times, philosophers thought that if you took a piece of metal and cut it in half, both pieces would still be the same metal. This was true. They thought that if you cut it in half again, you would still have the same metal. This was true. They also thought that if you kept cutting it into smaller and smaller pieces, you could keep this up forever and still have the same metal no matter how small those pieces became. This was false. A metal can be cut into smaller and smaller pieces. These pieces may be so small that a microscope is required to see them. Eventually, this process reaches a limit. This limit is the atom.

An *atom* is the smallest part of an element that retains all the properties of the element. If an atom is divided, the new pieces do not have the same properties as the original material. The word *atom* means "indivisible."

According to theory, an atom is made up of three types of particles, each having a different electrical charge. *Protons* have a positive charge, *neutrons* have a neutral charge, and *electrons* have a negative charge. Uncharged atoms have an equal number of protons and electrons, so their charges cancel. Atoms with an unequal number of protons and electrons are called *ions*.

The protons and neutrons form the nucleus of the atom, around which the electrons orbit. See Figure 2-1. There may be several rings of electrons, with a specific number in each ring. The first ring contains up to two

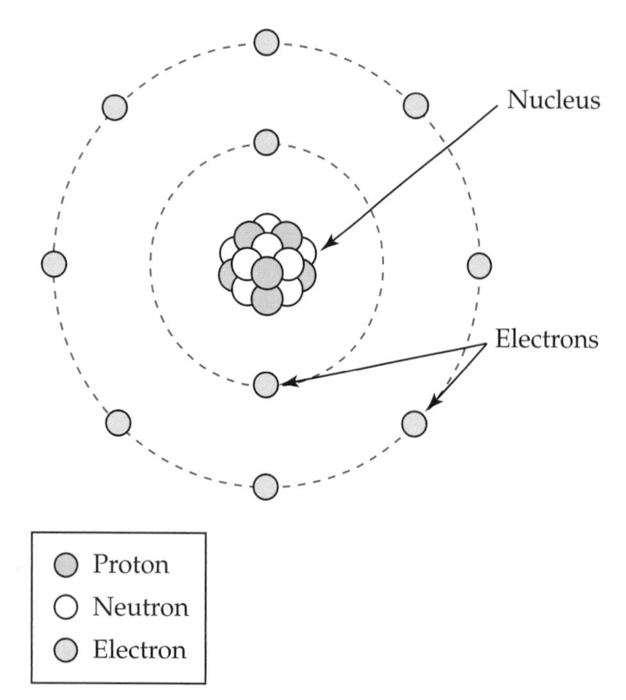

Figure 2-1. *An atom is composed of electrons, protons, and neutrons. The protons and neutrons form the nucleus.*

electrons. The second ring contains up to eight electrons. This third and subsequent rings contain up to eighteen electrons.

Elements

An *element* is a pure substance composed of a single material. It is as simple as a material can be. You cannot divide or separate an element into any other type of material. Whether you heat it, freeze it, machine it, break it, compress it, or use any other normal mechanical procedure, an element remains the same basic material that it was when you started.

There are over 100 known and universally established elements. See Figure 2-2. If everything on earth was broken down into its simplest form, all things could be separated into these elements. Some of the more common elements include oxygen, nitrogen, chlorine, hydrogen, gold, lead, copper, iron, silver, manganese, aluminum, magnesium, and sulfur.

The periodic table lists the elements in the order of their atomic numbers. The atomic number is the number of protons in the element's atoms. For example, the atomic number of carbon (C) is 6, so a carbon atom has six protons.

The hydrogen atom (H) contains only one electron. Hydrogen has one electron in its first ring, and one proton in its nucleus. Helium has two electrons and lithium has three electrons. Two electrons fill the first ring and the third one is left to travel in the second ring. Diagrams of these atoms are shown in Figure 2-3.

At room temperature, most elements are solid. Good examples are gold, iron, and lead. Several other elements are gases, such as oxygen and nitrogen. A few are normally liquids, such as bromine and mercury.

A *metal* is an element that has several of the following metallic properties:

- Ability to conduct electricity.
- Ability to conduct heat.
- Hardness.
- High density.
- Not transparent.

All metals possess some of these properties; many metals possess all these properties.

Molecules

A substance created by the chemical joining of two or more elements is called a compound. When two or more elements combine, their atoms join and form *molecules* of the compound. For example, oxygen atoms and hydrogen atoms combine to form water molecules.

An atom is the smallest part of an element, and a molecule is the smallest part of a compound. It takes two or more elements to make a compound, and it takes two or more atoms to make a molecule. See Figure 2-4.

The atoms in a molecule are joined together by chemical action. The atoms borrow, lend, or share the electrons in their outer ring. In a molecule of water, Figure 2-5, oxygen borrows the atoms of hydrogen to fill its outer ring.

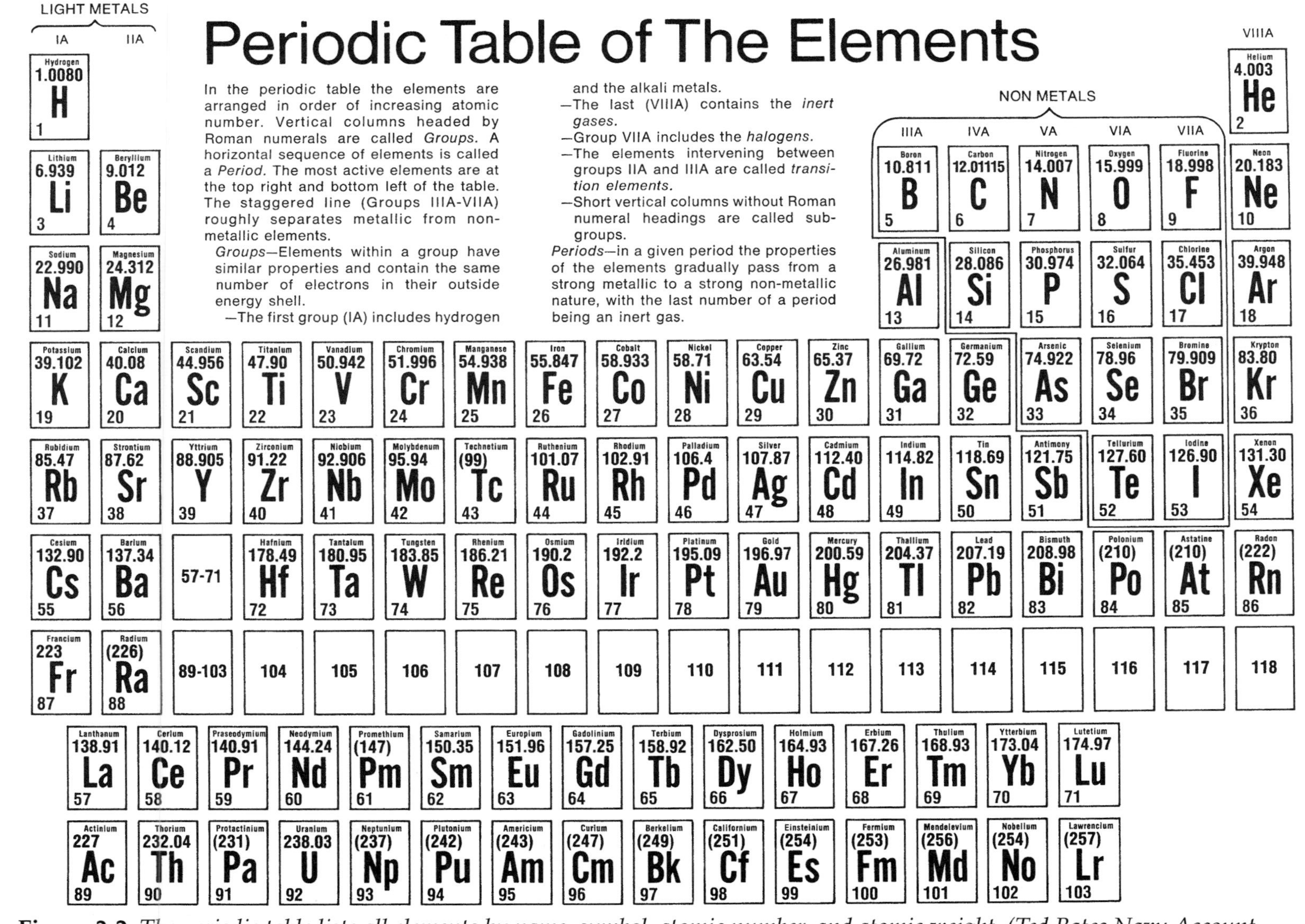

Figure 2-2. *The periodic table lists all elements by name, symbol, atomic number, and atomic weight. (Ted Bates Navy Account Group, Ted Bates & Co., Inc.)*

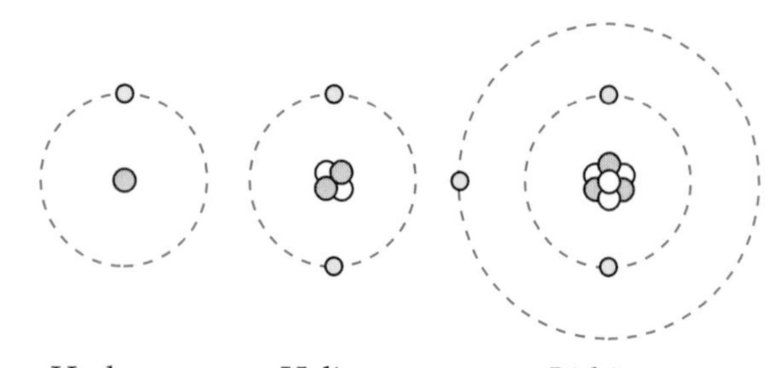

Figure 2-3. *Hydrogen has only one electron, helium has two electrons, and lithium has three electrons.*

Grains and Crystals

When a large group of atoms or molecules get together, they form a *family.* These families of atoms may be large enough to be seen by the naked eye. Such a family is known as a "grain" or "crystal." In a grain or crystal, all of the atoms orient themselves in neat, orderly formations.

Grains and crystals will be covered in detail in Chapter 7.

Compounds

A *compound* is a material composed of two or more chemically joined elements. A compound is not just a single element. In its simplest form, it is still made up of at least two elements. See Figure 2-4.

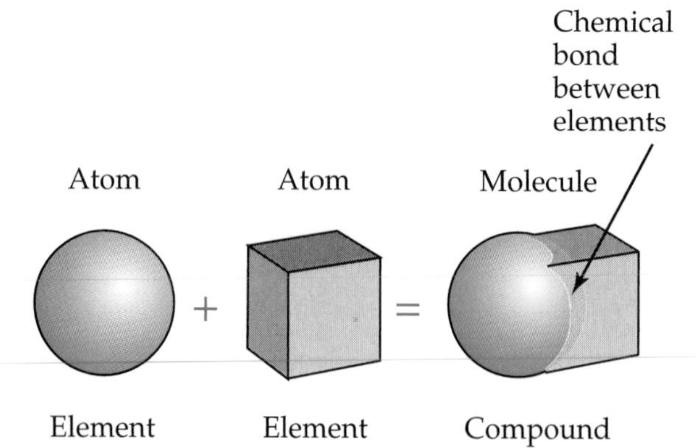

Figure 2-4. *A compound is made up of at least two different elements. The properties of the compound may be very different than the properties of its separate elements. (Note: The shapes used in this figure are not the actual shapes of atoms and elements.)*

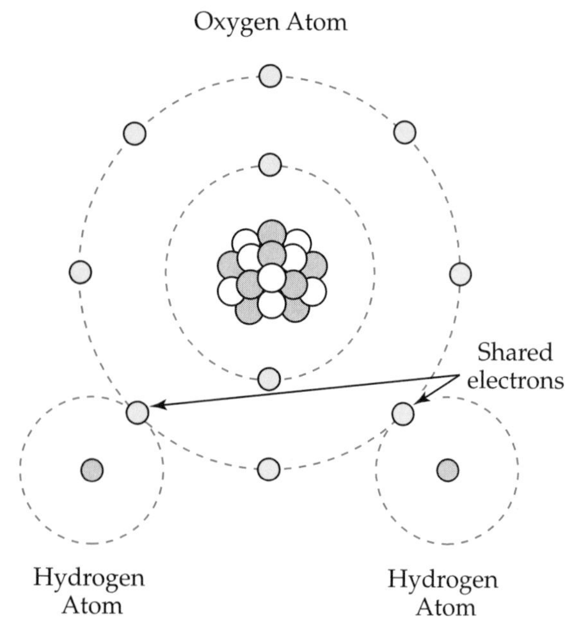

Figure 2-5. *A molecule of water is formed by chemical action between the oxygen and hydrogen atoms that make up the molecule.*

The elements in a compound are chemically joined and, therefore, very difficult to separate. The elements stay permanently joined unless special chemical action is taken to break down the compound.

One interesting feature of a compound is that its characteristics may be entirely different from the elements that make it up. Iron sulfide is made up of iron and sulfur. Iron is magnetic, but iron sulfide is not.

Water is composed of hydrogen and oxygen, both of which are gases. Hydrogen and oxygen are both flammable. However, when joined together, they become water, a compound that will put out fires. See Figure 2-6.

Sodium and chlorine can be chemically combined to produce table salt. Both sodium and chlorine are poisonous. Sodium is an innocent-looking, silvery metal that will burn your hand if you touch it. Chlorine is a greenish, poisonous gas that can kill you. Yet, when these two poisons are chemically combined, they become table salt, safe material to eat.

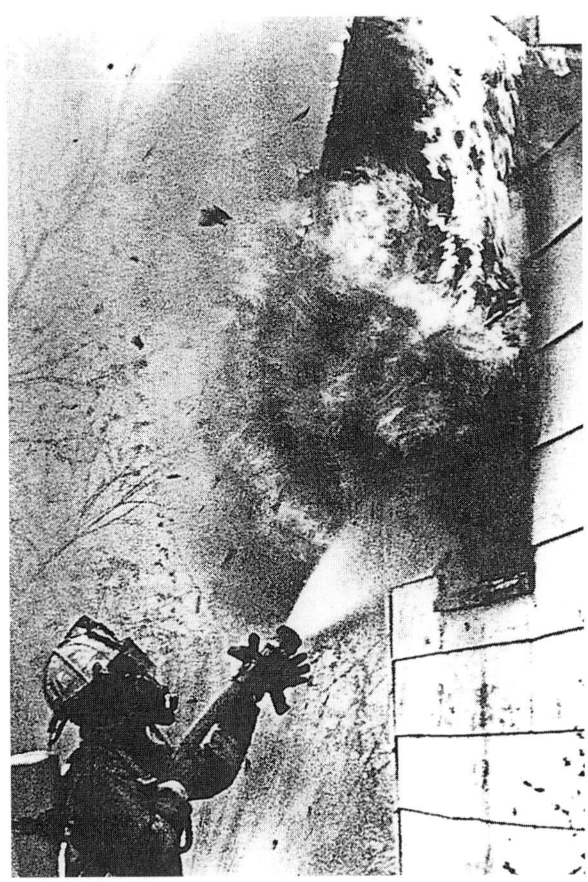

Figure 2-6. *The characteristics of a compound may be very different than the characteristics of the individual elements that form the compound. Hydrogen and oxygen are flammable compounds, but they combine to form a compound (water) that is used to extinguish fires. (Jack Klasey)*

Mixtures

A *mixture* is a material composed of two or more elements or compounds mixed together, but not chemically joined. A mixture is not just one material. In its simplest form, it is still made up of at least two elements or compounds. See Figure 2-7.

The difference between a mixture and a compound is the ease with which the elements can be separated. The elements in a mixture are not chemically joined, while the elements in a compound are chemically joined. Normally, filtering can be used to separate the components of a mixture.

In a mixture, no component completely loses its own identity. Therefore, the characteristics of a mixture are similar to the characteristics of the items that make it up. This is another way in which a compound and a mixture are different.

Iron-rich vitamin tablets contain a mixture of iron and other vitamins. The iron can be removed by grinding up the tablet, then using a magnet to collect the iron particles.

Muddy water is a mixture. In this case, a filter is not even necessary to separate the dirt from water. Just leaving the jar of muddy

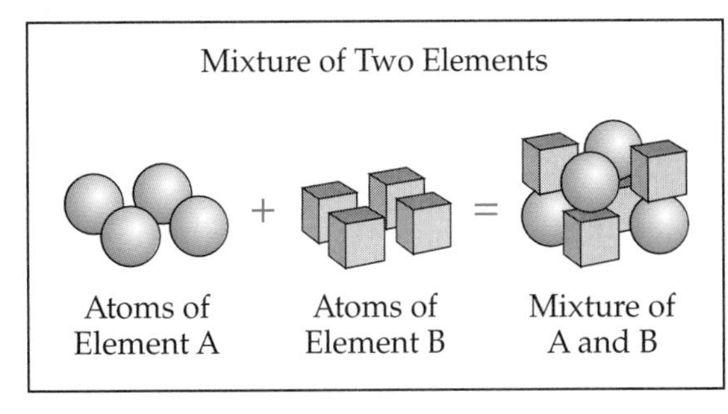

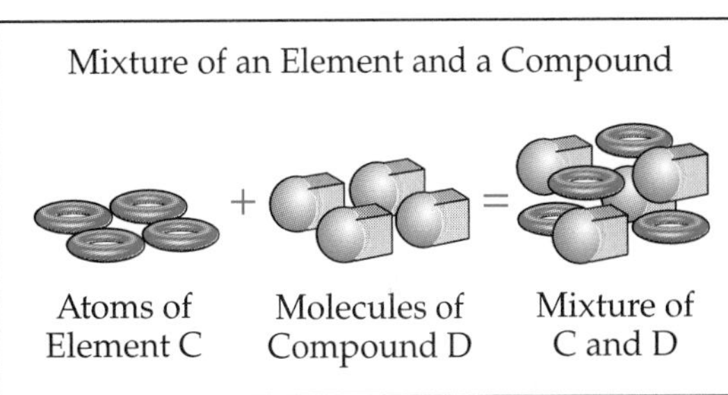

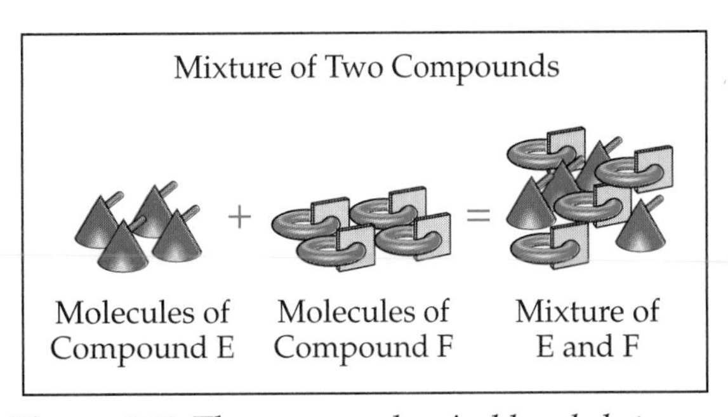

Figure 2-7. *There are no chemical bonds between the atoms or molecules combined in a mixture. The particles are combined physically, but not chemically.*

Self-Demonstration Magnetic Properties of Materials

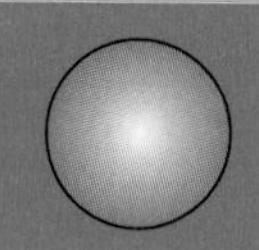

Obtain small pieces of aluminum, brass, plastic, gray cast iron, and several different types of steel, perhaps 1018, 1045, 1095, 4140, 52100, A6 tool steel, W1 tool steel, and 302 stainless steel. Other similar materials may be substituted if any of these are not available. Many of these same materials will be used for other Self-Demonstrations later in this text.

The exact size and shape of each piece is not important, as long as each piece has a flat surface (preferably two opposite flat surfaces).

Obtain a magnet. Line the material pieces in a neat row. Slowly bring the end of the magnet toward each material until it touches each piece. As the magnet nears each piece, the material will either be attracted to the magnet, repelled by the magnet (depending on which end of the magnet is approaching), or unaffected by the magnet. Note which materials are affected by the magnet's presence.

All ferrous materials will be affected by the magnet, including cast iron and all steel samples. Aluminum, brass, and any other nonferrous materials should not be affected. Plastic should not be affected.

water stand for a period of time will permit the items in the mixture to separate.

The oil in your automobile engine is a mixture of petroleum and additives. These can be separated. Homogenized milk is a mixture of milk and cream. These can be separated.

The iron and carbon atoms in steel do not chemically combine with each other. Compounds and molecules are not formed in steel. The atoms of iron and carbon are merely "mixed" together and become an alloy, or solid solution.

Solutions

A solution is a special kind of mixture. A *solution* is a mixture in which one substance is thoroughly dissolved in the other. When two materials combine and become a solution, one of the two is the "dictator" and the other is "submissive." The dictator dissolves the other substance. To look at a solution, you see only the dictator material, and not the dissolved material.

Generally, the dictator material in a solution is a liquid. The dissolved material generally is either a liquid or a solid. Examples are salt water or sugar water. After sugar is dissolved in water, it is difficult to recognize the difference between sugar water and regular water. Water is the dictator; it has totally overpowered the sugar.

The dictator material is known as the *solvent.* The dissolved material is known as the *solute.* Generally, there must be significantly more solvent than solute in order to perform the dissolving action.

The properties of a solution generally are very similar to the solvent. There will be some difference due to the influence of the solute, but not a great deal.

Solid Solutions

A *solid solution* is a solution in which both the solvent and solute are solids. At first, this sounds impossible. How can you mix a solid material into another solid material and cause dissolving to take place? Sugar cannot be dissolved in ice. If pieces of solid carbon are mixed up with pieces of solid iron, they will not dissolve.

The dissolving action can take place at elevated temperatures when both solids melt and become liquids. At these higher temperatures, iron dissolves many other elements, especially carbon. Iron becomes the solvent. Small amounts of carbon, phosphorus, or manganese become the solute.

At elevated temperatures, copper dissolves small amounts of zinc, lead, tin, or nickel. Many other materials behave this way, but iron and copper are two of the most common solvents.

Alloys

When two or more metals are dissolved together in a solid solution, the new material is known as an *alloy.* Steel is an alloy of iron and carbon. Bronze is an alloy of copper and tin. Brass is an alloy of copper and zinc.

The metals that are dissolved—the solutes—are also called *alloys* or *alloying elements.* Thus, the word *alloy* has two meanings.

- The dissolved metal material.
- The solid solution that is made up of alloys and solvent.

Applying Chemical Terms to Steel

The chemical and metallurgical terms from this chapter that appear most often in the study of metallurgy are crystal, atom, and alloy. Steel is an alloy (solid solution). Iron is the solvent, carbon is always one of the solutes.

Many other alloys are dissolved in iron to make up different types of steel. Some of these alloys include sulfur, manganese, aluminum, phosphorus, molybdenum, tungsten, and silicon. As these atoms collect in colonies and solidify, crystals are born. These terms will occur frequently throughout the study of metallurgy.

Test Your Knowledge

Write your answers on a separate sheet of paper. Do not write in this book.

1. What does the word *atom* mean?
2. What is the smallest part of a piece of silver that retains all the properties of silver?
3. List the three types of particles in an atom.
4. How are elements arranged in the periodic table?
5. List five general properties of metals.
6. What type of particle is formed when two or more atoms are chemically bonded?
7. Name an example of a compound with properties that are very different than the properties of the separate elements found in the compound.
8. List two major differences between compounds and mixtures.
9. Name an example of a method of separating the individual components of a mixture.
10. What are the two components of a solution called?
11. How are solid solutions created?
12. What is the solvent and the solute in steel?
13. List the two definitions of *alloy.*
14. Name five alloys that are used in various types of steel.
15. Carbon has six electrons. Sketch a diagram of carbon similar to Figure 2-3.
16. Fluorine has nine electrons. Sketch a diagram of fluorine.
17. Sketch a diagram of an element that has more than two electron rings. Use the periodic table (Figure 2-2) to select the element.

Section Two

Properties of Metals

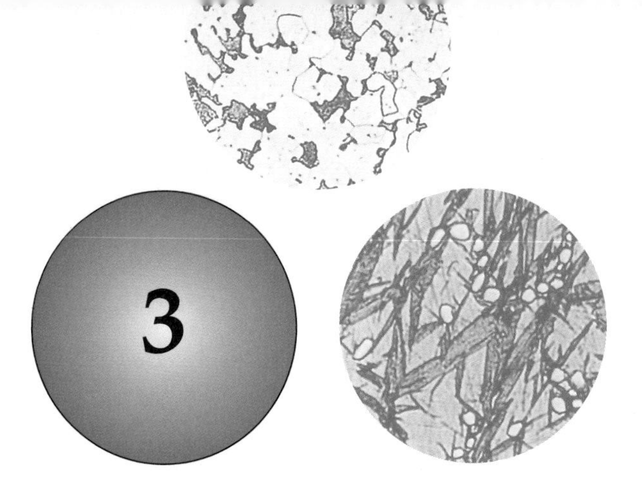

Hardness

After studying this chapter, you will be able to:

- Explain what hardness is.
- Describe how the hardness of metal is found.
- Compare different hardness testing methods.
- Discuss how each hardness testing method works.
- Convert between hardness scales.

What Is Hardness?

Hardness is perhaps the most important property of metals you will encounter during your study of metallurgy. It is a difficult word to define. However, one good definition of *hardness* is "a measure of resistance to deformation." Another is "a measure of resistance to penetration." Both of these definitions refer to the resistance of a metal surface to be damaged, dented, worn away, or deteriorated in any way as a result of a force or pressure against it.

Therefore, to invent a means of measuring hardness, you would have to create a machine with a penetrator or pointer that would try to dent or cut into a surface. A large force or weight would supply the power behind the penetrator. The size of the resulting dent or penetration in the sample would be the measure of the hardness of the material. See Figure 3-1.

The Relationship of Hardness to Other Properties

Hardness is important in the study of metallurgy because it relates to several other key properties of metal, especially strength, brittleness, and ductility. By measuring the hardness of a metal, you are also indirectly measuring the strength, brittleness, and ductility of the metal.

Hardness, then, is similar to a family's grocery list. From the size of the list, you can estimate the size of a family. From the items on the list, you can guess the family's financial

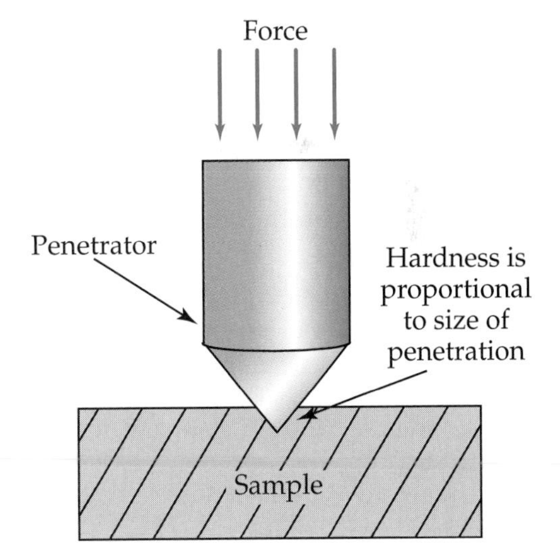

Figure 3-1. *Hardness is proportional to size of penetration. The harder the material, the smaller the penetration.*

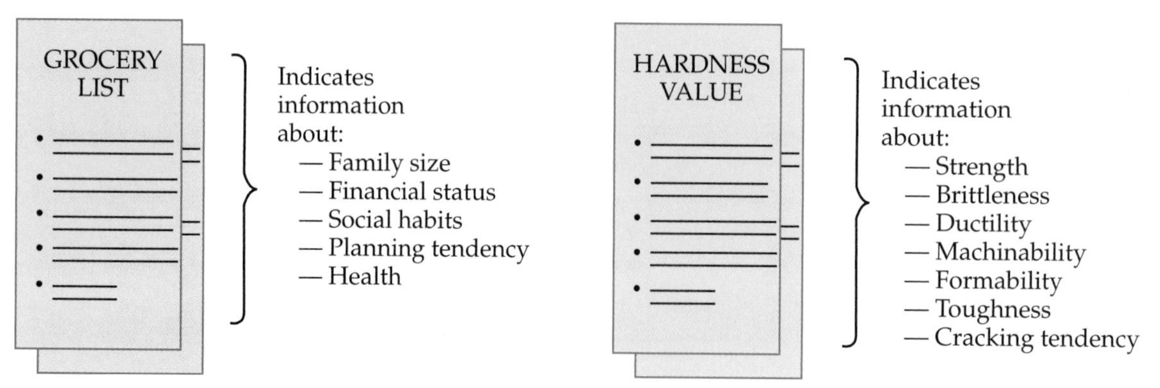

Figure 3-2. *Hardness is a gauge for other characteristics in the same way that a family's grocery list indicates family traits.*

status and social habits. Similarly, hardness is a gauge for many other characteristics of a metal's family traits. See Figure 3-2. These other traits or properties will be discussed in detail in Chapter 4, Material Properties.

Measuring Hardness

There are many different methods used to measure the hardness of a metal. Some of the most common methods will be discussed in this chapter. The two basic categories of hardness testing methods are penetration hardness and scratch hardness.

Penetration hardness is a very accurate measuring technique in which a precision machine is used. A penetrator on this machine is forced against the metal sample. The size of the resulting impression (dent) is measured, and the measurement is converted to a hardness number.

A *scratch hardness* test is very fast and crude. The metal sample is scratched by the edge of a tool or object. No numerical value of hardness is calculated. The sample is called either "hard" or "soft," depending on whether or not a scratch results.

Penetration hardness testing is a relatively expensive and accurate method compared to scratch hardness. See Figure 3-3. Most industrial companies concerned with hardness have some form of a penetration hardness tester.

Units of Hardness

Time is measured in seconds or hours. Weight is measured in units of pounds, ounces, or kilograms. Distance can be measured in units of feet, miles, or meters. All of these are rather obvious units of measurement. The units of hardness, however, are not so obvious.

Hardness is measured in many different units. Some examples are BHN, DPH, Shore units, Knoop units, R_C, and 15N. No single unit is used universally as the main unit of measure. Instead, each type of machine tends to have its own units. Therefore, many conversion charts are necessary to convert the units of one testing method to those of another hardness scale.

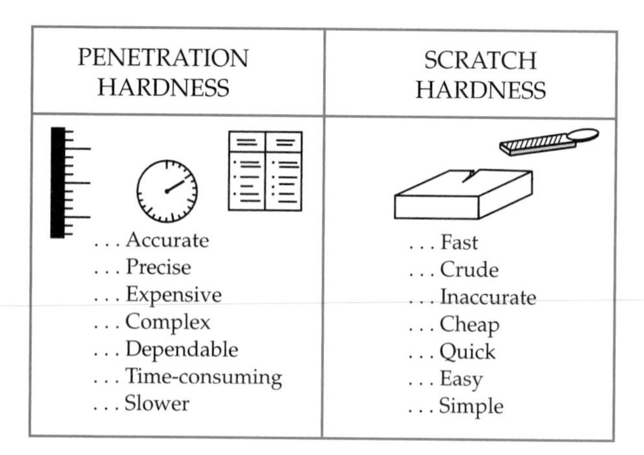

Figure 3-3. *There are two categories of hardness testing methods: penetration hardness and scratch hardness.*

Hardness Testing Methods

There are many different hardness testing methods and many different hardness testing machines. Nine of the most common ones will be discussed in detail in this chapter. Each method has advantages and disadvantages over the other methods. Therefore, determination of which method to use or which machine to buy depends on the individual application.

The following are nine of the most common hardness testing methods:

- Brinell hardness
- Vickers microhardness
- Knoop microhardness
- Rockwell hardness
- Rockwell superficial hardness
- Shore scleroscope hardness
- Sonodur hardness
- Mohs hardness
- File hardness

Brinell Hardness Testing Method

The *Brinell hardness testing method* is one of the oldest methods of hardness testing. A Brinell hardness tester is shown in Figure 3-4. The testing procedure is illustrated in Figure 3-5.

Brinell hardness testing procedure

The following steps are involved in a Brinell hardness test:

1. The metal sample is placed on the machine's anvil.
2. The hardened-steel penetrator (round ball) is slowly brought into contact with the test sample, Figure 3-6, either automatically or by manual operation. The contact pressure between the penetrator and the sample increases until a force of 3000 kilograms (3000 kg) is reached. The ball penetrator is harder than the sample, so a round dent is impressed onto the sample.
3. After the ball has made the round indenture in the sample, it is released and the sample is removed. See Figure 3-7.

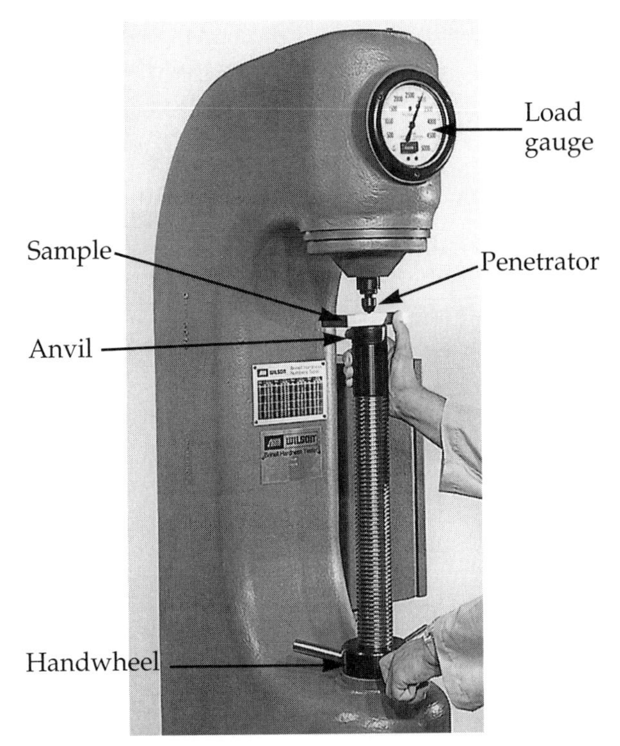

Figure 3-4. *A metal sample being tested on a Brinell hardness tester. (Instron Corporation)*

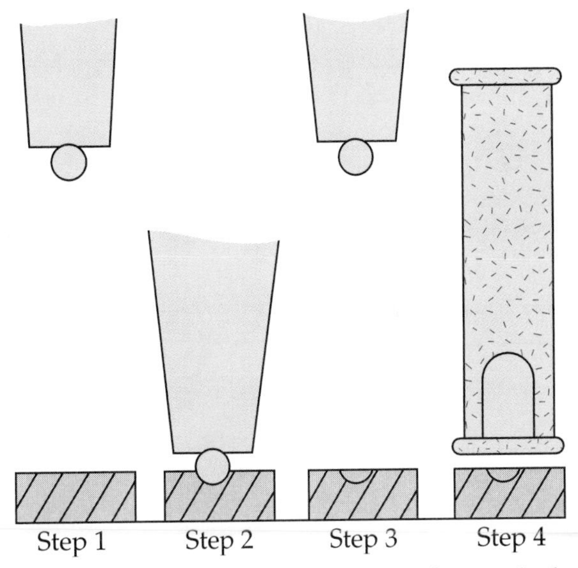

Figure 3-5. *The Brinell hardness testing method: Step 1—The sample is placed on the anvil. Step 2—The penetrator contacts and indents the sample. Step 3—The penetrator is released. Step 4—A microscope with a calibrated lens is used to measure the diameter of the dent.*

4. A small microscope with a calibrated lens is brought into contact with the sample, Figure 3-8. The diameter of the dent is measured in millimeters. The measurement is then converted to a hardness value. This conversion can be done by means of the formula shown in Figure 3-9 or, more commonly, by using a conversion chart.

The conversion chart shown in Figure 3-10 has two columns. The left column lists the diameter of the indenture in millimeters. The right column gives the hardness value in units of BHN (Brinell hardness number). These BHN values are derived from the formula shown in Figure 3-9.

The tungsten-carbide ball penetrator is approximately 10 millimeters (10 mm) in diameter. Generally, the force used in the test is 3000 kg. Sometimes, a second BHN scale is used in conjunction with a 500 kg force. This scale is used primarily for softer and thinner samples. In making the tests, loads should be applied against the sample for a minimum of 15 seconds. The sample should be relatively smooth, flat, clean, and horizontal.

The round ball makes a relatively large impression on the sample, compared to other hardness testing methods. Many of the other methods use a pointed penetrator, which will not deform the sample as much. Due to the greater area of penetration, Brinell samples are generally scrapped after being tested. Also, because of the large indenture, a Brinell hardness tester cannot be used to measure the hardness of very thin samples.

Figure 3-6. *The penetrator of a Brinell hardness tester contacting the surface of a test sample. (Instron Corporation)*

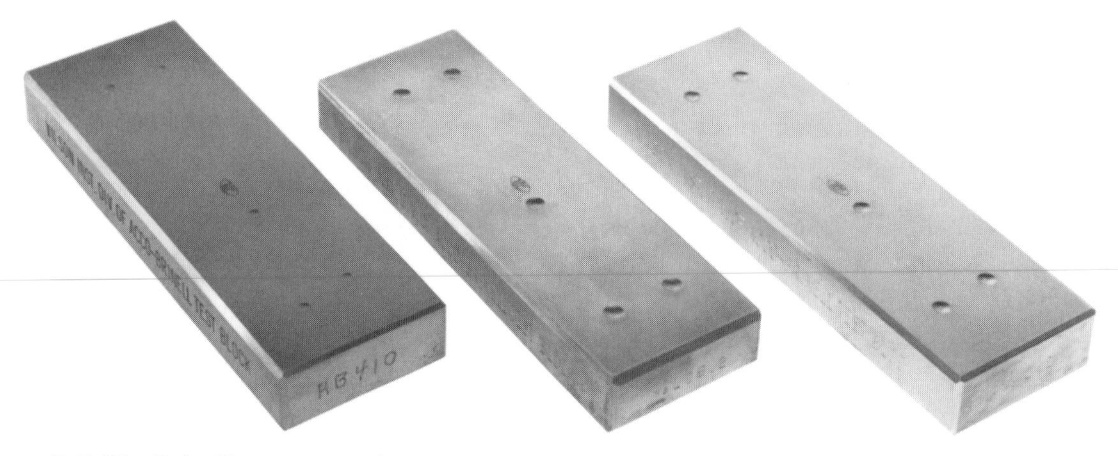

Figure 3-7. *The Brinell penetrator places a round dent on the sample. (Instron Corporation)*

Figure 3-8. *A Brinell microscope is used in conjunction with the hardness tester. (Left. Instron Corporation. Right. Engineering and Scientific Equipment Ltd.)*

$$\text{BHN} = \frac{\text{Load (kg)}}{\text{Surface Area (mm}^2\text{)}}$$

$$\text{BHN} = \frac{F}{\frac{\pi D}{2}\left(D - \sqrt{D^2 - d^2}\right)}$$

BHN = Brinell hardness number
F = Force or load in kilograms
D = Diameter of ball penetrator in millimeters

Figure 3-9. *This formula is used to calculate the Brinell hardness number from the diameter of the dent in the sample.*

The Brinell method generally is restricted to softer steels or other softer metals. Too much force is required on the penetrator to make a measurable dent on a very hard surface. However, Brinell is a very accurate hardness measuring method for soft materials.

Typical BHN values are shown in Figure 3-11.

Vickers Hardness Testing Method

Sometimes, the large indentation caused by the Brinell testing method makes the tested sample useless. If the tested part is to be used, it is critical that the size of indenture be as small as possible. This is particularly true for thin, brittle, or small parts.

In these applications, equipment known as *microhardness testers* are used. These testers use smaller loads and sharply pointed penetrators, so they make very small indentations. There are two common methods of microhardness testing: *Vickers hardness testing* and *Knoop hardness testing.*

Brinell Hardness Number Conversions			
Diameter of Ball Impression (mm)	BHN*	Diameter of Ball Impression (mm)	BHN*
2.25	745	3.60	285
2.30	710	3.65	277
2.35	682	3.70	269
2.40	653	3.75	262
2.45	627	3.80	255
		3.85	248
2.55	578	3.90	241
2.60	555	3.95	235
2.65	534	4.00	229
2.70	514	4.05	223
		4.10	217
2.75	495	4.15	212
2.80	477	4.25	203
2.85	461	4.35	192
2.90	444	4.40	187
2.95	432	4.50	179
3.00	415	4.60	170
3.05	401	4.65	166
3.10	388	4.80	156
3.15	375	4.90	149
3.20	363	5.00	143
3.25	352	5.10	137
3.30	341	5.20	131
3.35	331	5.30	126
3.40	321	5.40	121
3.45	311	5.50	116
3.50	302	5.60	112
3.55	293		
*3000 kg load, 10 mm ball			

Figure 3-10. *This chart can be used to determine the Brinell hardness number based on the diameter of the dent. (Teledyne Vasco)*

Applications of these microhardness testers include thin plates, metal foils, and fine wire. Glass and ceramics, which might fracture due to large indentations, can be tested with a microhardness testing method. The hardness of coatings and case-hardened parts can also be tested with microhardness testers. These machines are often used to test areas smaller than the size of a crystal or grain. These testers are more often found in research laboratories than on a manufacturing production floor.

A *Vickers hardness tester* is shown in Figure 3-12. The operation of this tester is similar to the operation of the Brinell tester. The differences between Brinell hardness testing and Vickers hardness testing are listed in Figure 3-13. The following are the three main differences:

- The penetrator has a different shape.
- The load (force) is less.
- The methods use different units.

Vickers hardness testing procedure

The testing procedure for Vickers hardness testing is nearly identical to the procedure for Brinell hardness testing. The following steps make up the procedure:

1. The sample to be tested is placed on the anvil of the tester, below a hardened steel penetrator with a diamond point.

Typical BHN Values	
Cold Rolled Steel (unquenched)	150 BHN
Quenched Steel	600 BHN
Stainless Steel (unquenched)	150 BHN
Cast Iron	200 BHN
Wrought Iron	100 BHN
Aluminum	100 BHN
Annealed Copper	45 BHN
Brass	120 BHN
Magnesium	60 BHN

Figure 3-11. *Brinell hardness values vary for different types of metal.*

2. This square diamond penetrator is slowly brought into contact with the sample. The contact pressure between the penetrator and sample is increased until 50 kg of force is reached.
3. The penetrator is retracted and the sample shows a small, pyramidal shaped hole, Figure 3-14.
4. The diagonal of the indentation is measured. The length of the diagonal is converted to a DPH (diamond-pyramid hardness) value by use of a formula, Figure 3-15, or a table, Figure 3-16.

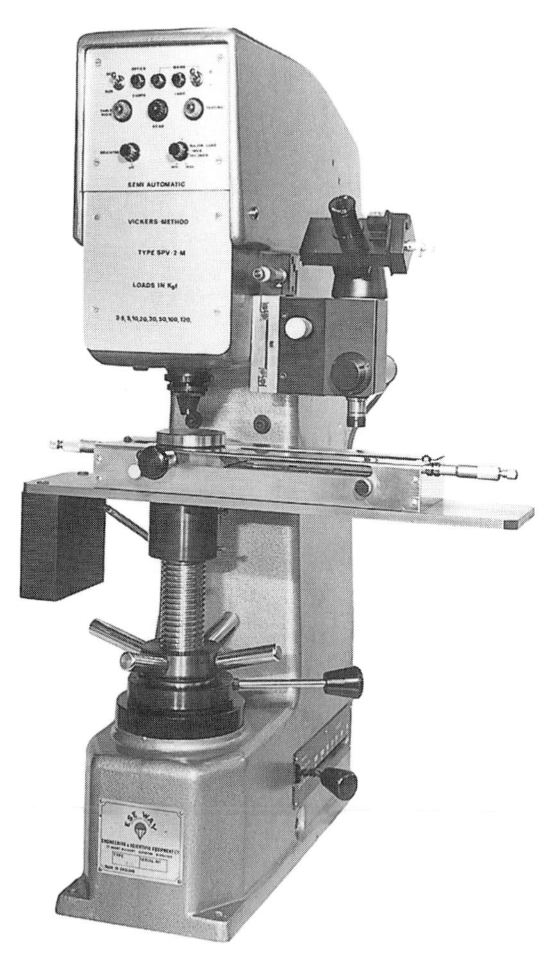

Figure 3-12. *This Vickers hardness tester is similar to a Brinell hardness tester. (Engineering and Scientific Equipment Ltd.)*

Comparison of Brinell and Vickers Hardness Testing

	Brinell	Vickers
Penetrator	10 mm Diameter Ball	Square Diamond
Load	500 kg or 3000 kg	50 kg
Units	BHN	DPH

Figure 3-13. *A chart comparing Brinell and Vickers hardness testing.*

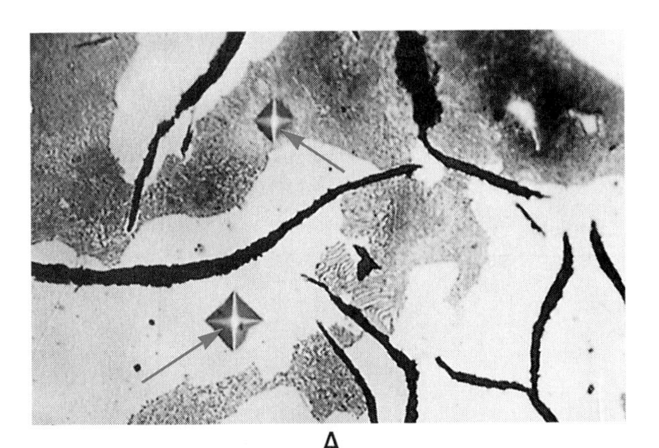

A

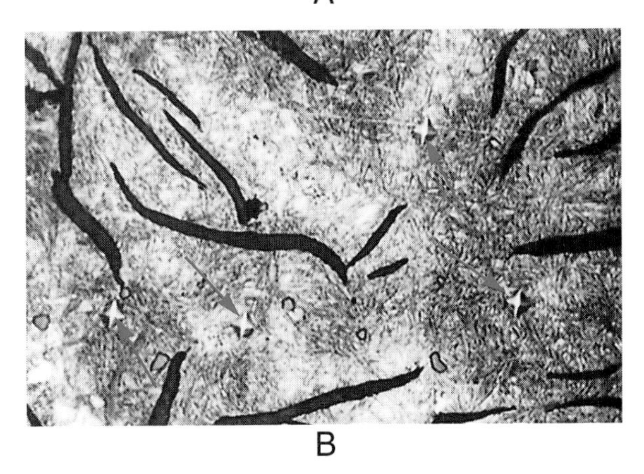

B

Figure 3-14. *Magnified Vickers hardness test samples. A—Cast iron part. The dent in the white area is larger than the dent in the grayish area because the iron structure is harder in the grayish area. B—This gray iron is much harder than the cast iron, so the indentations are much smaller. (Iron Castings Society)*

The length of the indentation diagonal can be measured in several ways:

- Using a microscope similar to the Brinell microscope.
- Using a calibrated micrometer barrel on the tester, Figure 3-17A.
- Using a digital readout.
- Using computer software to display an image of the indentation on a computer monitor, Figure 3-17B.

The angle of the penetrator head measures approximately 136°, Figure 3-18. The load most commonly applied is 50 kg. Some Vickers hardness testers occasionally use either 5, 10, 20, 30, or 100 kg loads. The load is held on the sample for about 30 seconds. The surface should be smooth, flat, clean, and horizontal before testing begins.

Note that the 50 kg load is considerably less than the 3000 kg load used with Brinell testing. Therefore, the sample is not damaged as severely as it is in the Brinell hardness test.

$$DPH = \frac{\text{Load (kg)}}{\text{Surface Area (mm}^2\text{)}}$$

$$DPH = \frac{1.85\,F}{d^2}$$

DPH = Vickers diamond-pyramid hardness
F = Force or load in kilograms
d = Diagonal length of indentation in millimeters

Figure 3-15. *This formula is used to determine the DPH number for Vickers hardness testing.*

Advantages of Vickers hardness testing

The following are some of the advantages of the Vickers hardness testing method over the Brinell hardness testing method:

- Vickers testing can be used on harder materials because the pointed penetrator can probe into a hard surface more easily than a ball penetrator can.
- Vickers testing can be used on smaller areas.
- Vickers testing requires a smaller load.

Vickers Hardness Conversion	
Diagonal Length of Impression (mm)	**DPH***
.30	1030
.35	757
.40	579
.45	458
.50	371
.55	306
.60	258
.65	219
.70	189
.75	165
.80	145
.85	128
*Based on 50 kg load	

Figure 3-16. *This table is used to determine the DPH number for Vickers hardness testing.*

Knoop Hardness Testing Method

The *Knoop hardness testing method* uses an even smaller load than the Vickers testing method. A load of less than 4 kg is used with the Knoop method. A load of merely 25 grams (25 g) can be used for extremely small areas. Most new microhardness testers are equipped for both Vickers and Knoop testing, Figure 3-19.

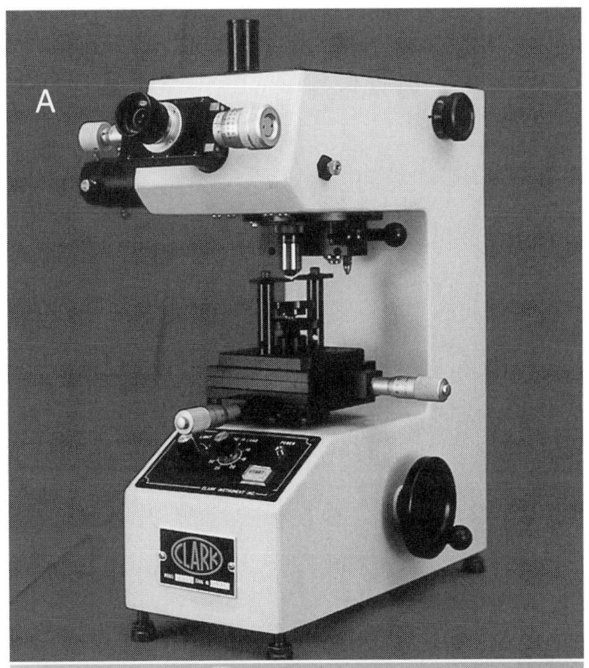

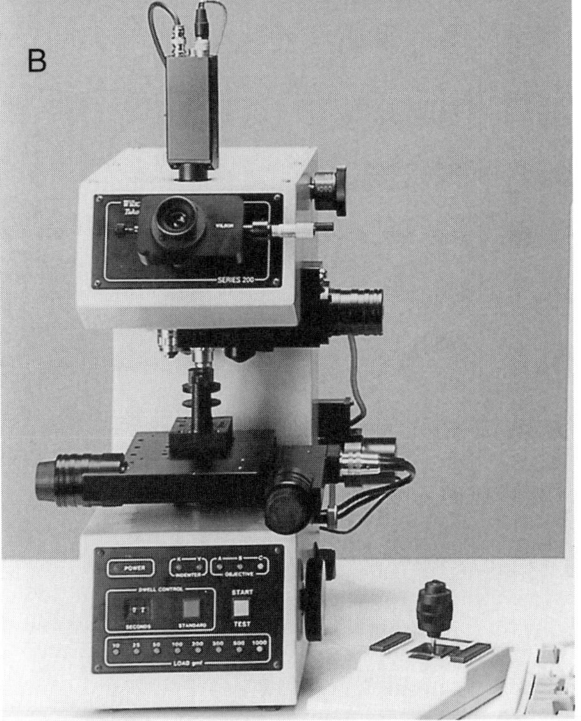

Figure 3-17. *A—This tester has a calibrated micrometer barrel for measuring the indentation. B—This tester has an automatic image analysis setup that shows an enlargement of the indentation on a monitor. (SUN-TEK Corporation and Instron Corporation)*

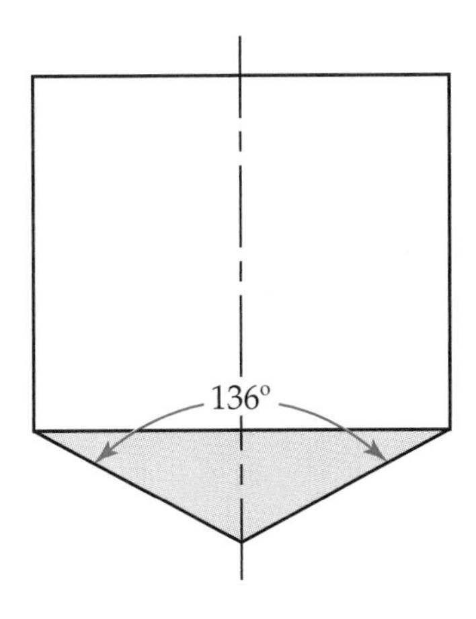

Figure 3-18. *The penetrator angle is approximately 136° for Vickers hardness testing.*

The Knoop penetrator has a diamond cross section known as an "elongated pyramid." The ratio of the diagonals is 7 to 1. See Figure 3-20.

The steps involved in making a Knoop hardness test are essentially identical to those of the Vickers test method. With the Knoop method, a load is applied and an indentation is made. The indentation is measured. A formula or chart is used to convert to a Knoop hardness value, or the value can be read directly from a digital readout on the tester. The hardness units are simply called "units Knoop."

Advantages of Knoop hardness testing

The following are the main advantages of Knoop hardness testing over the Vickers and Brinell testing methods:

- Knoop testing does essentially no damage to the specimen.
- Knoop testing can be used on very thin materials.
- Knoop testing can be used on very small surface areas.

As in the other two methods, the surface of the specimen should be smooth, flat, clean, and horizontal before testing.

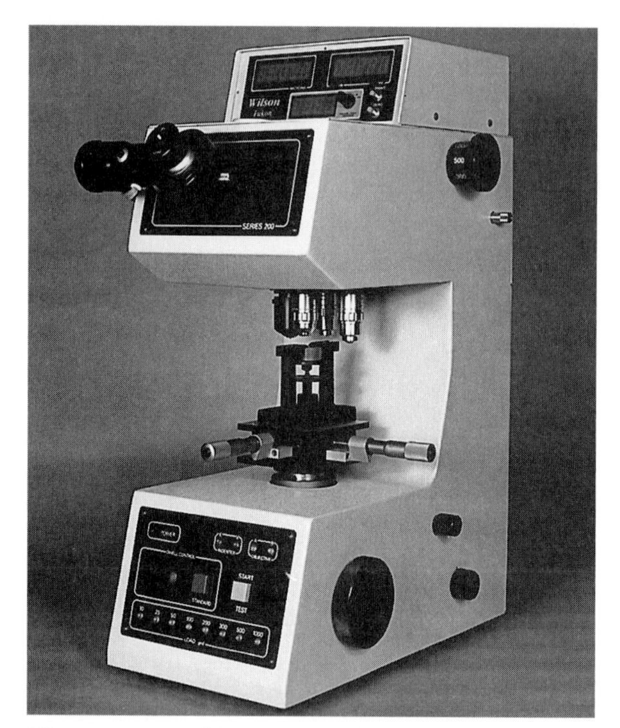

Figure 3-19. *This microhardness tester with digital output is used for Vickers and Knoop testing. (Instron Corporation)*

Rockwell Hardness Testing Method

The *Rockwell hardness testing method* is the most widely used of all metal hardness testing methods. The Rockwell testing method eliminates the effects of small surface imperfections by applying a preliminary load (minor load) to the sample before the hardness test is taken. Thus, Rockwell hardness values are very accurate. A Rockwell hardness tester is shown in Figure 3-21.

Rockwell hardness testing procedure

A Rockwell hardness test has two loading steps. A preliminary *minor load* is applied first. The *major load* follows and comprises the actual hardness test. A total of three separate steps are involved (see Figure 3-22):

1. The sample is placed on the anvil. The anvil is raised manually until the sample contacts the penetrator. Then, the sample is raised slightly higher until a minor load of about 10 kg is applied, Figure 3-23.

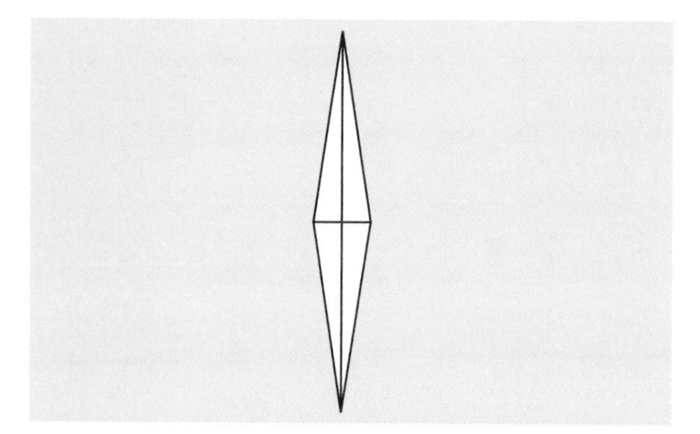

Figure 3-20. *The Knoop penetrator has an elongated pyramid shape.*

2. After the minor load is applied, the major load (60 kg, 100 kg, or 150 kg) is applied by actuating a handle or lever on the front of the machine, Figure 3-24. As the major load is applied, the penetrator moves deeper into the sample.

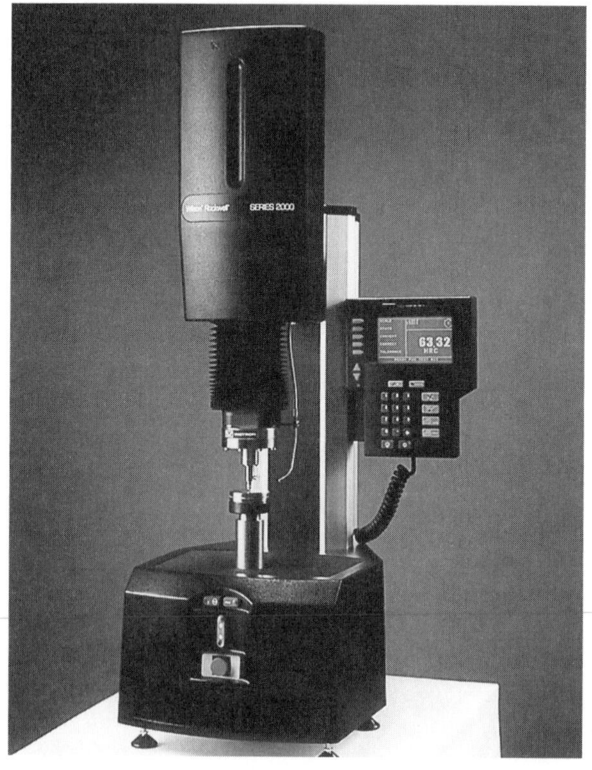

Figure 3-21. *This Rockwell hardness tester shows the Rockwell hardness number on a display screen. (Instron Corporation)*

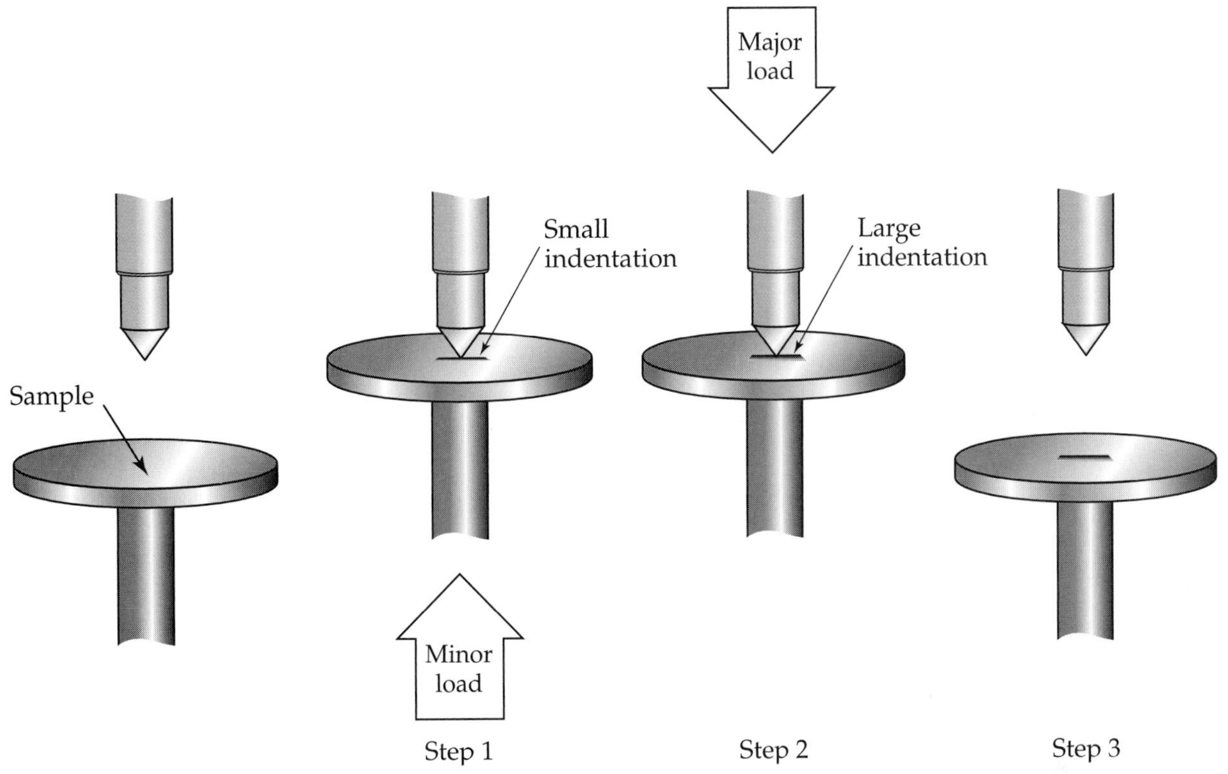

Figure 3-22. *Rockwell hardness test. Step 1—Sample is raised to penetrator after placement and minor load is applied. Step 2—Major load is applied and scale is read. Step 3—Sample is lowered.*

3. The hardness value is read directly from either a digital readout or from a rotary dial on the machine, Figure 3-25. No intermediate microscope or manual approximations need to be made. The scale reads in Rockwell hardness values, which are based on the depth of penetration. The machine automatically converts the depth reading to a Rockwell hardness value.

The minor load causes the penetrator tip to be slightly below the surface of the sample, rather than on the outside surface of the metal. The hardness measurement is then based on the major load. This eliminates the adverse effects of surface scale, surface roughness, and lack of flatness, smoothness, and cleanliness.

Three different penetrators are used for Rockwell testing, and three different loads are commonly applied. The three Rockwell penetrators are:

- A 1/8″ diameter tungsten-carbide ball.
- A 1/16″ diameter tungsten-carbide ball, Figure 3-26A.
- A conical-shaped, diamond-point penetrator, Figure 3-26B.

The following loads are used for Rockwell hardness tests:

- 60 kg
- 100 kg
- 150 kg

Therefore, there are nine possible combinations of penetrators and loads with the Rockwell system. Each combination has a unique hardness value scale. The relationship between the three penetrators, the three loads, and the nine Rockwell scales is shown in Figure 3-27. Which scale would be used for the hardest materials? Hard materials require the largest kilogram load and the sharpest penetrator. Therefore, the

Figure 3-23. *In a Rockwell hardness test, a minor load of about 10 kg is applied. (SUN-TEK Corporation)*

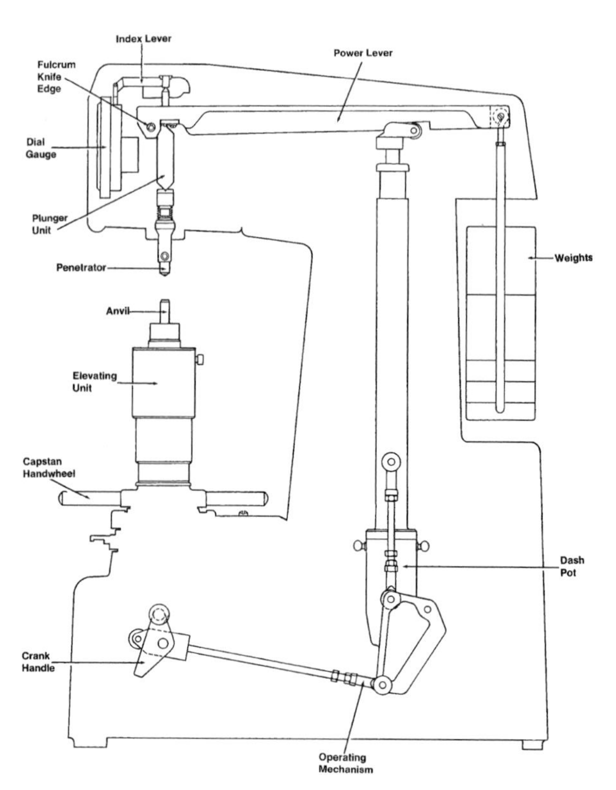

Figure 3-24. *This schematic diagram illustrates the operation of a Rockwell hardness tester. (Instron Corporation)*

Rockwell C scale is used for the hardest materials.

Can you understand why the Rockwell H scale (R_H) is used for the softest materials? This scale is so soft that it normally would not be used for any metals.

For testing steel, the two scales most often used are the Rockwell C scale (R_C) and the Rockwell B scale (R_B). The Rockwell C scale is used to test hard steels and the Rockwell B scale is used to test the softer, low-carbon steels, aluminum, and other soft nonferrous materials. The dial on most Rockwell testers has both a B and C scale. The hardest value that steel can attain is about 70 R_C (Rockwell C scale).

Testing accuracy

A load-cell tester may improve the accuracy of Rockwell hardness tests. This tester incorporates a closed-loop load-cell that eliminates friction in the Rockwell hardness tester. The penetrator and depth-measuring device are mounted directly on the load cell, which compensates for variations that may occur in the hardness reading. The friction caused by many pivot and slide points is eliminated with this method. While a load-cell equipped tester will deliver the most accurate hardness measurements, mechanical dead-weight testers are still widely used by manufacturers.

Advantages of Rockwell hardness testing

The Rockwell hardness testing method has the following two key advantages:

- The minor load greatly reduces or eliminates the effect of surface imperfections.
- Human error is reduced because the hardness value can be read directly from a scale.

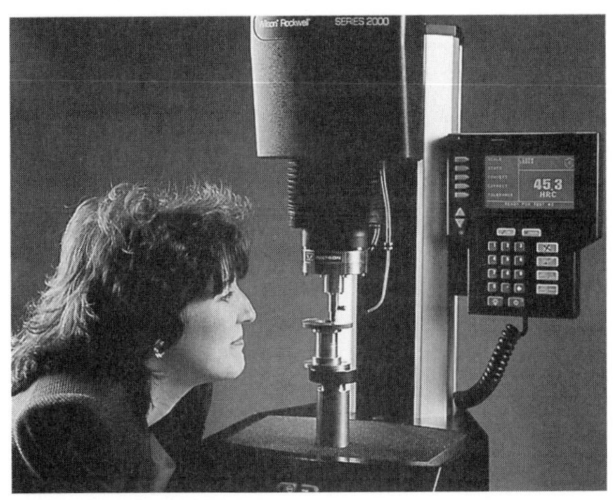

Figure 3-25. *The Rockwell hardness value can be read directly from a digital readout. (Instron Corporation)*

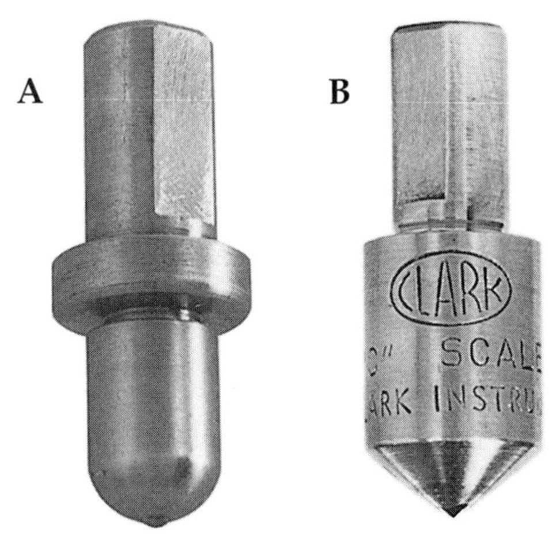

Figure 3-26. *Rockwell hardness tester penetrators. A—1/16″ diameter ball penetrator. B—Diamond-point penetrator. (Instron Corporation and SUN-TEK Corporation)*

Rockwell Superficial Hardness Testing Method

The *Rockwell superficial hardness testing method* is similar to the basic Rockwell hardness testing method. The difference is that the Rockwell superficial tester tests the hardness closer to the outside surface of the metal. See Figure 3-28. A tester that can perform both Rockwell hardness tests and Rockwell superficial tests is shown in Figure 3-29.

Rockwell superficial hardness testing procedure

The Rockwell superficial hardness testing method uses smaller loads than the Rockwell hardness testing method. The three common loads used on the superficial tester are 15 kg, 30 kg, and 45 kg. The same diamond-point penetrator and 1/16″ ball used in the Rockwell hardness test are also used in the Rockwell superficial hardness test.

There are two choices of penetrators and three choices of loading, so there are six total combinations available for use on the Rockwell superficial tester. Each combination uses one of the following scales: 15N, 30N, 45N, 15T, 30T, or 45T. See Figure 3-30.

For example, 55-15N would mean that a material has a hardness value of 55 units measured on a Rockwell superficial test machine using a diamond-point penetrator and a 15 kg load. A reading of 75-30T indicates a material hardness value of 75 units tested with the 1/16″ ball and a 30 kg load.

The Rockwell superficial tester penetrator-load combination for the hardest steels would be the one used with the 45N scale. This combination makes use of the diamond-point penetrator and the 45 kg load.

Advantages of Rockwell superficial hardness testing

The Rockwell superficial hardness testing method has the following advantages:

- Thin materials can be tested.
- Hardness near the surface can be tested.
- Case-hardened surfaces can be tested. (Case hardening is discussed in Chapter 15)

Many companies use Rockwell superficial testers for all hardness tests. Even though their intended purpose is testing hardness close to the outer surface, they are dependable for nearly all manufacturing applications.

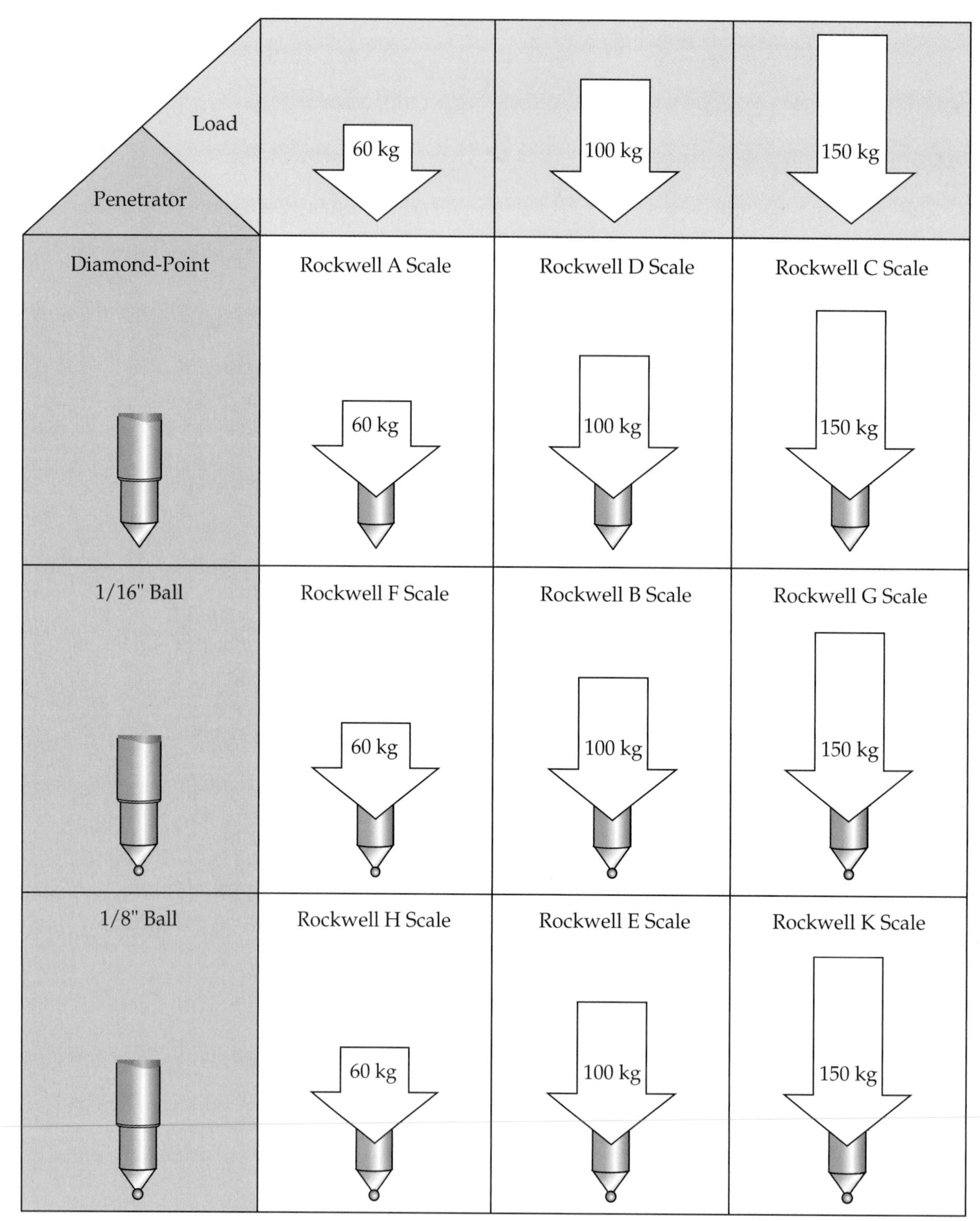

Figure 3-27. *There are nine combinations of penetrators and loads possible with the Rockwell hardness testing method. Each combination has a unique Rockwell scale.*

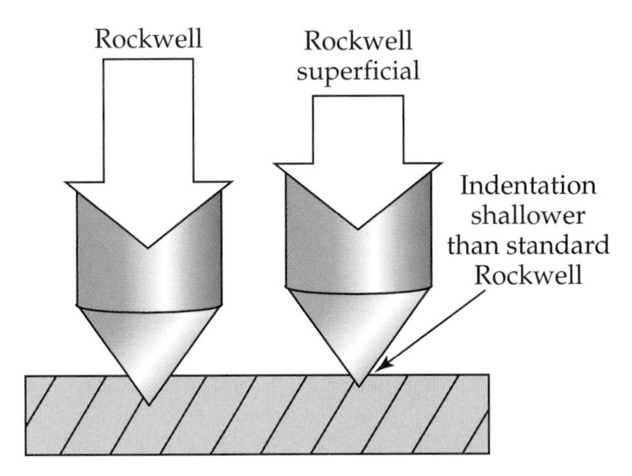

Figure 3-28. *The penetrator in a Rockwell superficial hardness tester does not penetrate as deeply as the penetrator in a Rockwell hardness tester.*

Shore Scleroscope Hardness Testing Method

The *Shore scleroscope hardness testing method* is entirely different than any other method discussed thus far. The surface is not penetrated by a diamond-point or ball penetrator. A ball or hammer is dropped onto the sample, and the hardness is proportional to the bounce of the object. Two Shore scleroscope models are shown in Figure 3-31.

Shore scleroscope hardness testing procedure

The sample to be tested rests on an anvil. A small metal ball or hammer drops from a height of 10″. The hammer strikes the test sample and rebounds. This hardness testing method assumes that the higher the hammer rebounds, the harder the material. The hammer weighs 40 grains. This is less than 0.001 pounds, which is less than 3 grams.

The height of the first bounce is monitored. On most Shore scleroscopes, the scale has a follower that follows the first bounce. The reading can be read directly off a dial, with little human reading error.

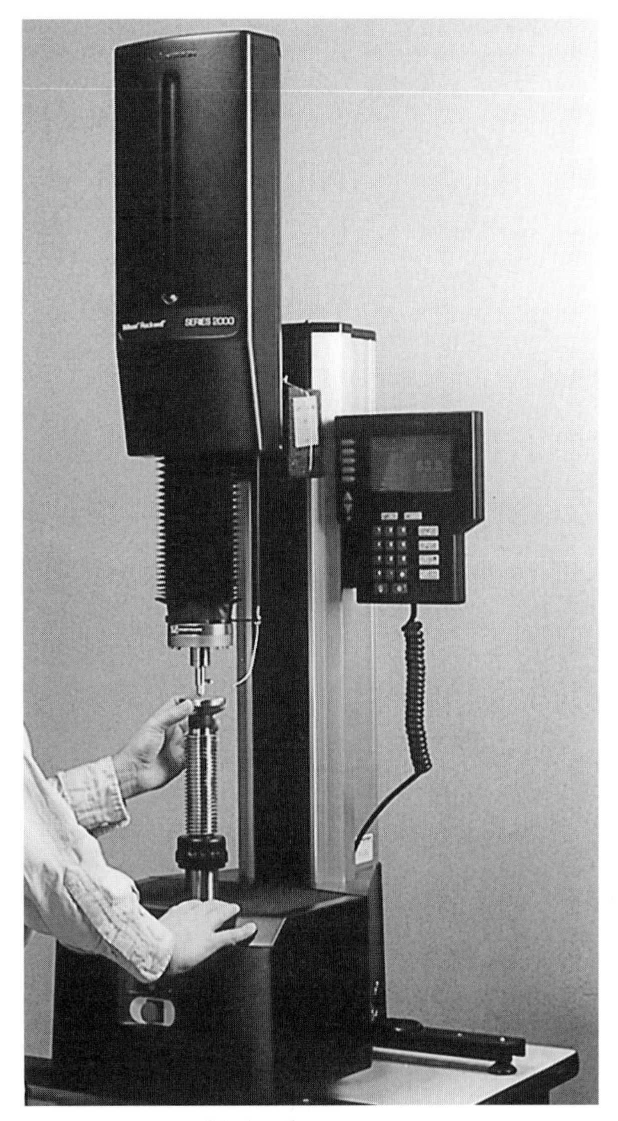

Figure 3-29. *This hardness tester can measure either Rockwell or Rockwell superficial hardness values. (Instron Corporation)*

This height is converted into a hardness value in units of *Shore*. For example, if the hammer bounces 6 1/4″ after it strikes the sample, the hardness is 100 units Shore. If the hammer bounces 3 1/8″ high, the hardness value is 50 units Shore. If the hammer strikes the sample and does not bounce, the hardness of the material is 0 units Shore, Figure 3-32.

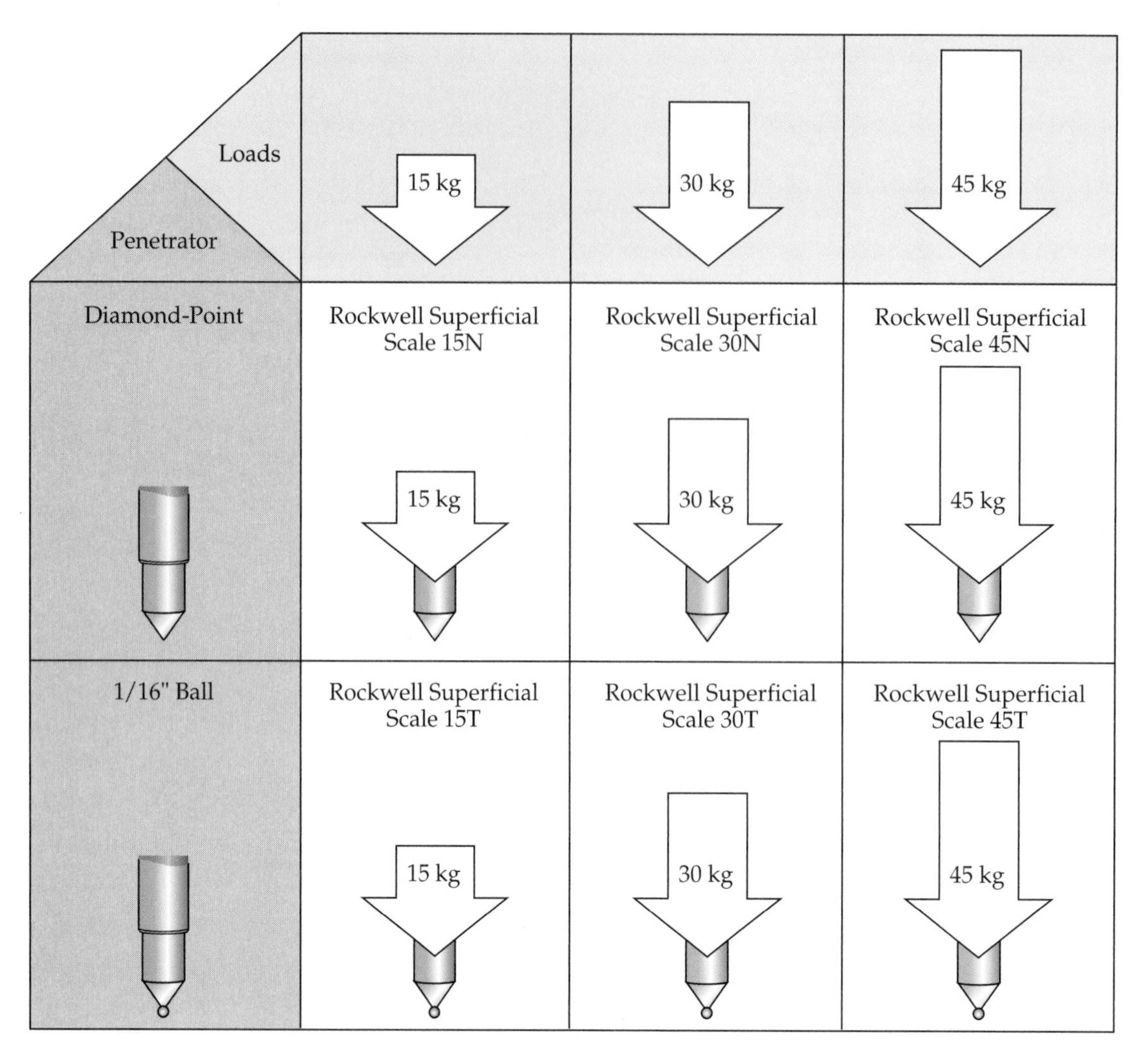

Figure 3-30. *This table relates the six Rockwell superficial scales to their penetrators and loads.*

This method may sound very unorthodox. However, the correlation between Shore values and Rockwell and Brinell values is very close.

An imperfection in the surface noticeably affects the movement of the hammer. Therefore, it is important to have a surface that is smooth, flat, clean, and horizontal when using the Shore scleroscope method. This procedure requires better surface quality than most other hardness testing methods.

Advantages of Shore scleroscope hardness testing

The following are some of the advantages of the Shore scleroscope hardness testing method:

- The impression made is negligible, because nothing actually penetrates the surface.
- The machine is small and portable; it can be carried around a factory. A conventional Rockwell or Brinell tester is too large to conveniently maneuver down the aisles of a manufacturing plant.

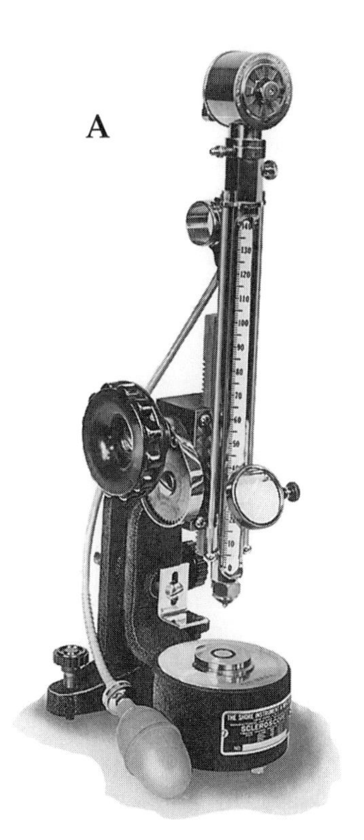

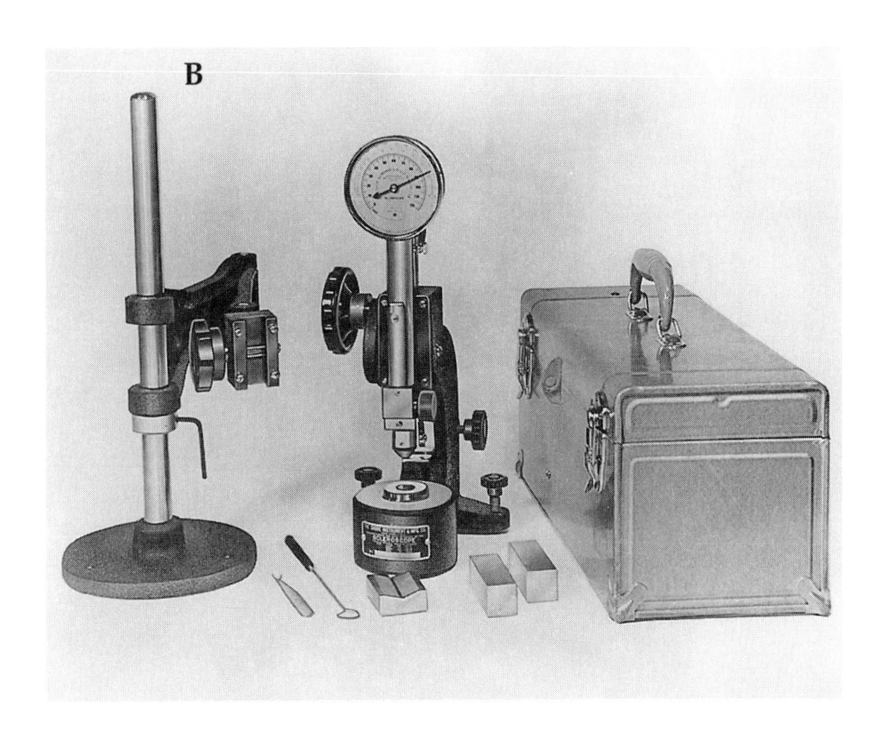

Figure 3-31. *The Shore scleroscope tests hardness by measuring the height of hammer bounce. A—Shore scleroscope. B—Shore scleroscope, clamping stand, and carrying case. One advantage of this tester is its portability. (The Shore Instrument & Manufacturing Company, Inc.)*

Sonodur Hardness Testing Method

The *Sonodur hardness testing method* is very different than the other hardness testing methods. This test method is based on the way hardness affects the natural resonant frequency of a piece of metal.

Sonodur hardness testing procedure

In making the Sonodur hardness test, a diamond-tipped magnetostrictive rod (0.75 mm in diameter) is pressed against the sample to be tested. An electrical coil is used to vibrate the rod, and the frequency at which it vibrates the greatest is determined in the electronic portion of the machine. This frequency is known as the *resonant frequency.* The harder the material, the higher the resonant frequency.

The resonant frequency value is then converted to a hardness number. The Sonodur machine reads in BHN units.

Advantages of Sonodur hardness testing

The Sonodur machine is small and portable. It gives a very quick response and does not damage the specimen. The Sonodur hardness testing method is newer than most other methods, and is considered very accurate.

Mohs Scale Hardness Testing Method

The *Mohs scale* was probably the first hardness testing method invented. In ancient times, philosophers discovered the need to devise a hardness scale. They selected ten

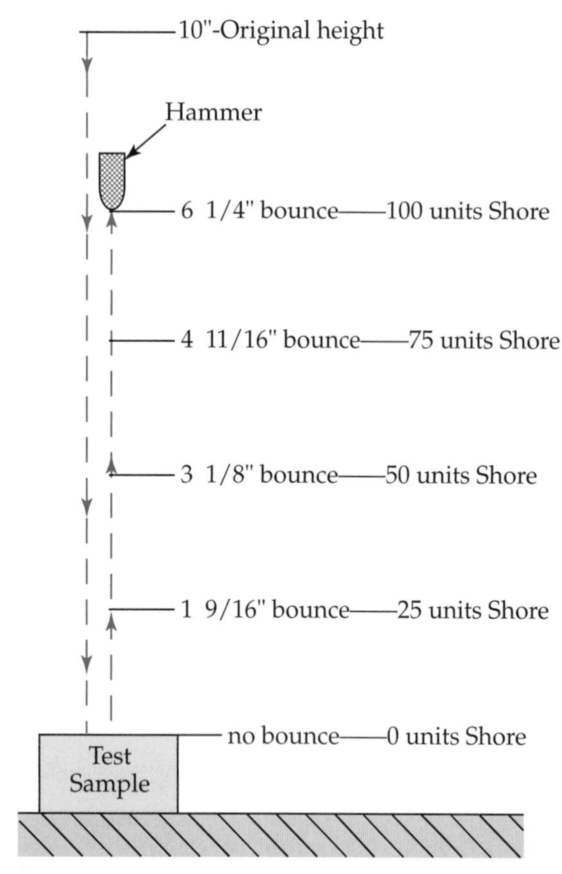

Figure 3-32. *This chart relates the height of the hammer bounce to units of Shore.*

stones of varying hardness. The softest stone was given a hardness value of 1. The hardest stone was given a hardness value of 10. The other eight stones were assigned the values of 2 through 9, based on increasing hardness. These ten stones are listed in the table in Figure 3-33.

To use the Mohs scale for hardness testing, the material sample is struck with one of the stones. If the sample is scratched, it is softer than the stone. Therefore, a softer stone is used to try to scratch the sample. This procedure continues until a stone that does not scratch the sample is found. For example, suppose topaz, quartz, and orthoclase scratch your sample but apatite and fluorspar do not. Then, the hardness of your sample is between

Mohs Scale	
Value	**Stone**
1	Talc
2	Gypsum
3	Calcite
4	Fluorspar
5	Apatite
6	Orthoclase
7	Quartz
8	Topaz
9	Sapphire
10	Diamond

Figure 3-33. *These ten stones make up the Mohs scale.*

5 and 6 units Mohs. This certainly is not an accurate method, but it was useful at the time it was developed. Today, this method is no longer applied, except for a minimal use in the field of geology.

File Hardness Testing Method

The *file hardness testing method* is fast, simple, and convenient—but inaccurate. However, it does provide a quick measurement, so it is used extensively in industry today. See Figure 3-34.

In the file hardness test, the inspector takes a file in one hand and the test material in the other. An attempt is made to scratch the test material by scraping it once with the edge of the file. If the material does not scratch, it is said to be *file hard.* If it does scratch, it is said to be *not file hard.*

This may sound very crude and useless as a hardness testing method. However, in industry, specific numerical hardness values are not always important. Often, an inspector merely wants to check if some parts have received a heat treatment or not. With one

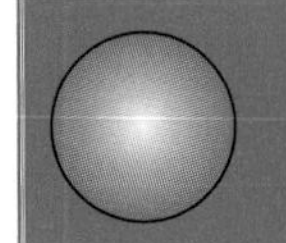

Self-Demonstration
File Hardness Testing

Obtain small pieces of aluminum, brass, plastic, gray cast iron, and several different types of steel, including 1018, 1045, 1095, W1 tool steel, and 302 stainless steel. Other similar materials may be substituted if these are not available. Many of these same materials will be used for other Self-Demonstrations later in this text.

Obtain a "sharp" file. Strike the edge of each sample with the edge of the file, using a smooth downward stroke. It is important that each sample receives the same strike intensity from your file. Try this several times on each sample, using different degrees of power.

Note the depth of the indenture on each sample due to the file strike. Is there a large variation in depth? The softer materials, like aluminum and plastic, should show a much deeper penetration.

Heat all of the steel samples to 1700°F. Let them soak at that temperature for at least half an hour, and then plunge them into a bucket of cold water. Perform file hardness tests on these samples again. Are they much more resistant to the file stroke? They should be! Does the high-carbon steel sample now have more hardness resistance than the low-carbon steel sample? It should!

File hardness testing is a very quick test and is therefore used extensively in industry, even though its results are not as accurate as those gained by using a Rockwell hardness tester. Some inspectors become very skilled at file hardness testing after doing it for many years. Did you notice that your technique improved as you tested more and more samples?

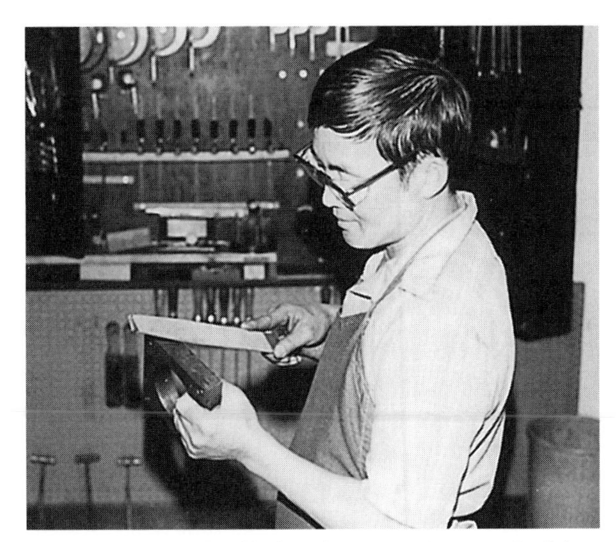

Figure 3-34. *The file hardness testing method is used extensively in industry because it gives quick results.*

quick stroke of a file, they have their answer.

This test method is not intended to give an accurate hardness value. However, many inspectors who have performed the file hardness test for years are very skilled at it and may argue that they can tell the numerical hardness value within a few points. One is inclined to believe these experienced people.

The file hardness test is dependent on the sensitivity of the hand of the inspector and on the sharpness of the file, so most companies rely on only one or two experienced people to run the test. Then, the results can be consistent and considered dependable. It is also important to use a relatively new file so that the hardness of the file does not cause a variation in the test.

Method	Basis	Penetrator	Load	Reading Symbol
Rockwell	Depth of Penetration	Diamond Point 1/16″ Ball or 1/8″ Ball	60-100-150 kg	R_C etc
Rockwell Superficial	Depth of Penetration	Diamond Point 1/16″ Ball or 1/8 Ball″	15-30-45 kg	15N 30T etc
Brinell	Area of Penetration	10mm Ball	500-3000 kg	BHN
File	Appearance of Scratch	File	Manual	None
Shore Scleroscope	Height of Bounce	40 Grain Weight	Gravity	Units Shore
Vickers	Area of Penetration	Pyramidal Diamond	5 to 120 kg	DPH
Knoop	Area of Penetration	Pyramidal Diamond	25-3600 grams	Units Knoop
Sonodur	Frequency of Vibration	Magnetostrictive Rod	N.A.	BHN
Mohs Scale	Appearance of Scratch	10 Stones	Manual	Units Mohs

Figure 3-35. *This chart compares various hardness testing methods.*

Comparing Hardness Testing Methods

All nine of the hardness testing methods are summarized in Figure 3-35. This chart will give you a quick check of the relative advantages and disadvantages of each technique.

Conversion Scales

Hardness can be measured in many different units, using different machines and scales. Therefore, there must be a simple way to convert one hardness scale value to another scale value. If company A has a Rockwell tester and company B has a Brinell tester and they are working together in manufacturing the same parts, they must be able to communicate with each other.

The answer lies in the use of *conversion charts* such as those shown in Figure 3-36 and Figure 3-37. With charts of this kind, hardness values can be converted to other scales that measure in the same hardness range. For example, Figure 3-38 shows that a 52 Rockwell C scale reading is equivalent to a 77 Rockwell A scale reading. If a blueprint calls for a hardness of 235 BHN (Brinell) and you have only a Rockwell testing machine, the material should have a 99 Rockwell B scale value, or a 22 Rockwell C scale value. See Figure 3-39. Also, a hardness of 321 BHN is

ROCKWELL				SUPERFICIAL ROCKWELL			BRINELL		Vickers of Firth Diamond Hardness Number	Sclero-scope	Tensile Strength
Diamond Brale			1 16" Ball	"N" Brale Penetrator			10 mm Ball 3000 kgm Load				
150 kgm C Scale	60 kgm A Scale	100 kgm D Scale	100 kgm B Scale	15 kg Load 15N	30 kg Load 30N	45 kg Load 45N	Diam. of Ball Im-pression in mm	Hardness Number			Equivalent 1000 Lb. Sq. In.
68	86	...	...	93	84	75	...	...	940	97	...
68	85	...	...	93	84	75	...	...	920	96	...
67	85	...	...	93	84	74	...	...	900	95	...
66	85	...	...	93	83	74	...	...	880	93	...
66	84	75	...	92	82	72	2.25	745	832	91	...
64	83	74	...	92	81	71	2.30	710	800	88	...
63	83	73	...	91	80	70	2.30	710	772	87	...
62	82	72	...	91	79	69	2.35	682	746	85	...
61	82	72	...	91	78	68	2.35	682	720	83	...
60	81	71	...	90	78	67	2.40	653	697	82	...
59	81	70	...	90	77	66	2.45	627	674	80	326
58	80	69	...	89	76	64	2.55	578	653	78	315
57	80	69	...	89	75	63	2.55	578	633	77	304
56	79	68	...	88	74	62	2.60	555	613	75	294
55	79	67	...	88	73	61	2.60	555	595	74	287
54	78	66	...	87	72	60	2.65	534	577	72	279
53	77	65	...	87	71	59	2.70	514	560	71	269
52	77	65	...	86	70	57	2.75	495	544	69	261
51	76	64	...	86	69	56	2.75	495	528	68	254
50	76	63	...	86	69	55	2.80	477	513	67	245
49	75	62	...	85	68	54	2.85	461	498	65	238
48	75	61	...	85	67	53	2.90	444	484	64	232
47	74	61	...	84	66	51	2.90	444	471	63	225
46	73	60	...	84	65	50	2.95	432	458	62	219
45	73	59	...	83	64	49	3.00	415	446	61	211
44	73	59	...	83	63	48	3.00	415	434	59	206
43	72	58	...	82	62	47	3.05	401	423	58	202
42	72	57	...	82	61	46	3.10	388	412	56	198
41	71	56	...	81	60	44	3.10	388	402	55	191
40	70	55	...	80	60	43	3.15	375	392	54	185
39	70	55	...	80	59	42	3.20	363	382	53	181
38	69	54	...	79	58	41	3.25	352	372	51	176
37	69	53	...	79	57	40	3.30	341	363	50	171
36	68	52	109*	78	56	39	3.35	331	354	49	168
35	68	52	109*	78	55	37	3.35	331	345	48	163
34	67	51	108*	77	54	36	3.40	321	336	46	159
33	67	50	108*	77	53	35	3.45	311	327	45	154
32	66	49	107*	76	52	34	3.50	302	318	44	150
31	66	48	106*	76	51	33	3.55	293	310	43	146
30	65	48	106*	75	50	32	3.60	285	302	42	142
29	65	47	105*	75	50	30	3.65	277	294	41	138
28	64	46	104*	74	49	29	3.70	269	286	40	134
27	64	45	103*	73	48	28	3.75	262	279	39	131
26	63	45	103*	73	47	27	3.80	255	272	38	126
25	63	44	102	72	46	26	3.80	255	266	37	124
24	62	43	101	72	45	24	3.85	248	260	37	122
23	62	42	100	71	44	23	3.90	241	254	36	118
22	62	42	99	71	43	22	3.95	235	248	35	116
21	61	41	99	70	42	21	4.00	229	243	35	113
20	61	40	98	69	42	20	4.05	223	238	34	111
18	...	...	97	...	...	...	4.10	217	230	33	107
16*	...	...	95	...	...	...	4.15	212	222	32	102
14*	...	...	94	...	...	...	4.25	203	213	31	98
12*	...	...	92	...	...	...	4.35	192	204	29	92
10*	...	...	90	...	...	...	4.40	187	195	28	90
8*	...	...	89	...	...	...	4.50	179	187	27	87
6*	...	...	87	...	...	...	4.60	170	180	26	83
4*	...	...	85	...	...	...	4.65	166	173	25	79
2*	...	...	84	...	...	...	4.80	156	166	24	77
0*	...	...	82	...	...	...	4.80	156	160	24	74
...	...	...	81	...	...	...	4.90	149	156	23	73
...	...	...	79	...	...	...	5.00	143	150	22	70
...	...	...	77	...	...	...	5.10	137	143	21	67
...	...	...	74	...	...	...	5.20	131	137	21	65
...	...	...	72	...	...	...	5.30	126	132	20	62
...	...	...	70	...	...	...	5.40	121	127	19	60
...	...	...	68	...	...	...	5.50	116	122	18	58
...	...	...	65	...	...	...	5.60	112	117	15	56

*Numbers are Rockwell "B" or "C" values not ordinarily determined.

Figure 3-36. *A hardness conversion chart permits hardness values to be converted to other scales. (Teledyne Vasco)*

Hardened steel and hard alloys

C 150 Kg. "Brale" Rockwell Hardness Tester	A 60 Kg. "Brale" Rockwell Hardness Tester	D 100 Kg. "Brale" Rockwell Hardness Tester	15-N 15 Kg. N "Brale" Rockwell Superficial Hardness Tester	30-N 30 Kg. N "Brale" Rockwell Superficial Hardness Tester	45-N 45 Kg. N "Brale" Rockwell Superficial Hardness Tester	Diamond Pyramid Hardness 10 Kg. (Square Base Diamond Pyramid - 136° Apex Angle)	Knoop Hardness 500 Gr. & Over (Knoop Elongated Diamond)	Brinell Hardness 3000 Kg. (Brinell Hultgren 10mm. Ball)	G 150 Kg. 1/16" Ball Rockwell Hardness Tester	Tensile Strength Approx. Only (Thousand lbs. per square inch)
80	92.0	86.5	96.5	92.0	87.0	1865	—	—	—	INEXACT AND ONLY FOR STEEL
79	91.5	85.5	—	91.5	86.5	1787	—	—	—	
78	91.0	84.5	96.0	91.0	85.5	1710	—	—	—	
77	90.5	84.0	—	90.5	84.5	1633	—	—	—	
76	90.0	83.0	95.5	90.0	83.5	1556	—	—	—	
75	89.5	82.5	—	89.0	82.5	1478	—	—	—	
74	89.0	81.5	95.0	88.5	81.5	1400	—	—	—	
73	88.5	81.0	—	88.0	80.5	1323	—	—	—	
72	88.0	80.0	94.5	87.0	79.5	1245	—	—	—	
71	87.0	79.5	—	86.5	78.5	1160	—	—	—	
70	86.5	78.5	94.0	86.0	77.5	1076	972	—	—	
69	86.0	78.0	93.5	85.0	76.5	1004	946	—	—	
68	85.5	77.0	—	84.5	75.5	942	920	—	—	
67	85.0	76.0	93.0	83.5	74.5	894	895	—	—	
66	84.5	75.5	92.5	83.0	73.0	854	870	—	—	
65	84.0	74.5	92.0	82.0	72.0	820	846	—	—	
64	83.5	74.0	—	81.0	71.0	789	822	—	—	
63	83.0	73.0	91.5	80.0	70.0	763	799	—	—	
62	82.5	72.5	91.0	79.0	69.0	739	776	—	—	—
61	81.5	71.5	90.5	78.5	67.5	716	754	—	—	—
60	81.0	71.0	90.0	77.5	66.5	695	732	614	—	—
59	80.5	70.0	89.5	76.5	65.5	675	710	600	—	—
58	80.0	69.0	—	75.5	64.0	655	690	587	—	—
57	79.5	68.5	89.0	75.0	63.0	636	670	573	—	—
56	79.0	67.5	88.5	74.0	62.0	617	650	560	—	—
55	78.5	67.0	88.0	73.0	61.0	598	630	547	—	301
54	78.0	66.0	87.5	72.0	59.5	580	612	534	—	291
53	77.5	65.5	87.0	71.0	58.5	562	594	522	—	282
52	77.0	64.5	86.5	70.5	57.5	545	576	509	—	273
51	76.5	64.0	86.0	69.5	56.0	528	558	496	—	264
50	76.0	63.0	85.5	68.5	55.0	513	542	484	—	255
49	75.5	62.0	85.0	67.5	54.0	498	526	472	—	246
48	74.5	61.5	84.5	66.5	52.5	485	510	460	—	237
47	74.0	60.5	84.0	66.0	51.5	471	495	448	—	229
46	73.5	60.0	83.5	65.0	50.0	458	480	437	—	221
45	73.0	59.0	83.0	64.0	49.0	446	466	426	—	214
44	72.5	58.5	82.5	63.0	48.0	435	452	415	—	207
43	72.0	57.5	82.0	62.0	46.5	424	438	404	—	200
42	71.5	57.0	81.5	61.5	45.5	413	426	393	—	194
41	71.0	56.0	81.0	60.5	44.5	403	414	382	—	188
40	70.5	55.5	80.5	59.5	43.0	393	402	372	—	182
39	70.0	54.5	80.0	58.5	42.0	383	391	362	—	177
38	69.5	54.0	79.5	57.5	41.0	373	380	352	—	171
37	69.0	53.0	79.0	56.5	39.5	363	370	342	—	166
36	68.5	52.5	78.5	56.0	38.5	353	360	332	—	162
35	68.0	51.5	78.0	55.0	37.0	343	351	322	—	157
34	67.5	50.5	77.0	54.0	36.0	334	342	313	—	153
33	67.0	50.0	76.5	53.0	35.0	325	334	305	—	148
32	66.5	49.0	76.0	52.0	33.5	317	326	297	—	144
31	66.0	48.5	75.5	51.5	32.5	309	318	290	—	140
30	65.5	47.5	75.0	50.5	31.5	301	311	283	92.0	136
29	65.0	47.0	74.5	49.5	30.0	293	304	276	91.0	132
28	64.5	46.0	74.0	48.5	29.0	285	297	270	90.0	129
27	64.0	45.5	73.5	47.5	28.0	278	290	265	89.0	126
26	63.5	44.5	72.5	47.0	26.5	271	284	260	88.0	123
25	63.0	44.0	72.0	46.0	25.5	264	278	255	87.0	120
24	62.5	43.0	71.5	45.0	24.0	257	272	250	86.0	117
23	62.0	42.5	71.0	44.0	23.0	251	266	245	84.5	115
22	61.5	41.5	70.5	43.0	22.0	246	261	240	83.5	112
21	61.0	41.0	70.0	42.5	20.5	241	256	235	82.5	110
20	60.5	40.0	69.5	41.5	19.5	236	251	230	81.0	108

Soft steel, grey and malleable cast iron and most non-ferrous metal.

B 100 Kg. 1/16" Ball Rockwell Hardness Tester	F 60 Kg. 1/16" Ball Rockwell Hardness Tester	G 150 Kg. 1/16" Ball Rockwell Hardness Tester	15-T 15 Kg. 1/16" Ball Rockwell Superficial Hardness Tester	30-T 30 Kg. 1/16" Ball Rockwell Superficial Hardness Tester	45-T 45 Kg. 1/16" Ball Rockwell Superficial Hardness Tester	E 100 Kg. 1/8" Ball Rockwell Hardness Tester	K 150 Kg. 1/8" Ball Rockwell Hardness Tester	A 60 Kg. "Brale" Rockwell Hardness Tester	Knoop Hardness 500 Gr. & Over (Knoop Elongated Diamond)	Brinell Hardness 500 Kg. 10mm. Ball (Standard Brinell)	Brinell Hardness 3000 Kg. D.P.H. 10 Kg. (Square Base Diamond Pyramid - 136° Apex Angle)	Tensile Strength Approx. Only (Thousand lbs. per square inch)
100	—	82.5	93.0	82.0	72.0	—	—	61.5	251	201	240	116
99	—	81.0	92.5	81.5	71.0	—	—	61.0	246	195	234	112
98	—	79.0	—	81.0	70.0	—	—	60.0	241	189	228	109
97	—	77.5	92.0	80.5	69.0	—	—	59.5	236	184	222	106
96	—	76.0	—	80.0	68.0	—	—	59.0	231	179	216	103
95	—	74.0	91.5	79.0	67.0	—	—	58.0	226	175	210	101
94	—	72.5	—	78.5	66.0	—	—	57.5	221	171	205	98
93	—	71.0	91.0	78.0	65.5	—	—	57.0	216	167	200	96
92	—	69.0	90.5	77.5	64.5	—	100	56.5	211	163	195	93
91	—	67.5	—	77.0	63.5	—	99.5	56.0	206	160	190	91
90	—	66.0	90.0	76.0	62.5	—	98.5	55.5	201	157	185	89
89	—	64.0	89.5	75.5	61.5	—	98.0	55.0	196	154	180	87
88	—	62.5	—	75.0	60.5	—	97.0	54.0	192	151	176	85
87	—	61.0	89.0	74.5	59.5	—	96.5	53.5	188	148	172	83
86	—	59.0	88.5	74.0	58.5	—	95.5	53.0	184	145	169	81
85	—	57.5	—	73.5	58.0	—	94.5	52.5	180	142	165	80
84	—	56.0	88.0	73.0	57.0	—	94.0	52.0	176	140	162	78
83	—	54.0	87.5	72.0	56.0	—	93.0	51.0	173	137	159	77
82	—	52.5	—	71.5	55.0	—	92.0	50.5	170	135	156	75
81	—	51.0	87.0	71.0	54.0	—	91.0	50.0	167	133	153	74
80	—	49.0	86.5	70.0	53.0	—	90.5	49.5	164	130	150	72
79	—	47.5	—	69.5	52.0	—	89.5	49.0	161	128	147	Even for steel, Tensile Strength relation to hardness is inexact, unless determined for specific material.
78	—	46.0	86.0	69.0	51.0	—	88.5	48.5	158	126	144	
77	—	44.0	85.5	68.0	50.0	—	88.0	48.0	155	124	141	
76	—	42.5	—	67.5	49.0	—	87.0	47.0	152	122	139	
75	99.5	41.0	85.0	67.0	48.5	—	86.0	46.5	150	120	137	
74	99.0	39.0	—	66.0	47.5	—	85.0	46.0	147	118	135	
73	98.5	37.5	84.5	65.5	46.5	—	84.5	45.5	145	116	132	
72	98.0	36.0	84.0	65.0	45.5	—	83.5	45.0	143	114	130	
71	97.5	34.5	—	64.0	44.5	100	82.5	44.5	141	112	127	
70	97.0	32.5	83.5	63.5	43.5	99.5	81.5	44.0	139	110	125	
69	96.0	31.0	83.0	62.5	42.5	99.0	81.0	43.5	137	109	123	
68	95.5	29.5	—	62.0	41.5	98.0	80.0	43.0	135	107	121	
67	95.0	28.0	82.5	61.5	40.5	97.5	79.0	42.5	133	106	119	
66	94.5	26.5	82.0	60.5	39.5	97.0	78.0	42.0	131	104	117	
65	94.0	25.0	—	60.0	38.5	96.0	77.5	—	129	102	116	
64	93.5	23.5	81.5	59.5	37.5	95.5	76.5	41.5	127	101	114	
63	93.0	22.0	81.0	58.5	36.5	95.0	75.5	41.0	125	99	112	
62	92.0	20.5	—	58.0	35.5	94.5	74.5	40.5	124	98	110	
61	91.5	19.0	80.5	57.0	34.5	93.5	74.0	40.0	122	96	108	
60	91.0	17.5	—	56.5	33.5	93.0	73.0	39.5	120	95	107	
59	90.5	16.0	80.0	56.0	32.0	92.5	72.0	39.0	118	94	106	
58	90.0	14.5	79.5	55.0	31.0	92.0	71.0	38.5	117	92	104	
57	89.5	13.0	—	54.5	30.0	91.0	70.5	38.0	115	91	103	
56	89.0	11.5	79.0	54.0	29.0	90.5	69.5	—	114	90	101	
55	88.0	10.0	78.5	53.0	28.0	90.0	68.5	37.5	112	89	100	
54	87.5	8.5	—	52.5	27.0	89.5	68.0	37.0	111	87	—	
53	87.0	7.0	78.0	51.5	26.0	89.0	67.0	36.5	110	86	—	
52	86.5	5.5	77.5	51.0	25.0	88.0	66.0	36.0	109	85	—	
51	86.0	4.0	—	50.5	24.0	87.5	65.0	35.5	108	84	—	
50	85.5	2.5	77.0	49.5	23.0	87.0	64.5	35.0	107	83	—	

B 100 Kg. 1/16" Ball Rockwell Hardness Tester	F 60 Kg. 1/16" Ball Rockwell Hardness Tester	15-T 15 Kg. 1/16" Ball Rockwell Superficial Hardness Tester	30-T 30 Kg. 1/16" Ball Rockwell Superficial Hardness Tester	45-T 45 Kg. 1/16" Ball Rockwell Superficial Hardness Tester	E 100 Kg. 1/8" Ball Rockwell Hardness Tester	H 60 Kg. 1/8" Ball Rockwell Hardness Tester	K 150 Kg. 1/8" Ball Rockwell Hardness Tester	A 60 Kg. "Brale" Rockwell Hardness Tester	Knoop Hardness 500 Gr. & Over (Knoop Elongated Diamond)	Brinell Hardness 500 Kg. 10mm. Ball (Standard Brinell)
50	85.5	77.0	49.5	23.0	87.0	—	64.5	35.0	107	83
49	85.0	76.5	49.0	22.0	86.5	—	63.5	—	106	82
48	84.5	—	48.5	20.5	85.5	—	62.5	34.5	105	81
47	84.0	76.0	47.5	19.5	85.0	—	61.5	34.0	104	80
46	83.0	75.5	47.0	18.5	84.5	—	61.0	33.5	103	—
45	82.5	—	46.0	17.5	84.0	—	60.0	33.0	102	79
44	82.0	75.0	45.5	16.5	83.5	—	59.0	32.5	101	78
43	81.5	74.5	45.0	15.5	82.5	—	58.0	32.0	100	77
42	81.0	—	44.0	14.5	82.0	—	57.5	31.5	99	76
41	80.5	74.0	43.5	13.5	81.5	—	56.5	31.0	98	75
40	79.5	73.5	43.0	12.5	81.0	—	55.5	—	97	—
39	79.0	—	42.0	11.0	80.0	—	54.5	30.5	96	74
38	78.5	73.0	41.5	10.0	79.5	—	54.0	30.0	95	73
37	78.0	72.5	40.5	9.0	79.0	—	53.0	29.5	94	72
36	77.5	—	40.0	8.0	78.5	100	52.0	29.0	93	—
35	77.0	72.0	39.5	7.0	78.0	99.5	51.5	28.5	92	71
34	76.5	71.5	38.5	6.0	77.0	99.0	50.5	28.0	91	70
33	75.5	—	38.0	5.0	76.5	—	49.5	—	90	69
32	75.0	71.0	37.5	4.0	76.0	98.5	48.5	27.5	89	—
31	74.5	—	36.5	3.0	75.5	98.0	48.0	27.0	88	68
30	74.0	70.5	36.0	2.0	75.0	—	47.0	26.5	—	67
29	73.5	70.0	35.5	1.0	74.0	97.5	46.0	26.0	—	—
28	73.0	—	34.5	—	73.5	97.0	45.0	25.5	—	66
27	72.5	69.5	34.0	—	73.0	96.5	44.5	25.0	85	—
26	72.0	69.0	33.0	—	72.5	—	43.5	24.5	—	65
25	71.0	—	32.5	—	72.0	96.0	42.5	—	—	64
24	70.5	68.5	32.0	—	71.0	95.5	41.5	24.0	—	—
23	70.0	68.0	31.0	—	70.5	—	41.0	23.5	82	63
22	69.5	—	30.5	—	70.0	95.0	40.0	23.0	—	—
21	69.0	67.5	29.5	—	69.5	94.5	39.0	22.5	—	62
20	68.5	—	29.0	—	68.5	—	38.0	22.0	—	—
19	68.0	67.0	28.5	—	68.0	94.0	37.5	21.5	79	61
18	67.0	66.5	27.5	—	67.5	93.5	36.5	—	—	—
17	66.5	—	27.0	—	67.0	93.0	35.5	21.0	—	60
16	66.0	66.0	26.0	—	66.5	—	35.0	20.5	—	—
15	65.5	65.5	25.5	—	65.5	92.5	34.0	20.0	76	59
14	65.0	—	25.0	—	65.0	92.0	33.0	—	—	—
13	64.5	65.0	24.0	—	64.5	—	32.0	—	—	58
12	64.0	64.5	23.5	—	64.0	91.5	31.5	—	—	—
11	63.5	—	23.0	—	63.5	91.0	30.5	—	73	—
10	63.0	64.0	22.0	—	62.5	90.5	29.5	—	—	57
9	62.0	—	21.5	—	62.0	—	29.0	—	—	—
8	61.5	63.5	20.5	—	61.5	90.0	28.0	—	71	—
7	61.0	63.0	20.0	—	61.0	89.5	27.0	—	—	56
6	60.5	—	19.5	—	60.5	—	26.0	—	—	—
5	60.0	62.5	18.5	—	60.0	89.0	25.5	—	69	55
4	59.5	62.0	18.0	—	59.0	88.5	24.5	—	—	—
3	59.0	—	17.0	—	58.5	88.0	23.5	—	—	—
2	58.0	61.5	16.5	—	58.0	—	23.0	—	68	54
1	57.5	61.0	16.0	—	57.5	87.5	22.0	—	—	—
0	57.0	—	15.0	—	57.0	87.0	21.0	—	67	53

Although conversion tables dealing with hardness can only be approximate and never mathematically exact, it is of considerable value to be able to compare different hardness scales.
This table is based on the assumption that the metal tested is homogeneous to a depth several times as great as the depth of the indentation.
In metal not homogeneous, different loads and different shapes of penetrators would penetrate, or at least meet the resistance of, metal of varying hardness, depending upon the depth of the indentation. Therefore, no recorded hardness value could be confirmed by another person unless shape of penetrator and actual load applied are both specified.

The indentation hardness values measured on the various scales depend on the work hardening behavior of the material during the test and this in turn depends on the degree of previous cold working of the material. The B-scale relationships in the table are based largely on annealed metals for the low values and cold worked metals for the higher values. Therefore, annealed metals of high B-scale hardness such as austenitic stainless steels, nickel and high nickel alloys do not conform closely to these general tables. Neither do cold-worked metals of low B-scale hardness such as aluminum and the softer alloys. Special correlations are needed for more exact relationships in these cases.

Figure 3-37. *This hardness conversion chart compares different hardness scales. (Instron Corporation)*

ROCKWELL				SUPERFICIAL ROCKWELL		
Diamond Brale			1 16" Ball	"N" Brale Penetrator		
150 kgm C Scale	60 kgm A Scale	100 kgm D Scale	100 kgm B Scale	15 kg Load 15N	30 kg Load 30N	45 kg Load 45N
68	86	...	...	93	84	75
68	85	...	...	93	84	75
67	85	...	...	93	84	74
66	85	...	...	93	83	74
66	84	75	...	92	82	72
64	83	74	...	92	81	71
63	83	73	...	91	80	70
62	82	72	...	91	79	69
61	82	72	...	91	78	68
60	81	71	...	90	78	67
59	81	70	...	90	77	66
58	80	69	...	89	76	64
57	80	69	...	89	75	63
56	79	68	...	88	74	62
55	79	67	...	88	73	61
54	78	66	...	87	72	60
53	77	65	...	87	71	59
52	77	65	...	86	70	57
51	76	64	...	86	69	56
50	76	63	...	86	69	55
49	75	62	...	85	68	54
48	75	61	...	85	67	53

Figure 3-38. *Converting from Rockwell C scale to Rockwell A scale. (Teledyne Vasco)*

the same hardness as 46 units Shore or 36-45N on the Rockwell superficial tester. See Figure 3-40.

The larger the numerical value of hardness, the harder the material. This is true of every hardness scale. Therefore, as we move toward the top of the charts, the hardness values represent harder material. If we move toward the bottom of the charts, the material is softer. For example, a hardness value of 28 Rockwell C scale would be slightly higher than 35 units Shore. A hardness of 432 BHN would be slightly higher than 434 DPH. Which would be harder, a value of 30 Rockwell C scale, 302 BHN, or 49 Shore? Many interesting problems can be presented to give you practice at reading these charts.

ROCKWELL				SUPERFICIAL ROCKWELL			BRINELL		Vickers of Firth Diamond Hardness Number
Diamond Brale			1 16" Ball	"N" Brale Penetrator			10 m/m Ball 3000 kgm Load		
150 kgm C Scale	60 kgm A Scale	100 kgm D Scale	100 kgm B Scale	15 kg Load 15N	30 kg Load 30N	45 kg Load 45N	Diam. of Ball Impression in mm	Hardness Number	
68	86	...	...	93	84	75	...	...	940
68	85	...	...	93	84	75	...	...	920
67	85	...	...	93	84	74	...	...	900
66	85	...	...	93	83	74	...	...	880
66	84	75	...	92	8[illegible]	72	2.25	745	832
27	64	45	103*	73	48	28	3.75	262	279
26	63	45	103*	73	47	27	3.80	255	272
25	63	44	102	72	46	26	3.80	255	266
24	62	43	101	72	45	24	3.85	248	260
23	62	42	100	71	44	23	3.90	241	254
22	62	42	99	71	43	22	3.95	235	248
21	61	41	99	70	42	21	4.00	229	243
20	61	40	98	69	42	20	4.05	223	238
18	...	...	97	...	...	...	4.10	217	230
16*	...	...	95	...	...	...	4.15	212	222
14*	...	...	94	...	...	...	4.25	203	213
12*	...	...	92	...	...	...	4.35	192	204

Figure 3-39. *Converting from BHN scale to Rockwell B and Rockwell C scale. (Teledyne Vasco)*

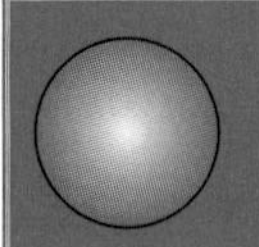

Self-Demonstration
Correlation of Hardness Scales

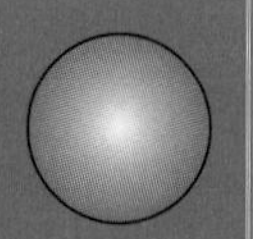

Obtain four small pieces of 1045 steel; 1095 or 1018 steel may be substituted if 1045 is not available. This material will be used for other Self-Demonstrations later in this text.

Test their hardness on the Rockwell C scale, using the diamond brale and a 150 kg load. Record these values. Next test the hardness of these same samples on the Rockwell A scale using the diamond brale and a 60 kg load. Record these values.

Test their hardness on the Rockwell D scale and the Rockwell B scale and record these values. Test their hardness on any other hardness tester that you have available to you. Do you have a Brinell hardness tester, a Shore scleroscope, a Rockwell superficial tester, or a microhardness tester? Record all these values also.

Now, use the hardness conversion charts in Figure 3-36 and Figure 3-37 to compare the hardness results.

Convert all the values to one of the Rockwell scales, such as the Rockwell C scale or the Rockwell A scale, and compare the values. How much do they vary?

It is expected that these hardness conversion values may deviate by a few hardness points. Do the hardness values of any of the materials consistently vary by more than the others?

SUPERFICIAL ROCKWELL "N" Brale Penetrator			BRINELL 10 mm Ball 3000 kgm Load		Vickers of Firth Diamond Hardness Number	Scleroscope	Tensile Strength Equivalent 1000 Lb. Sq. In.
15 kg Load 15N	30 kg Load 30N	45 kg Load 45N	Diam. of Ball Impression in mm	Hardness Number			
93	84	75	...	...	940	97	...
93	84	75	...	...	920	96	...
93	84	74	...	...	900	95	...
93	83	74	...	..	880	93	...
79	57	40	3.30	34[illegible]	363	50	171
78	56	39	3.35	331	354	49	168
78	55	37	3.35	331	345	48	163
77	54	36	3.40	321	336	46	159
77	53	35	3.45	311	327	45	154
76	52	34	3.50	302	318	44	150
76	51	33	3.55	293	310	43	146
75	50	32	3.60	285	302	42	142
75	50	30	3.65	277	294	41	138
74	49	29	3.70	269	286	40	134
73	48	28	3.75	262	279	39	131
73	47	27	3.80	255	272	38	126
72	46	26	3.80	[illegible]55	266	37	124

Figure 3-40. *Converting from BHN scale to Shore and 45N Rockwell superficial. (Teledyne Vasco)*

Test Your Knowledge

Write your answers on a separate sheet of paper. Do not write in this book.

1. In which hardness testing method is the hardness based on the diameter of an indenture?
2. In which hardness testing method does the impression look like a diamond, with one axis seven times as long than the other?
3. What hardness testing method is the most widely used of all methods?
4. In which hardness testing method is the hardness value dependent on the height of the bounce?
5. In which two hardness testing methods is the hardness dependent on the depth of penetration rather than the width of penetration?
6. In which hardness testing method is a diamond-tipped magnetostrictive rod used?
7. In which hardness testing methods is a minor load applied first to get through the outer surface of the metal before making the hardness test?
8. In most hardness testing methods, the surface should be as horizontal and as flat and smooth as possible. In which hardness testing method is this the most critical?
9. In which type of hardness testing method is the hardness dependent on the width of a square, diamond-shaped impression?
10. In which hardness testing method is a very small impression made by a diamond penetrator using a load so small it is measured in grams instead of kilograms?
11. What hardness testing method employs either a C, A, E, G, or K scale?
12. List the hardness testing methods that measure hardness in units of BHN.
13. Microhardness tester is another name for which hardness testing methods?
14. What hardness testing method uses a 10 mm diameter ball and a 3000 kg force?
15. The 30T scale refers to which hardness testing method?
16. Name the hardness testing method that measures in units of DPH.
17. Name the hardness testing method that involves scratching a surface with ten stones.
18. Name a hardness testing method that uses either a diamond-point penetrator or a ball penetrator, with loads of 15 kg, 30 kg, or 45 kg.
19. Which hardness testing method is the fastest to use, but does not give accurate numerical results?
20. In the 45T scale, what does the 45 stand for?
21. What is the greatest hardness value, in Rockwell units, that steel can attain?
22. Which Rockwell hardness scale is most commonly used for hard materials?
23. In the Rockwell hardness test, what scales use the diamond-point penetrator?
24. Which of the following values represents the hardest material: 35 Rockwell C, 69 Rockwell A, or 44 Rockwell D?
25. Which of the following values represents the hardest material: 60 BHN, 53 Rockwell C, or 271 DPH?
26. Which of the following values represents the hardest material: 38 Shore, 25 Rockwell C, or 97 Rockwell B?
27. Which of the following values represents the hardest material: 389 DPH, 61 Shore, or 389 BHN?
28. Which of the following values represents the hardest material: 426 Knoop, 30 Rockwell C, or 270 BHN?
29. Your firm has decided to invest in another hardness tester. If you already have a Rockwell hardness tester, what hardness testers would you consider for recommendation if your company uses primarily low-carbon steels?
30. Your firm is building a new plant in another section of the country. The budget for the new plant allows for the purchase of three new hardness tesers. If your company

uses all types of metal (including steel, cast iron, aluminum, brass, and also some plastic materials), which hardness testers would you consider and recommend?

31. The budget for a new plant allows for the purchase of two new hardness testers. If the company manufactures only high-alloy steel products, what factors should enter into the decision of which hardness testers to buy? Which hardness testers would you recommend?
32. If your firm is planning to get involved with products that will require a thin, hard outer surface on many of its metal parts, which hardness tester would you recommend?

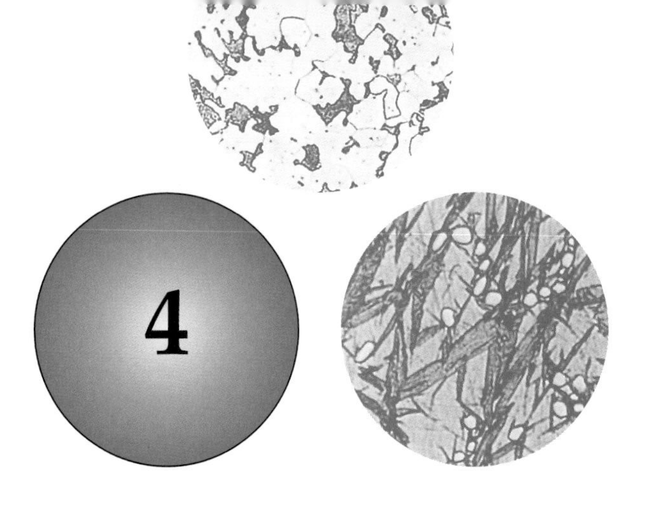

Material Properties

After studying this chapter, you will be able to:

- Explain the relationship between strength, hardness, and ductility.
- Calculate simple tensile, compressive, torsional, and flexural stresses.
- Compare the various types of stresses that materials must withstand.
- Define and calculate percent elongation and strain.
- Compare elastic and plastic deformation.
- Describe stress-strain diagrams for common materials.
- Explain modulus of elasticity.
- Determine lateral strain using Poisson's ratio.
- Describe types of corrosion.
- Explain thermal properties such as melting point, coefficient of thermal expansion, thermal conductivity, and specific heat.

Properties refer to the characteristics, abilities, special traits, strengths, advantages, disadvantages, and unusual features of a material. These material properties are used to compare different materials. Just as a human being may have the strength to lift heavy boxes or the ability to resist smoking cigarettes, similarly, a metal may exhibit strength to carry large loads or the ability to resist decay from corrosion.

Material properties can be divided into the following classes (Figure 4-1):

- Mechanical properties, such as strength, hardness, and modulus of elasticity.
- Chemical properties, such as corrosion resistance.
- Electrical properties, such as resistivity.
- Thermal properties, such as melting temperature.
- Other properties, such as density and wear.

Mechanical Properties

Hardness, ductility, and strength are the three properties discussed most often in metallurgy. These properties are related to one another. Generally, as strength and hardness increase, ductility decreases and the material becomes more brittle. As a material becomes more ductile, its strength and hardness are reduced.

Strength, hardness, and ductility are normally desirable properties in metal. Brittleness (the opposite of ductility) is generally a bad characteristic. Thus, a primary goal of metallurgical science is to find ways to increase the hardness and strength of a material without reducing its ductility.

Hardness

When certain alloys are added to metal, hardness and strength can be improved without decreasing the ductility. Millions of dollars of metallurgical research have gone into

Important Properties in Steel					
Mechanical Properties	**Stress/Strain Relationship Properties**	**Chemical Properties**	**Electrical Properties**	**Thermal Properties**	**Other Properties**
Hardness Strength Brittleness Ductility Stress Tensile strength Compressive strength Shear strength Torsional strength Flexure strength Fatigue strength Toughness Impact strength	Flexibility % elongation Strain Elasticity Plasticity Malleability Modulus of elasticity Stress-strain diagram Elastic range Plastic range Creep Poisson's ratio	Corrosion resistance Resistance to acids Resistance to alkali Resistance to other chemicals	Electrical conductivity Electrical resistance Dielectric strength Magnetic susceptibility	Coefficient of thermal expansion Melting temperature Thermal conductivity Heat capacity Specific heat	Weight Density Specific weight Wear Machinability Weldability

Figure 4-1. *General material properties can be grouped into several categories.*

the development of metals that can increase their hardness and strength without decreasing their ductility.

Ductility and Brittleness

The terms *ductile* and *brittle* are opposites. Both terms are used to describe a material's ductility. If a material stretches very much before it breaks, it is said to be *ductile,* or have high ductility. If a material stretches very little before it fractures, it is said to be *brittle,* or have little to no ductility, Figure 4-2.

In nearly all situations, ductility is more desirable than brittleness. Ductile materials resist shock better and absorb more energy before failure than brittle materials. However, in applications where no deformation is permitted, ductility is no longer an asset.

Brittle materials are usually stronger than ductile materials, but not always. Some ductile materials may resist high forces while stretching.

Low-carbon steel, aluminum, and rubber bands are ductile. Cast iron, glass, and uncooked spaghetti are brittle.

A common measure of ductility is percent elongation, which is discussed later in the chapter.

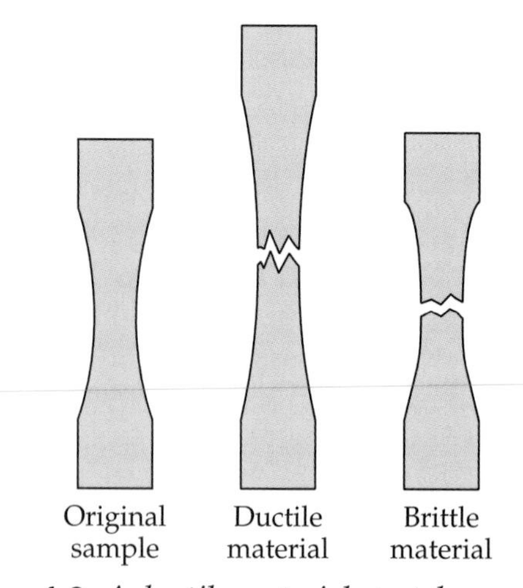

Figure 4-2. *A ductile material stretches much more before fracture than a brittle material does.*

Strength

There are many different types of strength, including tensile strength, compressive strength, shear strength, torsional strength, flexural strength, fatigue strength, and impact strength. Each type of strength is a measure of how a mate-rial reacts to a specific type of loading.

Stress

The amount of effort attempting to fracture an item is known as *stress.* The ability to resist that stress is known as *strength.* If the stress on a part exceeds its strength, the part will break.

Stress is mathematically equal to *force* or *load* divided by the cross-sectional area resisting that force. The units of stress are normally pounds per square inch (psi) or kips per square inch (ksi). One *kip* is equal to 1000 lb. Metric units are dynes per square centimeter.

For example, if a 1.2" diameter shaft is stretched by a uniformly applied force of 22,000 lb., the stress on the part is computed as follows:

$$\text{Cross-sectional area} = \frac{\pi \times D^2}{4}$$

$$= \frac{\pi \times (1.2")^2}{4}$$

$$= 1.131 \text{ in}^2$$

$$\text{Stress} = \frac{\text{Force}}{\text{Area}}$$

$$= \frac{22{,}000 \text{ lb.}}{1.131 \text{ in}^2}$$

$$= 19{,}450 \text{ psi}$$

Tensile strength is a material's ability to withstand stress in tension. Tension is a "pulling," Figure 4-3. This is perhaps the most important of all strengths. Most metals are very strong in tension.

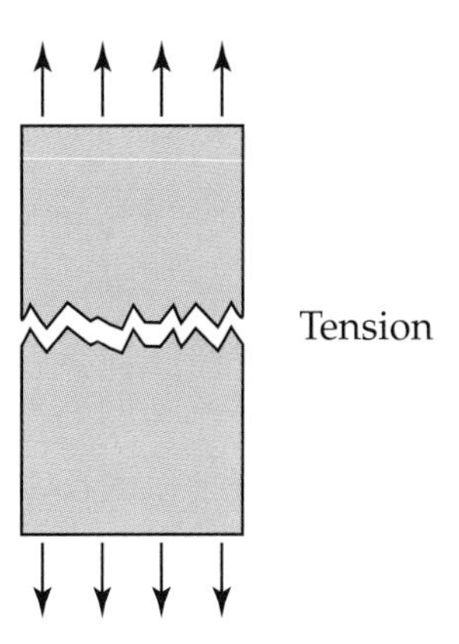

Figure 4-3. *A tensile load causes molecules in the material to be pulled away from one another.*

Tensile stress is equal to the applied force divided by the cross-sectional area that is resisting the tensile force. For example, if a cross-sectional area of 0.75 in^2 is resisting a force of 9900 pounds, the stress developed is:

$$\text{Tensile stress} = \frac{9900 \text{ lb.}}{0.75 \text{ in}^2}$$

$$= 13{,}200 \text{ psi}$$

Compressive strength is the ability to withstand "pressing" or "squeezing together," Figure 4-4. Cast iron has outstanding compressive strength.

A part loaded in compression must withstand a compressive stress equal to the compression force divided by the cross-sectional

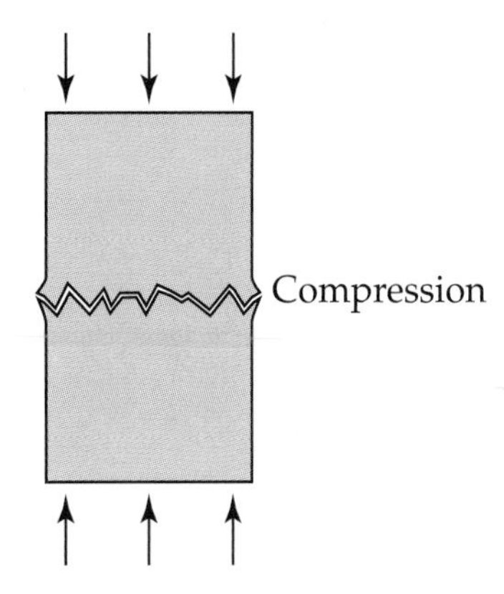

Figure 4-4. *Compression is a squeezing together.*

area withstanding the load. For example, if the compressive load is 555 lb. and the cross-sectional area is round with a diameter of 0.444", the compressive stress can be calculated as follows:

$$\text{Cross-sectional area} = \frac{\pi \times D^2}{4}$$

$$= \frac{\pi \times 0.444^2}{4}$$

$$= 0.155 \text{ in}^2$$

$$\text{Compressive stress} = \frac{\text{Load}}{\text{Area}}$$

$$= \frac{555 \text{ lb.}}{0.155 \text{ in}^2}$$

$$= 3580 \text{ psi}$$

Most materials have approximately equal abilities in resisting tension and compression. A few materials, such as cast iron and concrete, are able to take much higher compressive stresses than tensile stresses, Figure 4-5.

Shear strength is the ability to resist a "sliding past" type of action, Figure 4-6. A material's ability to withstand shear stress is normally less than its ability to resist tension or compression. However, the shear stresses that are built up in materials are also generally less.

Material	Tensile Strength (psi)	Compressive Strength (psi)
1025 Steel	70,000	70,000
1095 Steel	110,000	110,000
52100 Steel	140,000	140,000
Gray Cast Iron	35,000	110,000
Wrought Iron	40,000	40,000
Stainless Steel	95,000	95,000
Aluminum	40,000	40,000
Bronze	60,000	60,000
Zinc	20,000	20,000

Figure 4-5. *Cast iron has greater compression strength than tensile strength.*

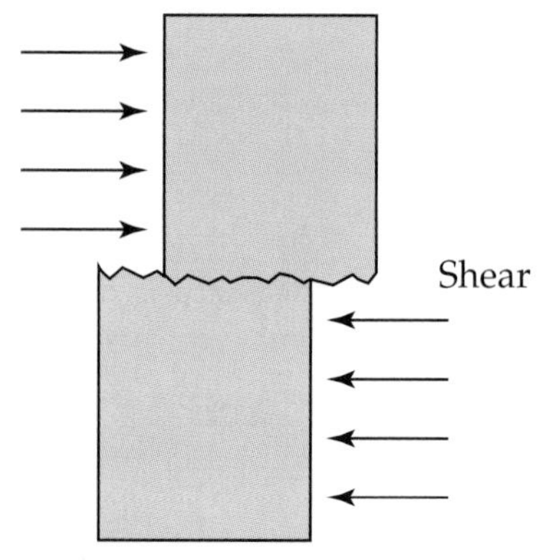

Figure 4-6. *Shear is a sliding past type of action.*

The shear stress developed in a part is equal to the shear force divided by the cross-sectional area withstanding the force. For example, in Figure 4-6, assume the shear force is 2220 pounds, the width of the bar is 0.75", and the depth into the page is 0.45". The shear stress is calculated as follows:

$$\text{Shear stress} = \frac{\text{Force}}{\text{Area}}$$

$$= \frac{2220 \text{ lb.}}{0.75" \times 0.45"}$$

$$= 6580 \text{ psi}$$

The resisting area for shear is not perpendicular to the line of action of the force, as is the case with tension and compression. This is a key difference in shear. The resisting area is always parallel to the line of action of the applied force.

Torsional strength is the ability to resist rotational shear, Figure 4-7. Torsion occurs in rotating machine parts. Round shafts that carry energy are common examples of this. When the stress becomes excessive, the crystals in the metal slide past each other and the cross section of the shaft fractures.

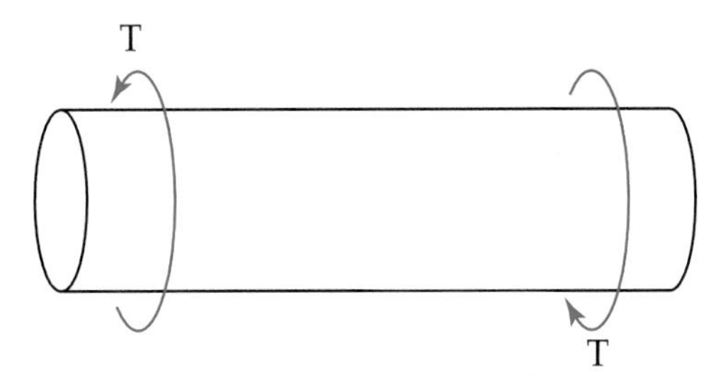

Figure 4-7. *Torsion occurs when a force causes rotation.*

The maximum torsional stress in a shaft or bar is determined from the following expression:

$$\text{Torsional stress} = \frac{T \times R}{J}$$

In this equation, "T" is the applied torque in inch-pounds, foot-pounds, etc., "R" is the distance from the center of the shaft to the outside surface, and "J" is the polar moment of inertia. For round shafts, the polar moment of inertia is determined using the following formula:

$$J = \frac{\pi \times D^4}{32}$$

In this equation, "D" is the diameter of the shaft. For example, if the applied torque is 555 in-lb. and the diameter of the shaft is 0.375", the torsional stress is calculated as follows:

$$J = \frac{\pi \times D^4}{32}$$

$$= \frac{\pi \times 0.375^4}{32}$$

$$= 0.00194 \text{ in}^4$$

$$\text{Torsional stress} = \frac{T \times R}{J}$$

$$= \frac{555 \text{ in-lb.} \times 0.1875''}{0.00194 \text{ in}^4}$$

$$= 53{,}640 \text{ psi}$$

Flexural strength is bending strength. It generally involves tension on one side of a material and compression on the opposite side, Figure 4-8. This is regularly encountered in beams and long parts in machines.

When flexure is applied to a part, there will be a *tensile stress* on one side of the part and a *compressive stress* on the opposite side. For a symmetrical beam, the maximum stress can be calculated using the following formula:

$$\text{Flexural stress} = \frac{M \times c}{I}$$

In this equation, "M" is the bending moment in inch-pounds, "c" is the distance from the neutral axis of the member in inches, and "I" is the moment of inertia. For a circular cross-section, the moment of inertia is calculated by the following:

$$I = \frac{\pi \times D^4}{64}$$

For example, if the diameter of the machine member is 0.75" and the applied moment

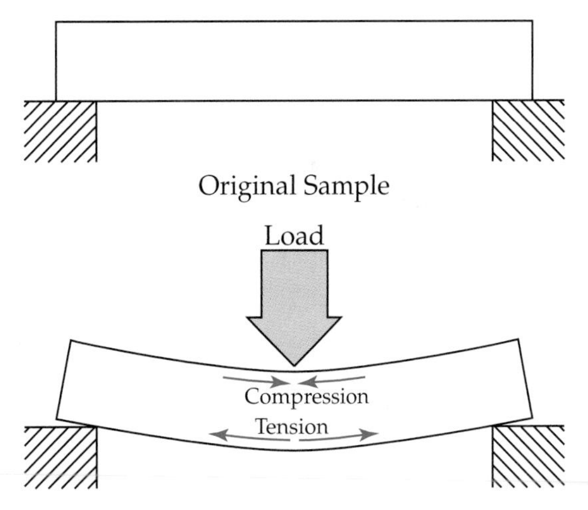

Figure 4-8. *A material loaded in flexure is subjected to both tensile and compressive stresses. The top of this sample is compressed, while the bottom is stretched in tension.*

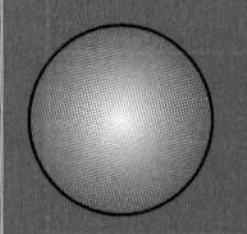

Self-Demonstration Flexural Strength

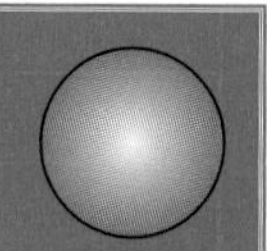

Obtain a flat piece of sheet metal, steel, or aluminum (1/32" or 1/16" thick). The piece should be 1" to 2" wide and at least 6" long. The exact dimensions are not critical. Sheet metal stock will likely be in inventory in your metallurgical laboratory stock room or machine shop stock room, or it may be purchased at a local hardware store.

Apply a coating of paint, glue, nail polish, or any other hard coating over a 1 in^2 section of the sheet metal near the middle of each side. Give the coating plenty of time to dry without disturbing the sample.

Carefully hold the sample at both ends with your hands and slowly bend it into a concave shape in one direction only.

As you slowly bend the sample, observe the coating. Flexural stress will put compression on one side of the sample and tension on the other side. Does the tension side begin to show small cracks? Does the compression side show signs of bulging?

at the break point is 8800 in-lb., the moment of inertia will be:

$$I = \frac{\pi \times D^4}{64}$$

$$= \frac{\pi \times 0.75^4}{64}$$

$$= 0.01553 \text{ in}^4$$

and the flexural stress will be:

$$\text{Flexural stress} = \frac{M \times c}{I}$$

$$= \frac{8800 \times 0.375}{0.01553}$$

$$= 212{,}500 \text{ psi}$$

In this example, there would be a tensile stress of 212,500 psi on the bottom of the shaft and a compressive stress of 212,500 on the top of the shaft. The two stresses together are called "flexure."

Comparing Types of Stress

The different types of stresses are the result of various types of loading. The type of loading determines the stress to which the material is subjected. A comparison of the main types of strengths and stresses are shown in Figure 4-9.

Fatigue strength or *endurance strength* refers to the ability of a material to resist repeated loading. A machine part may fail at a lower stress level if a force is continually applied and withdrawn. Vibration produces fatigue stress.

Some materials will hold up to a constant tensile stress or compressive stress without breaking, but will fail quickly under a smaller, repetitive fatigue stress. Slight cracks in the surface tend to grow and propagate across a piece of metal when it lacks good fatigue strength, Figure 4-10. Similarly, if you repeatedly fold and crease a piece of paper before trying to rip it, it will "give up" much more easily.

Toughness and *impact strength* measure the ability of a material to resist shock. A material

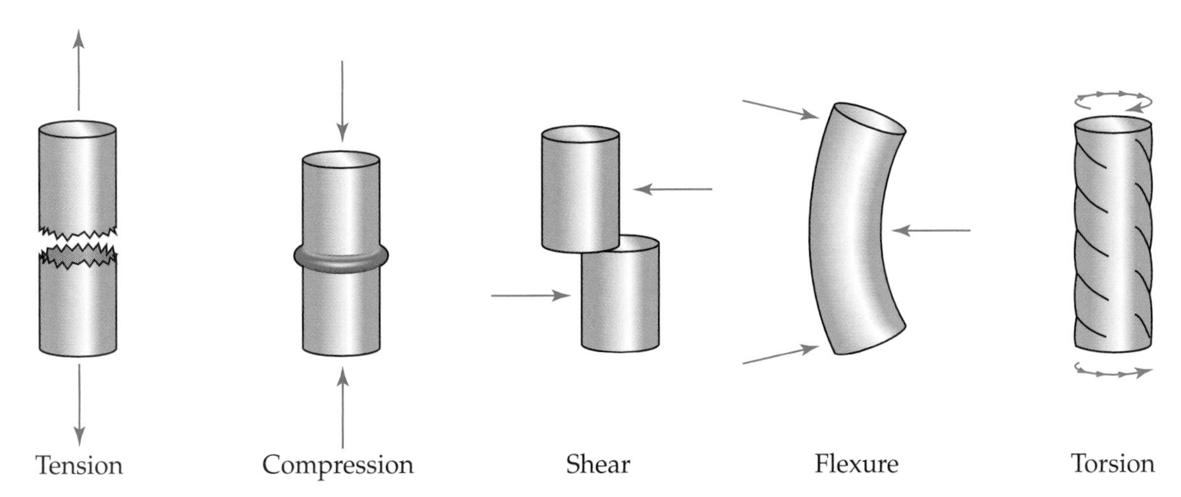

Figure 4-9. *Materials are subjected to many types of stresses. The forces acting on the object produce the stress. These stresses are resisted by the strength of the material.*

must have a good combination of both strength and ductility to resist shock. Air hammers, connecting rods in engines, and impact wrenches all must resist shock. Therefore, an important property in them is toughness.

Some materials can resist high forces or loads if the loads are applied gradually and gently. But some of these same materials cannot tolerate even a small force if it is suddenly applied. The science of karate illustrates this.

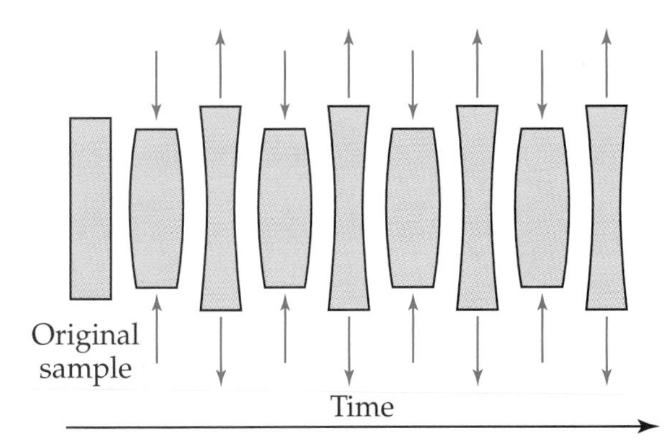

Figure 4-10. *Fatigue strength deals with repeated loading and unloading. In this figure, a sample is subjected to periodic compressive and tensile loading over time. Deformation of the part is exaggerated in this drawing.*

Strong materials can be broken with surprising ease if the speed of the blow is quite rapid.

To cite some examples, a material like cast iron, which has good compressive strength but poor ductility, does not have good shock resistance. Medium-carbon steel has fairly good strength and ductility. Therefore, it has good toughness and shock resistance.

Ductility

Ductility is the ability of a material to bend, stretch, or distort without breaking. A ductile material is flexible. A brittle material is not flexible.

When a metal is stretched in tension, the amount that its length increases can be referred to as *elongation, deformation, change of length,* or simply *stretch.* When a metal is reduced in length due to compression, the length reduction can be referred to as *deformation, contraction,* or *change of length.* Change of length resulting from tension or compression is measured in units of length.

A common measure of ductility is percent elongation at fracture. Other measures of ductility include strain and elasticity.

Percent elongation is the percentage that a material stretches before breaking. Mathematically, this is equal to the maximum amount of deformation divided by the original length. This value is then converted to a percentage by multiplying by 100:

$$\text{\% elongation} = \frac{\text{Deformation}}{\text{Original length}} \times 100\%$$

Consider the long bar in Figure 4-11 that stretched considerably before breaking. Assume that the original length of the bar was 33" and it stretched 2.22" before breaking. The percent elongation is calculated as follows:

$$\text{\% elongation} = \frac{\text{Deformation}}{\text{Original length}} \times 100\%$$

$$= \frac{2.22''}{33'' \times 100\%}$$

$$= 6.7\%$$

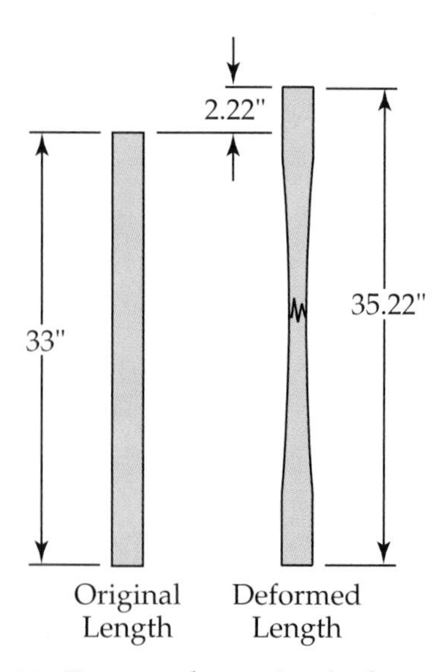

Figure 4-11. *Percent elongation is the percent of the original length a material stretches before it breaks.*

Strain is the ratio of deformation (or change in length) to original length. The only numerical difference between strain and percent elongation is a factor of 100. However, percent elongation generally refers to the stretch ratio at failure, while strain can refer to either the stretch ratio during a test before failure, or the maximum stretch ratio of the material.

Thus, strain is found by:

$$\text{Strain} = \frac{\text{Deformation}}{\text{Original length}}$$

The units of strain are commonly referred to as inches per inch, although strain can be considered as unitless since deformation and original length normally have the same units.

To determine the strain at failure of the bar in Figure 4-11,

$$\text{Strain} = \frac{\text{Deformation}}{\text{Original length}}$$

$$= \frac{2.22''}{33''}$$

$$= 0.0673 \text{ inches/inch}$$

Suppose you apply a load to a part and it stretches. Then, you gradually remove the load. Will the part snap back to its original length? If it does, the material is said to be *elastic*. If it doesn't, it is said to be *plastic*.

The terms elastic and plastic are also used to describe materials. An elastic material returns to its original shape after loading. A plastic material retains some of the deformation caused by the load. Elastic and plastic behavior is illustrated in Figure 4-12.

This ability of a material to return to its original length and shape after being stretched without any permanent deformation is referred to as *elasticity*. *Plasticity* is the opposite. It is the ability of a material to permanently deform and to retain its new shape without breaking.

Malleability is a form of plasticity. It is the ability of a material to permanently change to a new useful shape after being hammered, forged, pressed, or rolled. Malleability is needed for operations such as forging, drawing, extruding, or forming in a press. Most ductile materials are malleable.

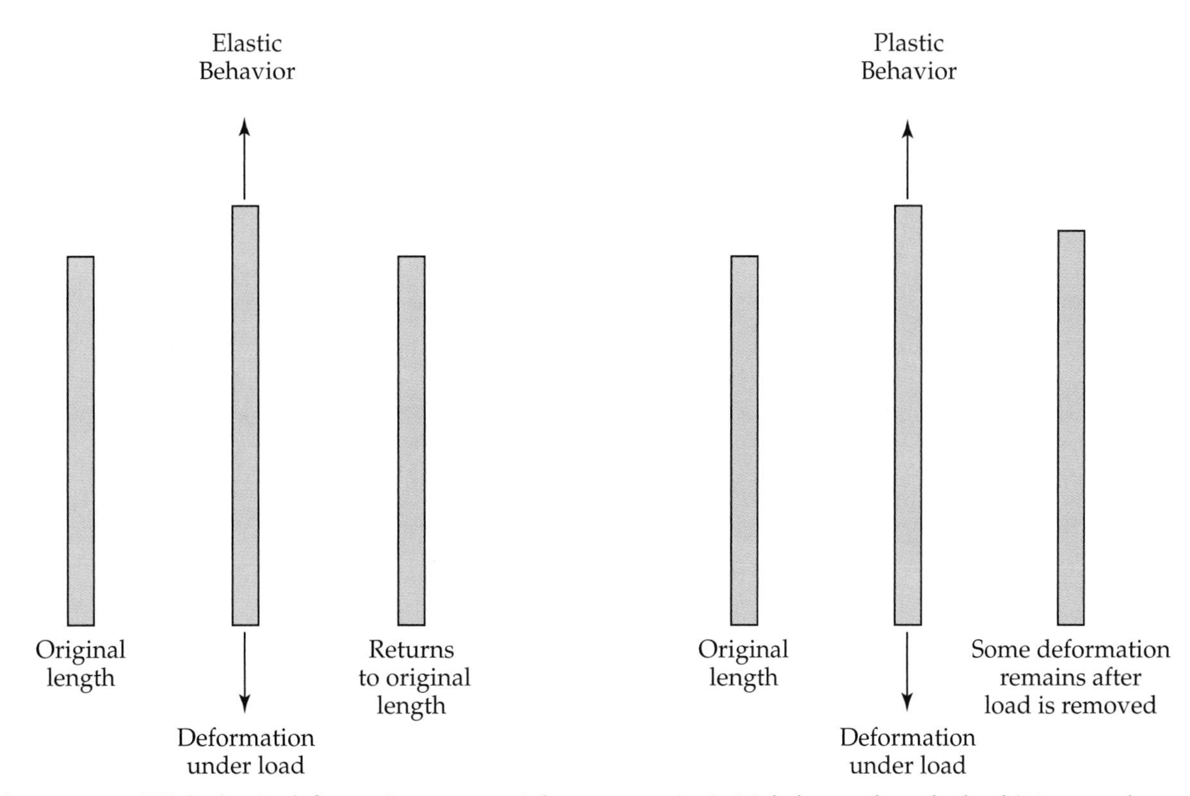

Figure 4-12. *With elastic deformation, a material returns to its initial shape when the load is removed. With plastic deformation, the material does not return to its original shape when the load is removed: some of the deformation remains.*

Relationship of Stress and Strain

There is a special relationship between stress and strain that is significant in the study of metallurgy. Strain increases as stress increases. For elastic deformation, strain increases in direct proportion to the increase in stress in many materials. For a given part, stress is proportional to force or load, and strain is proportional to elongation or change of length.

The ratio of stress to strain for a material is called *modulus of elasticity* or *Young's Modulus* (named after Thomas Young). The letter E is used to represent the modulus of elasticity in equations. It has the same units as stress (psi) and is calculated by dividing a stress by the strain it produces:

$$\text{Modulus of Elasticity} = \frac{\text{Stress}}{\text{Strain}}$$

This ratio remains constant for low levels of stress, i.e., stresses below the proportional limit stress, which will be discussed later in this chapter. The table in Figure 4-13 shows the modulus of elasticity for several materials.

We can use the relationship between stress and strain to accurately predict deformation for various loads. For example, a shaft that is 0.375" in diameter and 5.5" long stretches 0.0073" when subjected to a longitudinal force of 4400 lb. This is illustrated in Figure 4-14. In the following calculations, we will use this information to calculate the modulus of elasticity for the material. We will then use the modulus of elasticity to calculate the strain that would occur for different loads.

Material	Modulus of Elasticity (psi $\times 10^3$)
Aluminum	10,600
Gray cast iron	12,000
Malleable cast iron	26,000
Steel	30,000
Magnesium	6500
Titanium	16,500
Concrete	3000–8000
Wood	1500–2000

Figure 4-13. *This table lists the modulus of elasticity for several materials, both metallic and nonmetallic.*

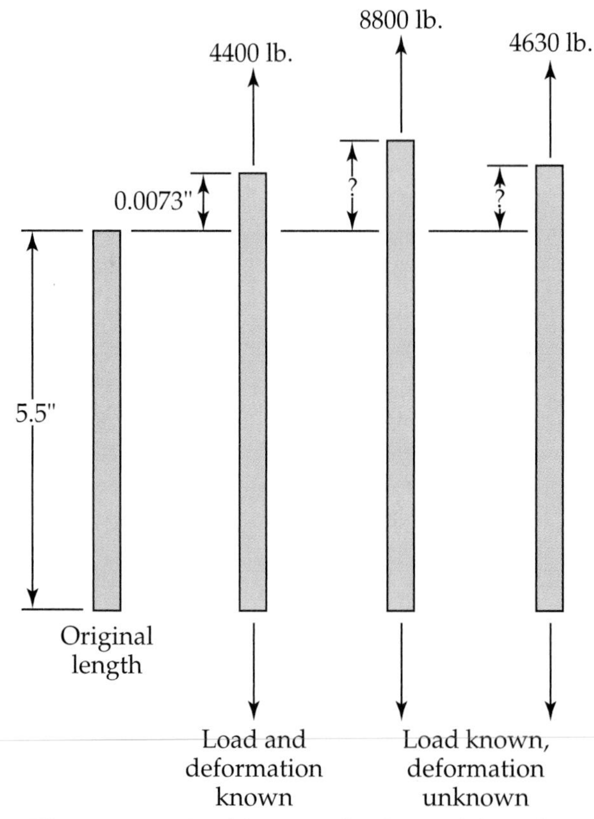

Figure 4-14. *In this example, the modulus of elasticity is calculated from the load and deformation shown. This information is then used to determine the strain under two different loads.*

The stress is calculated as discussed previously:

$$\text{Cross-sectional area} = \frac{\pi \times D^2}{4}$$

$$= \frac{\pi \times (0.375")^2}{4}$$

$$= 0.110 \text{ in}^2$$

$$\text{Stress} = \frac{\text{Force}}{\text{Area}}$$

$$= \frac{4400 \text{ lb.}}{0.110 \text{ in}^2}$$

$$= 40{,}000 \text{ psi}$$

After determining the stress, we can determine the strain:

$$\text{Strain} = \frac{\text{Deformation}}{\text{Original length}}$$

$$= \frac{0.0073"}{5.5"}$$

$$= 0.00133 \text{ inches/inch}$$

Using the stress and corresponding strain value, we can determine the modulus of elasticity for this material:

$$E = \frac{\text{Stress}}{\text{Strain}}$$

$$= \frac{40{,}000 \text{ psi}}{0.00133 \text{ inches/inch}}$$

$$= 30{,}000 \text{ psi}$$

Now that we know the modulus of elasticity, we can determine the strain resulting from any load in the elastic range. For example, we can calculate the strain for a force of 2000 lb. First, calculate the stress:

$$\text{Stress} = \frac{\text{Force}}{\text{Area}}$$

$$= \frac{2000 \text{ lb.}}{0.110 \text{ in}^2}$$

$$= 18{,}200 \text{ psi}$$

Then use the equation for the modulus of elasticity to determine the strain:

$$\text{Strain} = \frac{\text{Stress}}{\text{E}}$$

$$= \frac{18{,}200 \text{ psi}}{30{,}000 \text{ psi}}$$

$$= 0.607 \text{ inches/inch}$$

We could then calculate the expected deformation based on this strain.

Stress-Strain Diagrams

A graph of stress versus strain is known as a stress-strain diagram and describes the behavior of materials. A typical stress-strain diagram for low-carbon steel is shown in Figure 4-15. The proportional limit stress and elastic limit stress are shown in the diagram.

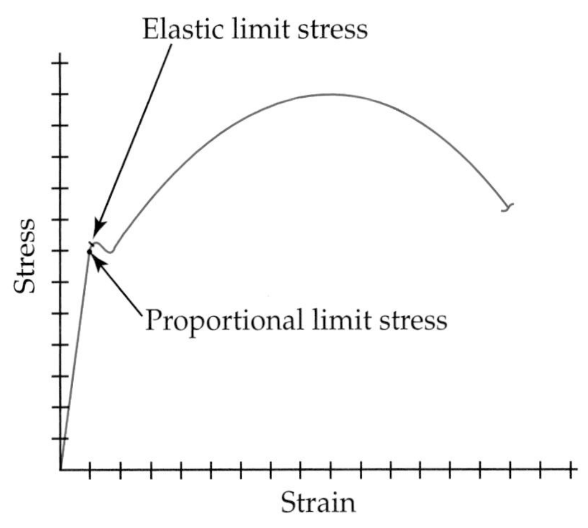

Figure 4-15. *A stress-strain diagram illustrates a material's tendency to deform under load. This is the stress-strain diagram for low-carbon steel.*

Proportional limit

As long as both stress and strain increase at a constant rate, the stress-strain diagram will be linear, i.e. a straight line. Note that in Figure 4-15, the straight line continues until it reaches the *proportional limit*. Beyond this point, strain increases at a faster rate than stress and the modulus of elasticity no longer applies.

Elastic and plastic range

If the material is stressed slightly beyond the proportional limit, it reaches a point known as the *elastic limit*. See Figure 4-15. If the material is stressed beyond this point, the material will not return to its original length when the load is released. It will be permanently deformed. If the load is released before the material reaches this elastic limit, it will return to its original length.

The portion of the stress-strain curve between zero and the elastic limit is known as the *elastic range*. Beyond the elastic limit, the ratio of stress to strain is no longer linear and the material will not return to its original length if the load is released. Strain and deformation grow more rapidly. This region is known as the *plastic range.*

Stress-strain diagrams have different shapes for different materials. See Figure 4-16. Aluminum and most nonferrous metals do not show the distinguishable proportional limit that ferrous metals do. Cast iron normally fractures very close to the proportional limit because it is so brittle.

Creep

When a material is loaded to a specific stress-strain level and held there for a long period of time, a phenomenon known as *creep* occurs. Plastic flow occurs and the material continues to stretch even though the stress is not increased. Creep is a slow plastic flow process and is more pronounced at higher temperatures. When it occurs, the function of a part may be jeopardized due to its dimensional change. At lower temperatures, creep may take months or years to affect the functioning of a part. However, at elevated temperatures, creep may be serious problem.

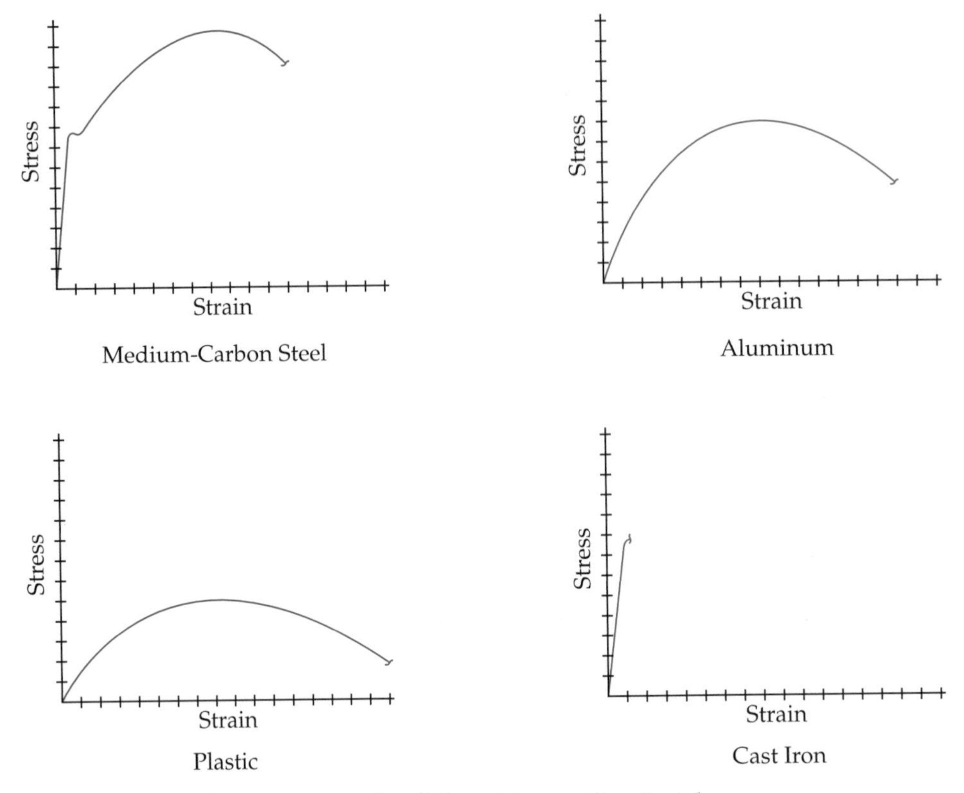

Figure 4-16. *Stress-strain diagrams vary for different types of materials.*

Poisson's Ratio

As a material is strained and elongated in one direction, the material decreases dimensionally in the other two directional planes. See Figure 4-17. The decrease in strain in the two perpendicular planes is less than the strain in the loaded direction. The relationship between these strain values is defined by *Poisson's ratio.* Poisson's ratio is the ratio of the lateral strain to the strain in the loaded direction. Poisson's ratio includes a negative sign, to illustrate that the lateral strain and axial strain have opposite directions.

$$\text{Poisson's ratio (m)} = -\frac{\text{Lateral strain}}{\text{Axial strain}}$$

The following example illustrates how Poisson's ratio is used to determine dimensional changes in cross-section.

A thin machine member is 77" long, and has a 0.55" × 0.33" rectangular cross section. The material has a Poisson's ratio of 0.3. If it is stretched 1.5" along its length, we can determine the

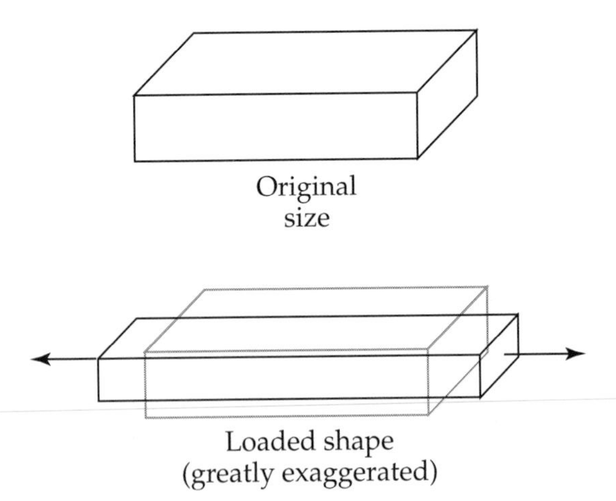

Figure 4-17. *An object elongated along one axis will contract along the axes perpendicular to the axis of loading. Poisson's ratio is the ratio of the transverse strain to the strain in the loaded direction.*

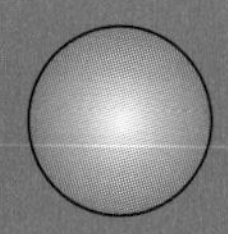

Self-Demonstration
Poisson's Ratio

Obtain some modeling clay. You will need a few cubes that are 1 to 2 in^3 in size.

Mold a sample into a cube with your hands. Do not apply a great deal of pressure or "squeezing" to the sample in making it take shape.

Measure the sample very accurately in each of the three dimensions. Calculate the volume (formula: length × width × height). Record these values.

Set the bottom of the cube on a flat surface. Set another flat object on the top of the sample.

Slowly apply pressure to the upper surface of the flat object as uniformly as possible until the clay cube begins to change shape. The clay should bulge at the sides. Before the clay changes dimensions very much, remove the upper object and accurately measure all three dimensions. If the bulging is not uniform throughout the side surfaces, you will need to average and estimate the width change. Did your side dimensional values (length and width) both increase?

Multiply the new length times the new width times the new height and compare to the original volume. Which volume product is greater? The original volume should be greater because the squeezing action should have compressed the sample slightly.

Calculate the Poisson's ratio value for the clay. Does your Poisson's ratio seem reasonable?

change in the cross section using Poisson's ratio. First, we must calculate the axial strain:

$$\text{Strain} = \frac{\text{Deformation}}{\text{Original length}} = \frac{1.5''}{77''} = 0.0195 \text{ inches/inch}$$

Once the axial strain is known, we can use Poisson's ratio to calculate the lateral strain:

$$\text{Lateral strain} = \text{Axial strain} \times -\text{Poisson's ratio} = 0.0195 \times (-0.3) = -0.00585$$

This lateral strain value is then used to calculate the deformation in the two cross-sectional dimensions:

$$\text{Deformation} = \text{Strain} \times \text{Original length} = -0.00585 \times 0.55'' = -0.0032''$$

$$\text{Deformation} = -0.00585 \times 0.33'' = -0.0019''$$

Chemical Properties

Corrosion resistance is perhaps the most important chemical property of a metal. A metal that has good corrosion resistance is able to protect itself against chemical attack by the environment. A corrosion resistant material can resist humidity without deteriorating. It can also resist sunlight, water, and heat.

Corrosion resistance may be the most important factor in selecting a material to be located in a corrosive environment. Materials are subjected to many types of corrosion:

- Oxidation. Oxidation involves a chemical reaction between a material and oxygen. Iron and steel rust when the iron chemically bonds with oxygen to form iron oxide, which is better known as rust.

- Galvanic corrosion. This type of corrosion is also referred to as electrochemical corrosion. In order for galvanic corrosion to occur, two dissimilar metals must be in contact with each other and both metals must also be connected by an *electrolyte,* a liquid that ionizes to bond with other ions. One metal acts as an anode, losing positive ions to the electrolyte and positive ions to the other metal (cathode). Pitting occurs on the anode as it loses these ions. The metal serving as the cathode is not corroded. See Figure 4-18.
- Pitting. This type of corrosion causes small pits to form on the surface of a material. Pitting is caused due to inconsistencies within the molecular and atomic composition of the material. This lack of consistency can result from many causes, including residual stresses, cracks, and processing procedures. Some materials, such as stainless steel, are more likely to be corroded than other materials. Pitting is also more likely to occur in specific environments, such as salt water or chlorine bleach. Figure 4-19 illustrates pitting.
- Intergranular corrosion. This type of corrosion occurs when the molecular composition of a material differs slightly at the grain boundaries. This condition can be caused by improper heat treatment or by improper chemical compositions in an alloy. See Figure 4-20.
- Stress corrosion cracking. This type of corrosion occurs in many types of metals. Residual stresses caused by improper processing produce cracks. See Figure 4-21.

Electrical Properties

If electricity can flow freely through a material, the material has high *electrical conductivity.* If the material refuses to let electricity flow through it, it has high *electrical resistance.* Steel has a very high electrical conductivity and a low resistance to electrical flow.

Dielectric strength is another popular electrical property. A material with good dielectric

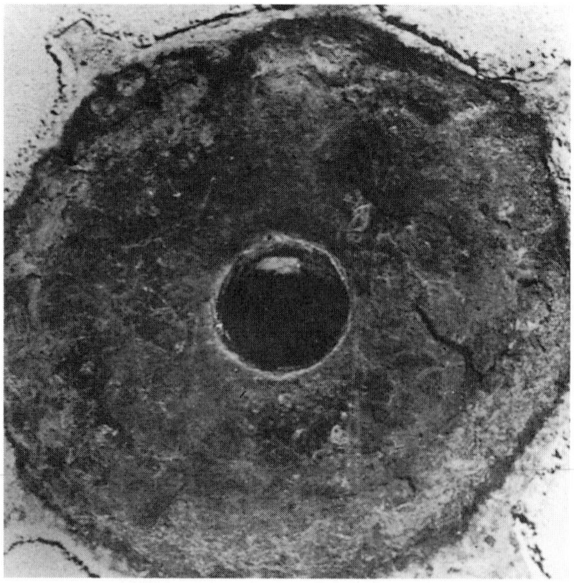

Figure 4-18. *In this example of galvanic corrosion, magnesium reacted with the steel core. (The International Nickel Company, Inc.)*

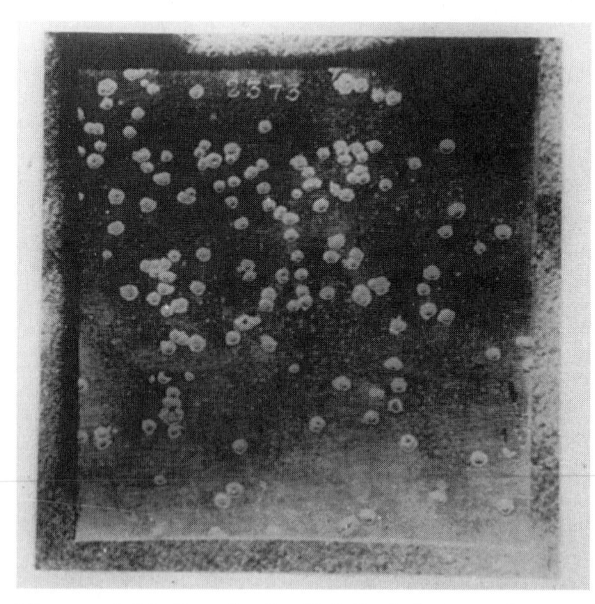

Figure 4-19. *Small pits distributed at random are evidence of pitting corrosion. (The International Nickel Company, Inc.)*

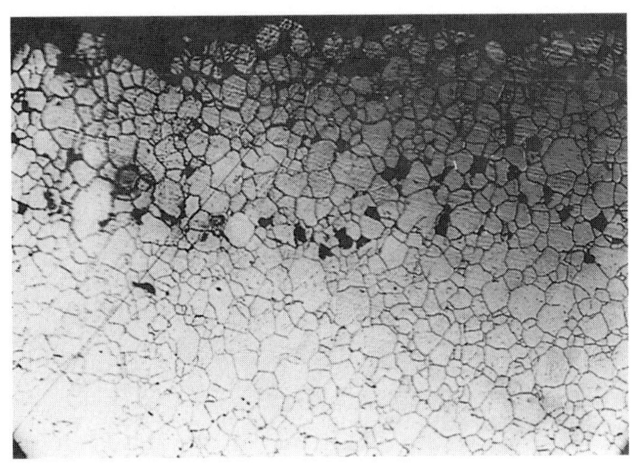

Figure 4-20. *Intergranular corrosion of steel under 100X magnification. (The International Nickel Company, Inc.)*

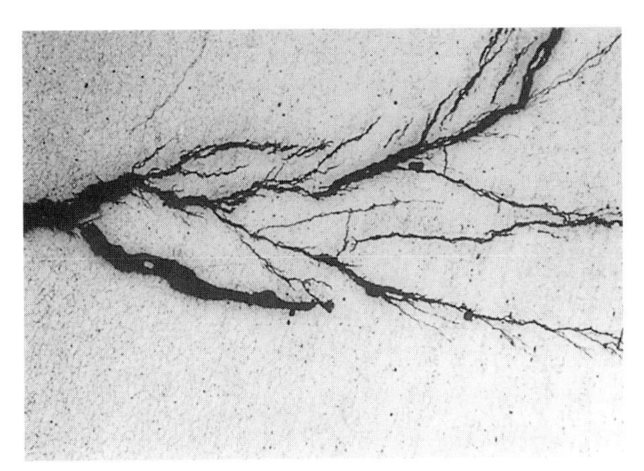

Figure 4-21. *Stress corrosion cracking of stainless steel under 100X magnification. (The International Nickel Company, Inc.)*

strength is able to withstand a large voltage over a prolonged time period without passing current or breaking down. A type of device used to test dielectric strength is shown in Figure 4-22.

Magnetic Properties

Some metals are magnetic. Some are not. The most recognized examples of magnetic materials are iron and steel at room temperature. It is interesting to note that iron and steel

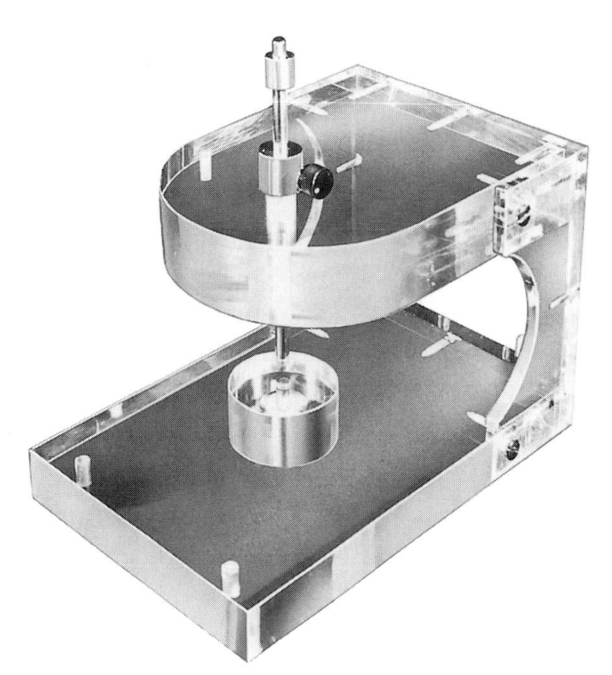

Figure 4-22. *This electrode assembly is used to test dielectric strength.*

at elevated temperatures are no longer magnetic. Most nonferrous metals, such as aluminum, magnesium, copper, and zinc, are not magnetic. The ability to be magnetized is referred to as *magnetic susceptibility*.

Thermal Properties

As the temperature changes, mechanical properties also change. The strength, hardness, ductility, and modulus of elasticity are normally affected. Some metals are affected more by temperature change than others. The effect of temperature on the properties of most metals is not linear. A loss of ductility, tensile strength, or impact strength as the temperature increases occurs suddenly as the temperature reaches a thermal transition level.

When a metal is heated, it expands. Some metals expand more rapidly than others, Figure 4-23. The *coefficient of thermal expansion* describes how fast a material expands when subjected to heat. In Figure 4-24, the coefficient of thermal expansion is compared for different materials. The greater the coefficient, the more a material expands.

The magnitude of the change of length for a material due to thermal expansion is given by the expression:

$$\Delta L = \alpha \times L \times \Delta T$$

where "α" is the coefficient of thermal expansion (in units of in/in/°F), "ΔL" is the change in length (in inches), "L" is the original length (in inches), and "ΔT" is the change in temperature (in °F).

For example, consider a long, thin aluminum connecting link in a business machine. It measures 0.062" by 0.125" by 7.7" long. How much would it increase in length due to an increase from room temperature (72°F) to 110°F?

First, calculate the change in temperature:

$$\Delta L = 110°F - 72°F$$
$$= 38°F$$

From Figure 4-24, the coefficient of thermal expansion for aluminum is 12.8×10^{-6} per °F. The increase in length would then be:

$$\Delta L = \alpha \times L \times \Delta T = (12.8 \times 10^{-6}/°F) \times 7.7" \times 38°F$$
$$= 0.00375"$$

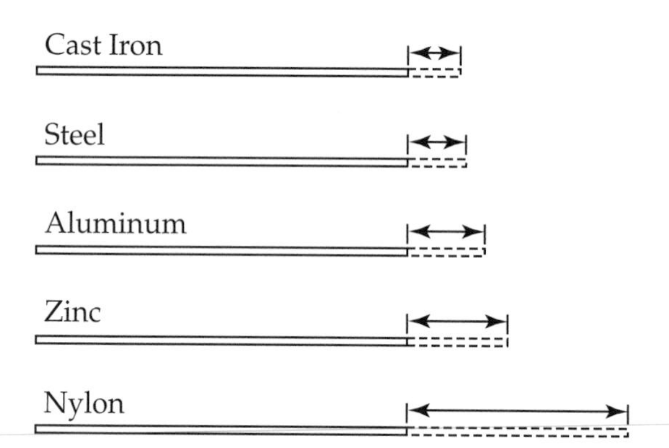

Figure 4-23. *Different materials increase in length more rapidly than others as the temperature increases. The arrows compare the thermal expansion of five different materials exposed to the same temperature increase.*

Melting point is another important thermal property. This is the temperature at which the material will change from a solid to a liquid. For steel, this temperature is in the vicinity of 3000°F. The temperatures at which selected materials melt are listed in the table in Figure 4-25.

If heat can travel rapidly across a material, it has high *thermal conductivity.* A material with a high degree of thermal conductivity transmits heat rapidly. If you are trying to get rid of heat, this property is an advantage. If you are trying to retain heat, this trait is a disadvantage.

Aluminum and copper both have high degrees of thermal conductivity. Steel is about average for a metal. The table in Figure 4-26 lists the thermal conductivity of several materials.

Heat capacity of a material is the amount of heat required to raise the temperature of a material one degree. This amount varies from material to material. It is measured in Btu (British thermal unit) per pound per °F or in Joules per kilogram per degree K

Material	Coefficient of Thermal Expansion (per °F)
Gray iron	6.0×10^{-6}
Steel	6.5×10^{-6}
Nickel	7.3×10^{-6}
Copper	9.2×10^{-6}
Bronze	10.0×10^{-6}
Brass	10.3×10^{-6}
Aluminum	12.8×10^{-6}
Magnesium	14.4×10^{-6}
Zinc	17×10^{-6}
Nylon	50×10^{-6}
Polystyrene	100×10^{-6}

Figure 4-24. *The table compares the coefficient of thermal expansion for different materials.*

(Kelvin). It is used to determine the specific heat of a material.

Specific heat is the ratio of the heat capacity of a material to the heat capacity of water. Thus, the specific heat of iron is the ratio of the amount of heat required to raise a certain mass of iron one degree to that required to raise the same mass of water one degree. The values of specific heat for some materials are shown in Figure 4-27.

Material	Melting Temperature (°F)
Gray iron	2400
Steel	2700
Nickel	2650
Copper	1980
Aluminum	1220
Magnesium	1200
Zinc	790
Lead	620
Nylon	300
Polystyrene	250

Figure 4-25. *This table compares the melting temperatures of different materials.*

Material	Thermal Conductivity (cal/cm²/°C/sec/cm)
Gray iron	0.11
Steel	0.11
Nickel	0.22
Copper	0.94
Aluminum	0.45
Magnesium	0.37
Zinc	0.27
Lead	0.08
Nylon	0.0006
Polystyrene	0.001

Figure 4-26. *This table compares how rapidly heat transfers across each material.*

Weight

Weight is a property that often is important. Often, a lighter weight material, such as aluminum or magnesium, has an advantage over steel. Less frequently, heavier materials have an advantage for certain applications.

The ratio of the weight of a material to its volume is often referred to as *density* or *specific weight*. Technically, specific weight is the correct term, but since the term density is so often used in actual practice, we will consider these two terms to be synonymous. The table in Figure 4-28 lists the density of several materials.

Wear

The ability of a material to withstand wearing away is a very important material

Material	Specific Heat Capacity
Gray iron	0.130
Steel	0.116
Nickel	0.109
Copper	0.093
Brass	0.093
Aluminum	0.217
Magnesium	0.222
Zinc	0.095
Lead	0.031
Nylon	0.400
Polystyrene	0.320
Water	1.000

Figure 4-27. *This table compares specific heat capacity for different materials.*

Material	Density (lb./ft^3)
Gray iron	482
Steel	490
Nickel	550
Copper	555
Aluminum	170
Magnesium	109
Zinc	440
Lead	710
Nylon	70
Polystyrene	60

Figure 4-28. *This table compares the densities of different materials.*

property. In many applications, a few thousandths of an inch of wear can cause an entire machine to fail.

Wear is the ability of a metal to resist a slow deterioration, usually over a long period of time. This deterioration may be caused by frictional scratching, scoring, galling, scuffing, or seizing. Wear is also caused by pitting or fretting. Figure 4-29 illustrates several different types of wear.

The ability to resist wear is highly dependent on the hardness of the material. The harder a material becomes generally determines how great its ability will be to resist wear.

Types of Wear

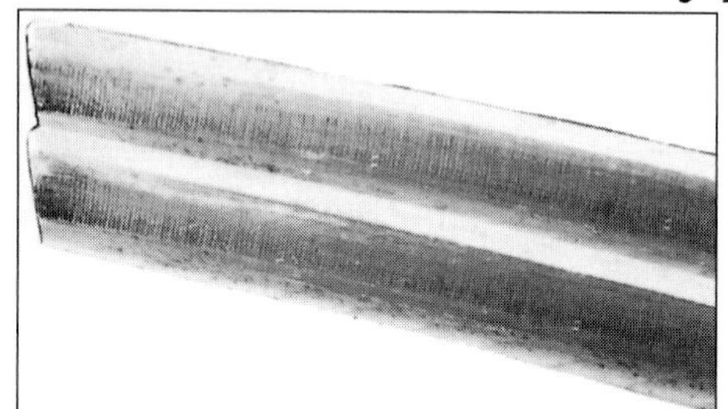

Normal Wear (Surface Polishing)

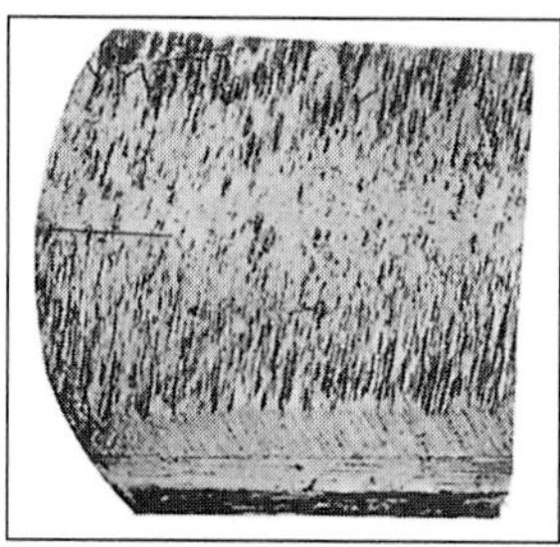

Abrasive Wear (Scratching)

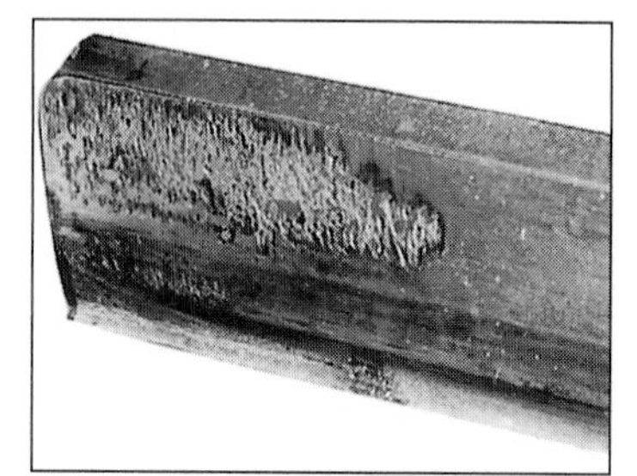

Adhesive Wear (Scoring, galling, scuffing, seizing)

Pitting Wear

Fretting Wear

Figure 4-29. *Wear can result from many causes. Some typical types of wear are illustrated in these photographs. (The Falk Corporation, subsidiary of Sundstrand Corporation)*

Proper Lubrication

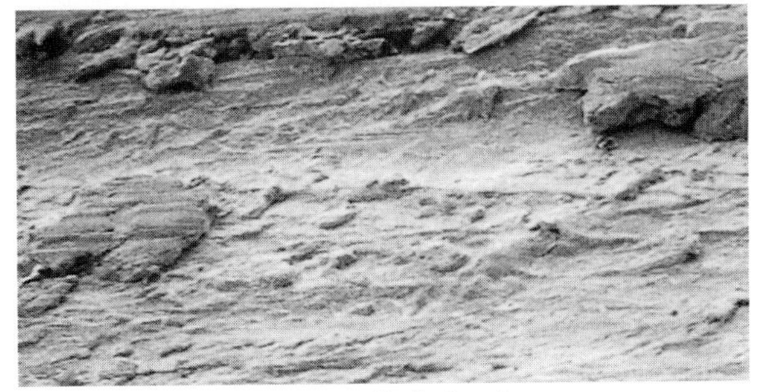
Improper Lubrication

Figure 4-30. *Close-ups of used engine bearing inserts magnified at 25X. The sample on the top was properly lubricated. The sample on the bottom is an example of improper lubrication. (Texaco Inc.)*

Proper lubrication is effective in reducing wear. Figure 4-30 shows the effects of improper and proper lubrication on a typical engine bearing insert.

Machinability

The ease with which a material can be cut is an important characteristic of a metal. This is known as *machinability*. This property refers to the convenience with which a material may be cut by turning, drilling, milling, boring and other cutting tools. If a part needs to be machined, it should be made from material with good machinability characteristics.

Weldability

The ability of a material to be fabricated satisfactorily by one of the common welding processes is known as *weldability.* Common welding processes today include arc welding, MIG welding, TIG welding, and gas welding.

Comparison Charts of Metal Properties

It is interesting to compare the properties of different metals. The table in Figure 4-31 compares most of these properties for steel, cast iron, wrought iron, aluminum, copper, bronze, brass, zinc, lead, nickel, tin, titanium, and tungsten.

Test Your Knowledge

Write your answers on a separate sheet of paper. Do not write in this book.

1. Name three important mechanical properties of a metal.
2. As hardness of a material increases, how is the strength of the material usually affected?
3. What type of strength indicates a material's ability to resist being "squeezed" together?
4. What type of strength indicates a material's ability to resist shock?
5. What type of strength indicates a material's ability to resist "pulling apart?"
6. What type of strength indicates a material's ability to resist repeated loading?
7. What type of strength indicates a material's ability to resist bending?
8. What type of strength indicates a material's ability to resist "sliding past?"
9. If a material does not stretch before it breaks, it is described as _____.
10. Percent elongation is a good measure of what other mechanical property?
11. Calculate the percent elongation and strain for a material with an initial length of 62" that stretched 2.04" when subjected to a tensile load.
12. Define *modulus of elasticity.*
13. What point on a stress-strain diagram occurs at the end of the elastic region?
14. How does temperature affect creep?
15. What type of strain is calculated using Poisson's ratio?

Material	Tensile Strength (psi x 10^3)	Compressive Strength (psi x 10^3)	Hardness (BHN)	Modulus of Elasticity (psi x 10^6)	Weight (lbs/ft^3)
Steel	60–200	Same as Tensile	150–620	30	490
Cast Iron	20–100	80–180	140–325	15	482
Wrought Iron	40	Same as Tensile	NA	28	490
Aluminum	20–60	Same as Tensile	50–110	10	170
Copper	30–60	Same as Tensile	40	17	555
Bronze	65–130	Same as Tensile	100–200	17	550
Brass	30–100	Same as Tensile	50–160	15	520
Zinc	20–30	NA	80–90	NA	440
Lead	2–5	NA	5–12	20	710
Nickel	45–60	NA	80–380	30	550
Tin	3–9	NA	7	6	450
Titanium	50–135	NA	NA	16	280
Tungsten	220	NA	NA	59	1180

Material	Electric Conductivity (100 Silver Comparison Scale)	Electric Resistance (Micro-OHM/CM)	Coefficient of Thermal Expansion (in/in°F x 10^{-6})	Melting Temperature (°F)	Thermal Conductivity (cal/cm^2/ °C/sec/cm)
Steel	12	17	6.3	2700	.11
Cast Iron	5	30	6.5	2400	.11
Wrought Iron	15	NA	6.6	2800	.18
Aluminum	63	3	12	1220	.45
Copper	98	2	9	1980	.94
Bronze	36	7	10	1840	.29
Brass	28	6.5	10	1700	.28
Zinc	30	6	17	790	.27
Lead	8.4	23	16	620	.08
Nickel	13	20	7	2650	.22
Tin	14	11.5	15	450	.16
Titanium	14	120	5	3300	.03
Tungsten	14	5.5	2.4	6100	.48

Figure 4-31. *This table compares many properties of different metals. Values listed may vary for different alloys and compositions of each material. Cost is not included because it fluctuates rapidly for some of the materials listed.*

16. What is dielectric strength?
17. A material that does not permit electricity to flow through it is said to have a high electrical _____.
18. What is thermal conductivity?
19. Which properties of metal would be most critical when selecting a material for a robot arm for the space shuttle that has to lift heavy satellites?
20. Which properties of metal would be most critical when selecting a material for a part in a toy metal truck?
21. Which properties of metal would be most critical when selecting a material for an 8' cross to be attached to the front of a church?
22. Which properties of metal would be most critical when selecting a material for a leg of a card table?

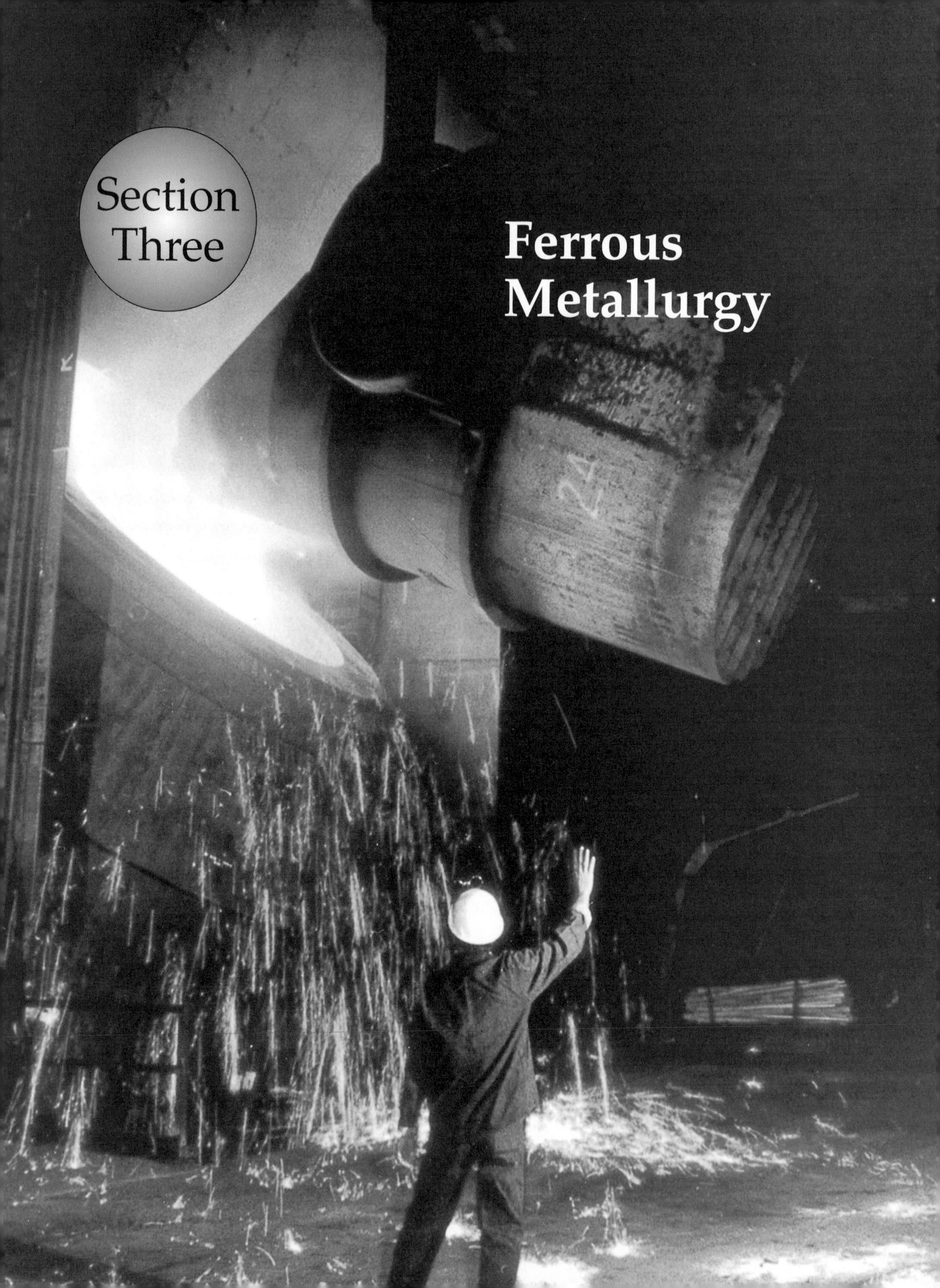
Section Three
Ferrous Metallurgy

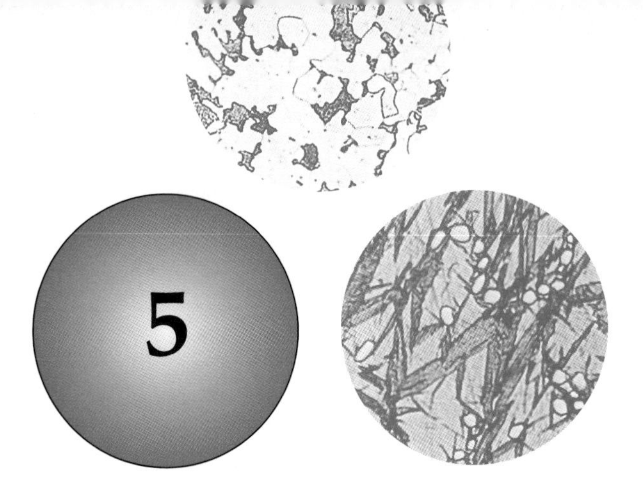

5 What Is Steel?

After studying this chapter, you will be able to:
- Describe the composition of steel.
- Identify the differences between steel and iron.
- Use the steel numbering system to identify various types of steel.
- Identify the effects of different alloying elements on steel.
- Describe various kinds of carbon steel and alloy steel.
- Compare different types of cast iron.

Steel is one of the most widely used materials in the world. It has high strength. It can be machined and formed easily. Also, steel is readily available and reasonably priced, as compared to other materials having similar physical properties.

Composition of Steel

Steel is a material composed primarily of iron. Most steel contains more than 90% iron. Many types of carbon steel contain more than 99% iron.

All types of steel contain a second element—carbon. Many other alloying elements are used in most steel, but iron and carbon are the only elements found in *all* steel. The percentage of carbon in steel ranges from just above 0% to approximately 2%. Most steel has between 0.15% and 1.0% carbon.

Steel with less carbon is more flexible (ductile) than high-carbon steel, but it is also weaker. As the carbon content increases, so do strength, hardness, and brittleness.

When steel is made, the iron dissolves the carbon. When there is too much carbon for the iron to "digest," the resulting alloy is no longer called steel. The carbon precipitates out and remains in the form of flakes or other shapes, Figure 5-1. Approximately 2% carbon is the most that can be dissolved in the iron.

Steel Numbering System

A steel numbering system is used to identify the many types of steel. A steel's numerical name usually consists of four numbers or digits. The first two digits refer to the alloy content. The last two digits (or three digits, in the case of a five-digit number) refer to the percentage of carbon in the steel. See Figure 5-2.

The steel numbering system is shown in Figure 5-3. In 5147 steel, for example, the "51" identifies chromium as a key alloying element. In 2517 steel, the "25" indicates that there is an unusual amount of nickel in this steel. In 4718 steel, the "47" indicates that the amount of chromium, nickel, and molybdenum in the steel is higher than average. In 1040 steel, the "10" means that the steel has very little alloy content except carbon. As shown by these examples, the first two numbers always give an indication of the alloy content in the steel.

The last two digits (or three digits) of a numerical name for steel indicate the percentage of carbon in the steel. A two-digit number

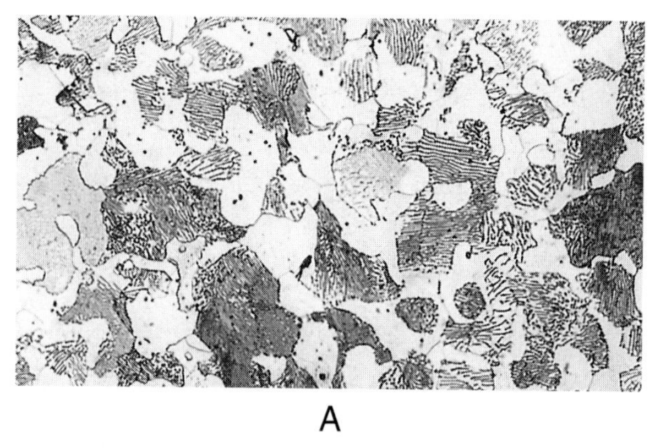

A

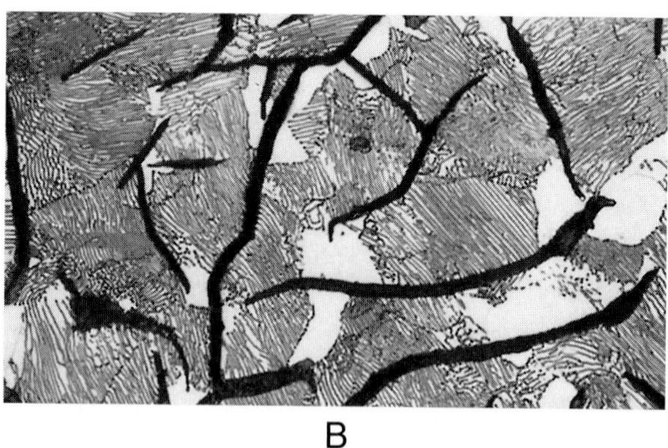

B

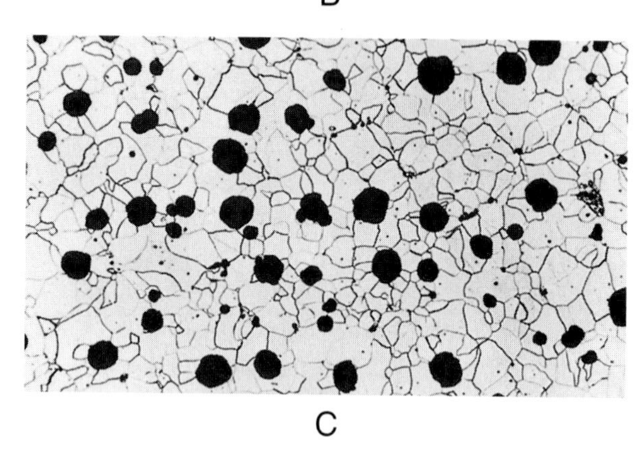

C

Figure 5-1. *Microstructural views of steel and cast iron. A—In steel, the iron dissolves the carbon. (Buehler Ltd.) B—In gray cast iron, the carbon precipitates out as carbon flakes. (Buehler Ltd.) C—In ductile cast iron, the carbon precipitates out as small round nodules. (Iron Castings Society)*

represents hundredths of a percent. In 1040 steel, for example, the "40" means there is 0.40% carbon in the steel. In 1018 steel, the "18" indicates that the steel has only 0.18% carbon, thus it is a low-carbon steel. There is approximately 0.60% carbon in 8660 steel, which would make it a medium-carbon steel.

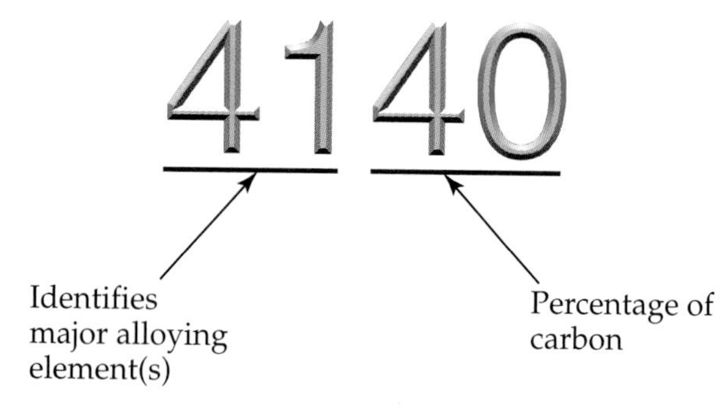

Figure 5-2. *A steel's name usually consists of four digits. It supplies information about the alloy content and percentage of carbon.*

When the carbon content of the steel is 1% or more, three digits are needed to describe the content. For example, 50100 steel contains 1.00% carbon. See Figure 5-4.

A steel numerical name provides much information about the alloy content of the steel. Indirectly, the number tells you about the quality, strength, and corrosion resistance of the steel. For example, 8622 steel has 0.22% carbon, 0.20% molybdenum, 0.50% chromium, and 0.55% nickel. Molybdenum makes this steel strong at high temperatures. Chromium improves its corrosion resistance, while nickel improves the steel's toughness.

Comparing Steel and Iron

The relationship of steel to cast iron and wrought iron is shown in Figure 5-5. The difference in the three materials is primarily based on the carbon content. Steel ranges from just above 0% carbon to approximately 2% carbon. Most types of cast iron contain 2% to 4% carbon. Wrought iron contains essentially no carbon. At approximately 6% carbon, the material becomes so brittle that it is relatively useless.

Steel Numbering System	
Steel Numerical Name	Key Alloys
10XX	Carbon only
11XX	Carbon only (free cutting)
13XX	Manganese
23XX	Nickel
25XX	Nickel
31XX	Nickel-Chromium
33XX	Nickel-Chromium
303XX	Nickel-Chromium
40XX	Molybdenum
41XX	Chromium-Molybdenum
43XX	Nickel-Chromium-Molybdenum
44XX	Manganese-Molybdenum
46XX	Nickel-Molybdenum
47XX	Nickel-Chromium-Molybdenum
48XX	Nickel-Molybdenum
50XX	Chromium
51XX	Chromium
501XX	Chromium
511XX	Chromium
521XX	Chromium
514XX	Chromium
515XX	Chromium
61XX	Chromium-Vanadium
81XX	Nickel-Chromium-Molybdenum
86XX	Nickel-Chromium-Molybdenum
87XX	Nickel-Chromium-Molybdenum
88XX	Nickel-Chromium-Molybdenum
92XX	Silicon-Manganese
93XX	Nickel-Chromium-Molybdenum
94XX	Nickel-Chromium-Molybdenum-Manganese
98XX	Nickel-Chromium-Molybdenum
XXBXX	Boron
XXLXX	Lead

Figure 5-3. *This table relates the alloy content in steel to the first two digits in its name.*

Examples of Steel Names and Carbon Content	
Identification Number	Carbon Content
1040	0.40%
1018	0.18%
8660	0.60%
50100	1.00%

Figure 5-4. *The last two or three digits in a numerical name for steel represent the carbon content in hundredths of a percent. Three digits are used when the carbon content is 1% or more.*

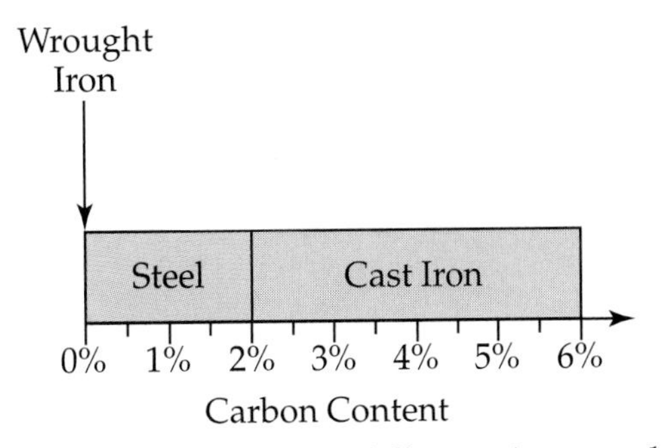

Figure 5-5. *The basic difference in wrought iron, steel, and cast iron is their percent carbon content.*

Alloying Elements

Most steel contains other ingredients in addition to iron and carbon. These ingredients are commonly called *alloying elements.* Most of the alloying elements in steel are present in small amounts, but they have a great effect on the properties of the steel. Some of these alloying elements and their effects are summarized in Figure 5-6.

Carbon, manganese, and nickel are added to steel to increase strength. To obtain better corrosion resistance or resistance to atmospheric conditions, chromium or copper may be added. If lead or sulfur is added, the steel will have better machinability. To obtain better physical properties at high temperature, tungsten or molybdenum are recommended.

The greater the amount (percentage) of the alloying elements, the more profound their effect on the steel. However, it is

Effects of Alloying Elements on Steel	
Alloying Element	**Effect on Steel**
Carbon	Hardness, strength, wear
Chromium	Corrosion resistance, hardenability
Lead	Machinability
Manganese	Strength, hardenability, more response to heat treatment
Aluminum	Deoxidization
Nickel	Toughness, strength
Silicon	Deoxidization, hardenability
Tungsten	High-temperature strength, wear
Molybdenum	High-temperature strength, hardenability
Sulfur	Machinability
Titanium	Elimination of carbide precipitation
Vanadium	Fine grain, toughness
Boron	Hardenability
Copper	Corrosion resistance, strength
Columbium	Elimination of carbide precipitation
Phosphorus	Strength
Tellurium	Machinability
Cobalt	Hardness, wear

Figure 5-6. *This table lists the effects of common alloying elements.*

unusual for steel to have more than 2% of any single alloying element. For example, phosphorous and sulfur are added to most types of steel, but the amount added rarely exceeds 0.05%. The percentages of alloying elements for some common types of steel are listed in Figure 5-7.

Types of Steel

There are many different categories and types of steel. However, most steel is classified as either *carbon steel* or *alloy steel,* Figure 5-8. Carbon steel contains relatively few alloying elements. Therefore, it is less expensive than alloy steel. Alloy steel has special qualities, such as increased strength, corrosion resistance, and the ability to resist wear at high temperatures.

Carbon steel is classified as low-carbon steel, medium-carbon steel, or high-carbon steel. There are many types of alloy steel, such as structural steel and maraging steel. See Figure 5-9. Stainless steel and tool steel are so widely used that they can be considered separate types of steel in themselves.

Carbon Steel

Carbon steel is the most common type of steel. About 90% of all steel made is carbon steel. It is sometimes called "plain carbon steel."

There are relatively few alloying elements in carbon steel; carbon is the dominant alloy. Most carbon steel is considerably less expensive than alloy steel. The three basic types of carbon steel are low-carbon steel, medium-carbon steel, and high-carbon steel.

Low-carbon steel

The largest percentage of all carbon steel is *low-carbon steel.* It contains between 0.05% and 0.35% carbon. Low-carbon steel lacks the ability to become as hard and strong as other steel. However, because it does not become very hard, it is easier to machine and work with in the manufacturing plant. Some characteristics of low-carbon steel are listed in Figure 5-10.

Low-carbon steel is the least expensive type of steel. For this reason, it has many uses. Applications include fence wire, auto bodies, galvanized sheets, storage tanks, large pipe, and various parts in buildings, bridges, and ships.

Low-carbon steel is not as strong and hard as some of the more expensive grades of alloy steel, but it is not weak or low in quality. All steel—even low-carbon steel—is very strong and can be trusted to support a great deal of force.

Examples of Alloying Elements in Steel											
Steel	**Type of Steel**	**Tensile Strength (× 1000 psi)**	**C**	**Mn**	**P**	**S**	**Si**	**Ni**	**Cr**	**Mo**	**V**
1025	Plain Carbon	60–103	0.22–0.28	0.30–0.60	0.04 max	0.05 max					
1045	Plain Carbon	80–182	0.43–0.50	0.60–0.90	0.04 max	0.05 max					
1095	Plain Carbon	90–213	0.90–1.03	0.30–0.50	0.04 max	0.05 max					
1112	Free Cutting Carbon	60–100	0.13 max	0.70–1.00	0.07–0.12	0.16–0.23					
1330	Manganese	90–162	0.28–0.33	1.60–1.90	0.035	0.040	0.20–0.35				
2517	Nickel	88–190	0.15–0.20	0.45–0.60	0.025	0.025	0.20–0.35	4.75–5.25			
3310	Nickel-Chromium	104–172	0.08–0.13	0.45–0.60	0.025	0.025	0.20–0.35	3.25–3.75	1.40–1.75		
4023	Molybdenum	105–170	0.20–0.25	0.70–0.90	0.035	0.040	0.20–0.35			0.20–0.30	
52100	Chromium	100–240	0.98–1.10	0.25–0.45	0.035	0.040	0.20–0.35		1.30–1.60		
6150	Chromium-Vanadium	96–230	0.48–0.53	0.70–0.90	0.035	0.040	0.20–0.35		0.80–1.10		0.15 min
9840	Nickel-Chromium Molybdenum	120–280	0.38–0.43	0.70–0.90	0.040	0.040	0.20–0.35	0.85–1.15	0.70–0.90	0.20–0.30	
4140	Chromium-Molybdenum	95–125	0.38–0.43	0.75–1.00	0.035	0.040	0.20–0.35		0.80–1.10	0.15–0.25	

Figure 5-7. *This table shows the alloy content of common types of steel.*

Steel	
Carbon Steel	**Alloy Steel**
• Less Expensive • Fewer Alloys • Few Special Properties	• More Expensive • More Alloys • Special Properties

Figure 5-8. *Most types of steel fall into one of two categories, carbon steel or alloy steel.*

Low-Carbon Steel
• 0.05–0.35% Carbon • Comparatively less strength • Comparatively less hardness • Easy machining & forming • Least expensive • Largest quantity produced

Figure 5-10. *Common characteristics of low-carbon steel.*

Medium-carbon steel

Medium-carbon steel contains 0.35% to 0.50% carbon. It can be heat treated. If it is heat treated properly, medium-carbon steel can become quite hard and strong. It is frequently used in forgings and high-strength castings. The characteristics of medium-carbon steel are shown in Figure 5-11.

Applications of medium-carbon steel include wheels, axles, crankshafts, and gears.

High-carbon steel

The carbon content in *high-carbon steel* is over 0.50%, and it may be over 1%. This type of steel can be readily heat treated to obtain high strength and high hardness.

The disadvantage of high hardness in steel is a relatively high rate of distortion and the potential of cracking or becoming very brittle during the hardening process. Nevertheless, high-carbon steel can be safely used for making tools, dies, knives, railroad wheels, and for many other applications requiring high strength. The characteristics of high-carbon steel are listed in Figure 5-12.

Alloy Steel

Alloy steel is a grade of steel in which one or more alloying elements have been added in larger amounts to produce special

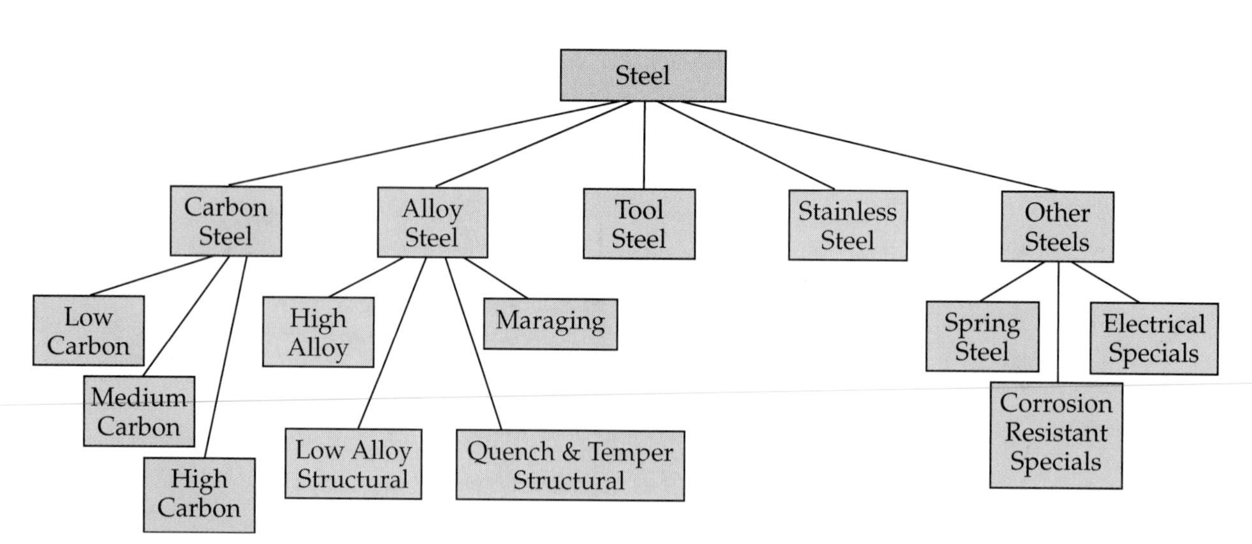

Figure 5-9. *Several types of alloy steel are so widely used that they are considered as separate classifications. These include stainless steel and tool steel.*

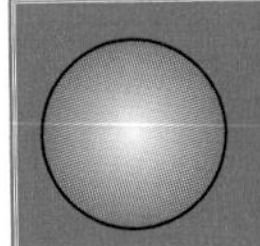

Self-Demonstration
Comparing Hardness of Different Types of Steel

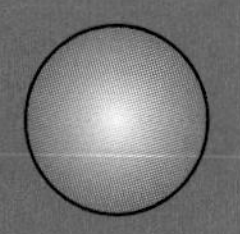

Obtain a small piece of several different types of steel, including 1018, 1045, 1095, 4140, and 52100 steel. Other similar types may be substituted if any of these are not available. If other types are substituted, try to obtain an assortment that includes low-carbon, medium-carbon, and high-carbon steel. Many of these same types will be used for other Self-Demonstrations later in this text.

Test the hardness of each of the five samples and record the hardness values. The Rockwell C scale is recommended for this test. Other testing methods may be used, as long as the method is consistent for each type of steel.

Heat the samples to 1700°F in an oven. After the samples have soaked at this temperature for at least one half hour, plunge them into a bucket of cold water.

After the samples are cool to the touch, check their hardness values again. It may be necessary to file off scale from some of the samples so that they rest flat on the testing surface.

Did the hardness value increase for each of the samples? (It should.)

Did the hardness values of the samples of high-carbon steel increase to higher values than the values for the samples of low-carbon steel? (They should. Most of the hardness values should increase considerably, especially the high-carbon steels.)

Medium-Carbon Steel
• 0.35–0.50% Carbon • Hard & strong after heat treating • More expensive than low-carbon steel

Figure 5-11. *Medium-carbon steel is harder, stronger after heat treating, and more expensive than low-carbon steel.*

High-Carbon Steel
• 0.50–1.00+% Carbon • Hard and strong after heat treating • More expensive than low- & medium-carbon steels

Figure 5-12. *High-carbon steel has a high level of hardness and strength.*

Carbon Steel	Alloy steel
• Lower Cost • Greater Availability	• Higher Strength • Better Wear • Toughness • Special High Temperature Behavior • Better Corrosion Resistance • Special Electrical Properties

Figure 5-13. *A comparison of the advantages offered by carbon steel and alloy steel.*

properties. Alloy steel has properties that cannot ordinarily be obtained with carbon steel. See Figure 5-13. Some common alloying elements for alloy steel are listed in Figure 5-14.

Alloying Elements That Define Alloy Steels
• Aluminum
• Chromium
• Cobalt
• Columbium
• Copper (Over 0.6%)
• Manganese (Over 1.6%)
• Molybdenum
• Nickel
• Silicon (Over 0.6%)
• Titanium
• Tungsten
• Vanadium
• Zirconium

Figure 5-14. *Common alloying elements used in alloy steel.*

Alloy steel is more expensive than carbon steel; it should be used only when a special property is needed. Some of the special properties that alloys may be used for include corrosion resistance, high temperature capability, electromagnetic behavior, high strength, and wear resistance.

Low-alloy structural steel

Low-alloy structural steel has less alloy content than other types of alloy steel. Generally, it is stronger than carbon steel and less expensive than high-alloy steel. The characteristics of low-alloy structural steel are listed in Figure 5-15.

Low-alloy structural steel contains more than an average amount of manganese, silicon, columbium, vanadium, and copper in its composition. This steel is primarily used for structural applications where light weight is important. Because it is stronger than carbon steel, low-alloy structural steel can be used to make smaller, lighter parts. Applications of different types of structural steel are shown in Figures 5-16, 5-17, and 5-18.

Low-Alloy Structural Steel
• Less alloys than other alloy steel
• More alloys than carbon steel
• Less expensive than other alloy steel
• More expensive than carbon steel
• Weldable
• Good corrosion resistance
• Structural applications
• Alloys include: Manganese-Silicon-Columbium-Vanadium-Copper

Figure 5-15. *Low-alloy structural steel is stronger than carbon steel and less expensive than other types of alloy steel.*

Figure 5-16. *More than 22,000 tons of structural steel were used during construction of the Pacific Gas and Electric Company Headquarters Building in San Francisco. This building is 34 stories high, it contains more than 1,000,000 ft² of floor space, and it rises 530′ above the street. (Bethlehem Steel Corporation)*

Figure 5-17. *Construction of the Madison Square Garden Sports and Entertainment Center in New York City required large amounts of structural steel. The structure's 404′ diameter roof is supported by 48 assemblies of 3 3/4″ steel strand cable. The cable-supported roof is one of the largest permanent suspension roofs in the United States. (Bethlehem Steel Corporation)*

Quench and temper structural steel

Quench and temper structural steel is stronger than low-alloy structural steel and has better impact resistance at lower temperatures. This type of steel also has more corrosion resistance and better physical properties in general than low-alloy structural steel. See Figure 5-19.

Quench and temper structural steel is often used in pressure vessels, submarine bodies, and other applications where additional cost is justified in order to obtain greater strength and corrosion resistance.

Maraging steel

Maraging steel contains large amounts of nickel and small amounts of carbon. Generally, maraging steel contains 18%–25% nickel. It has very high strength (up to 250,000 psi) and is able to maintain good ductility and good toughness. Some parts made of maraging steel can be stretched by 11% of the original size before breaking. The characteristics of maraging steel are listed in Figure 5-20.

Maraging steel is used in rocket motor cases and other aerospace applications where high strength, toughness, and ductility are necessary. A typical alloy composition of maraging steel is shown in Figure 5-21.

Tool Steel

Tool steel is so widely used that it is grouped separately from other types of steel, Figure 5-9. There are many different types of

Figure 5-18. *Different types of structural steel were used in construction of the Golden Gate Bridge. (American Iron and Steel Institute)*

tool steel. Some are used in cutting tools, molds, and dies, Figure 5-22. Tool steel is also used for general machine parts, where high strength, wear resistance, and dimensional stability are required.

Different categories of tool steel are listed in Figure 5-23. Each of these 11 categories denotes a type of tool steel used for a specific purpose. The "S" category of tool steel, for example, is used for applications requiring extreme shock resistance, such as air hammers or stamping dies.

Tool steels in the "A" series have special alloys that make them capable of becoming hard when quenched in air. Air quenching is a less violent quenching method than water quenching or oil quenching, so the steel is less likely to crack or distort during the quenching process. Tool steels in the "A" series are expensive, and are only used where dimensional accuracy is extremely important.

Quench & Temper Structural Steel
• Stronger than low-alloy structural steel • Better properties than low-alloy structural steel • More expensive than low-alloy structural steel • Structural applications

Figure 5-19. *Quench and temper structural steel is used for applications where strength and resistance to corrosion are important.*

Maraging Steel
• 18-25% Nickel • Low carbon content • Very high strength • Good ductility • Good toughness

Figure 5-20. *Large amounts of nickel are used in maraging steel.*

Typical Composition of Maraging Steels	
Alloying Element	**Amount**
Carbon	0.03%
Nickel	18.5%
Cobalt	7.5%
Molybdenum	4.8%
Titanium	0.40%
Zirconium	0.01%
Aluminum	0.10%
Silicon	0.10% max
Manganese	0.10% max
Sulfur	0.01% max
Phosphorus	0.01% max

Figure 5-21. *A typical composition of alloying elements for maraging steel.*

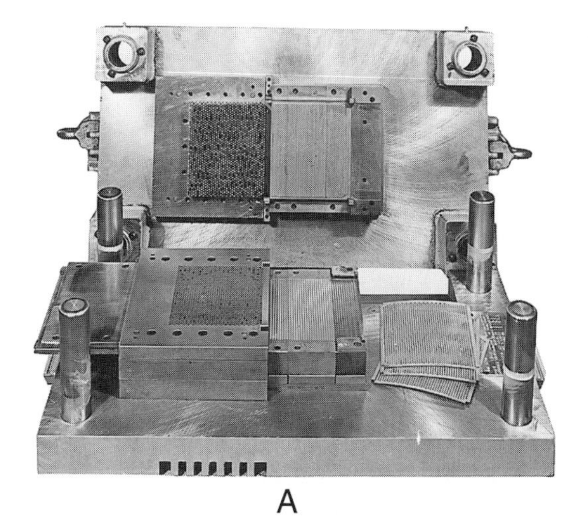

A

B

Figure 5-22. *Applications for tool steel. A—This piercing die is made from Carpenter Vega® a commercial brand of A6 air-hardening tool steel. (Carpenter Technology Corporation) B—This tool and die maker is polishing a slide insert of a Ford transmission case die. It is made of Crucible Nu-Die® V Densified™, a commercial brand of H13 tool steel. (Crucible Specialty Metals Division of Colt Industries)*

Tool Steel Categories		
Category	**Example**	**Description**
W	W1 & W5	Water Hardening
O	O1 & O6	Oil Hardening
A	A2 & AS	Air Hardening
D	D1 & D2	Oil or Air Hardening
S	S2 & S4	Shock Resisting
H	H10 & H41	Hot Working
M	M1 & M34	High Speed (Molybdenum)
T	T2 & T15	High Speed (Tungsten)
L	L1 & L2	Special Purpose
F	F1 & F3	Special Purpose
P	P5 & P20	Mold Making

Figure 5-23. *Common categories of tool steel.*

The "M" and "T" categories of tool steel are designed for high-speed work. These steels contain higher quantities of molybdenum and tungsten.

Tool steels in the "H" category have good strength at high temperature. They are used for hot-working processes, such as forging and die casting.

Stainless Steel

Stainless steel is a classification of special alloy steel that is used extensively, Figure 5-9. As the name implies, stainless steel is extremely resistant to corrosion. It costs more than carbon steel and is harder to cut and machine. However, if a machined part is expected to come into contact with a corrosive atmosphere, the extra cost is justified.

Stainless steel is commonly used to meet special sanitation requirements. It is designed for such applications as food processing and the transfer of chemicals through pipes.

All stainless steel contains high quantities of chromium alloy, and many types contain high quantities of nickel alloy. These alloys give stainless steel its superior ability to resist corrosion. See Figure 5-24.

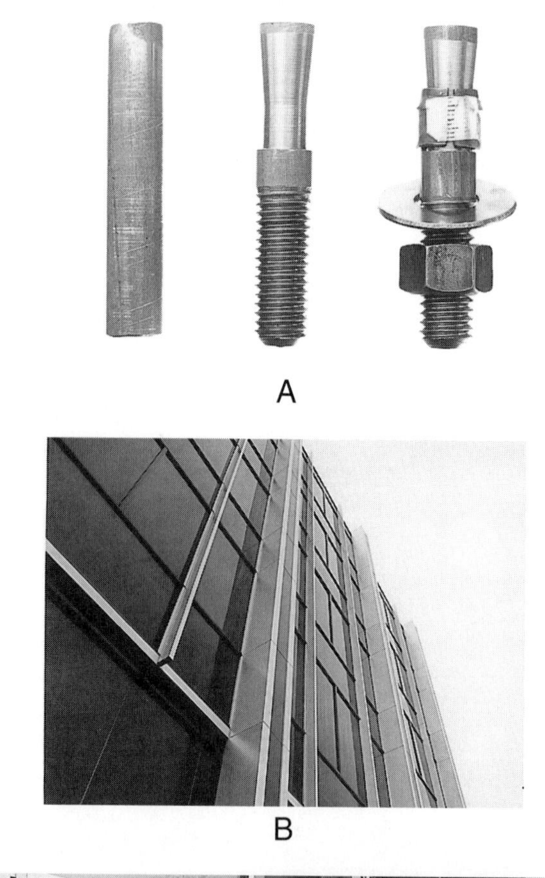

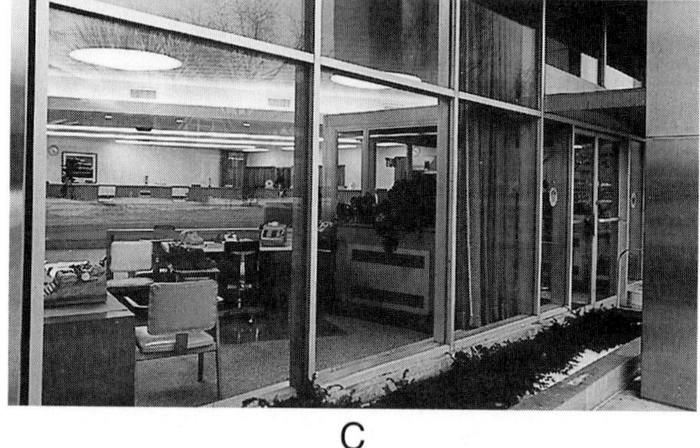

Figure 5-24. *Stainless steel applications. A—This stainless steel KWIK-BOLT® Concrete Anchor is made in three production stages. 1—1/2" round bar stock of 303 PLUS-X stainless steel. 2—Part has been machined and degreased. 3—KWIK-BOLT® complete with floating wedges, thermoplastic sleeve, and foil retainer. (Crucible Specialty Metals Division of Colt Industries) B—Stainless steel is used for the building trim on this structure. (Armco Inc.) C—Stainless steel is used as trim around these windows. (Armco Inc.)*

Spring Steel

Spring steel is a special classification of steel that has great hardness, strength, and elasticity. It is used in leaf springs, clock springs, knife blades, and golf club shafts.

Spring steel is normally purchased as a thin, flat material. The carbon content can vary anywhere from 0.35% to 1.4%.

Compared to carbon steel, spring steel is relatively expensive. However, it has special strength and ductility characteristics that make it worth the extra cost. Additional alloying elements in spring steel include manganese (up to 0.8%) and chromium, silicon, vanadium, or molybdenum.

Applications of spring steel include camera shutters, circular cutters in machine tool industry, fishing rods, putty knives, manicure files, bread knives, measuring tapes, steel rules, trowels for concrete work, feeler inspection gauges, and small springs for clocks and other devices. See Figure 5-25.

Other Types of High-Alloy Steel

There are many types of special purpose steels that have unusual alloy contents:

- In the electrical industry, special steels with high silicon content are used in generators and transformers.
- In rockets, missiles, jet aircraft, and nuclear power devices (where operation at high temperature is important), special alloy steels that contain titanium, columbium, nickel, and chromium are used.
- Special cobalt steels containing over 30% cobalt are used for electromagnets.

Many of these special purpose steels are extremely expensive and can be justified for use only when their unusual properties are required for safety or special needs.

Figure 5-25. *Spring steel is made into many different shapes for a variety of applications other than conventional springs. (Wallace Barnes Steel, Barnes Group Inc.)*

Cast Iron

Cast iron is a material that uses iron as its primary ingredient. It contains 2% to 6% carbon and small amounts of silicon. Other alloying elements are also used. Bear in mind that "cast iron" and "iron" are two very different terms. Consider it a coincidence that the word *iron* is used in both names.

How Steel and Cast Iron Differ

It is important to understand the difference between steel and cast iron. The difference can be described as follows. When steel is produced, the carbon that is added to the iron dissolves and disappears. This is similar to the dissolving of sugar in water. If you add a small amount of sugar to a glass of water, the sugar immediately dissolves and becomes invisible. If you add a little more sugar, the same thing happens. Eventually, if you add enough sugar, the water does not dissolve all of it. Some sugar precipitates out, and it becomes visible.

The same situation exists with iron and carbon in steel and cast iron. When carbon is added to iron, the carbon dissolves and disappears. Eventually, if enough carbon is added, it precipitates out. Steel is iron with the carbon in solution, which occurs below 1.6 to 2% (the percentage varies in different types of steel). Cast iron is iron in which some of the carbon has precipitated out and appears as flakes (as in gray cast iron, the most common iron of the cast iron family) or little spheres (nodular iron). See Figure 5-26.

The effects of the precipitated carbon are both good and bad. The flakes provide a cushioning effect for iron when it receives high compression loads, Figure 5-27. This makes cast iron a very good material when vibration is present along with large compressive loads. See Figure 5-28.

On the other hand, gray cast iron is more brittle than steel, and it has very poor strength when stretched in tension, Figure 5-29. The precipitated flakes encourage the formation of cracks and cause breakage.

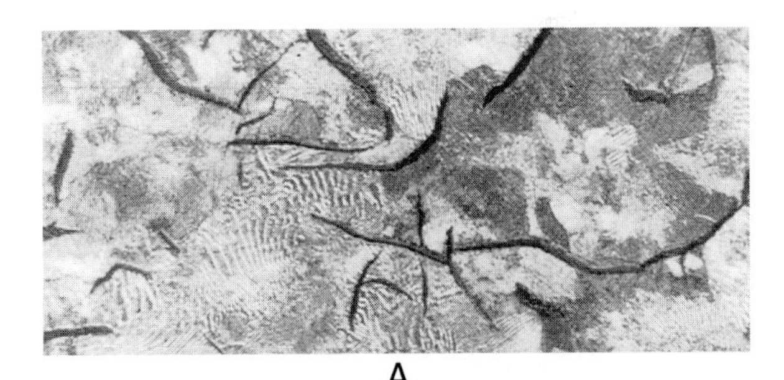

A

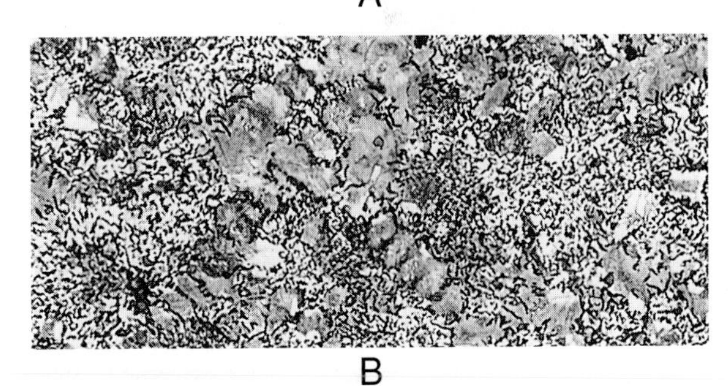

B

Figure 5-26. *Cast iron crystals. A—In gray cast iron, carbon flakes tend to crystallize out of the iron. (General Motors Powertrain Group) B—The graphite flakes in this cast iron sample are very fine and can barely be recognized. The flakes are the tiny black lines that appear throughout the photograph. (Buehler Ltd.)*

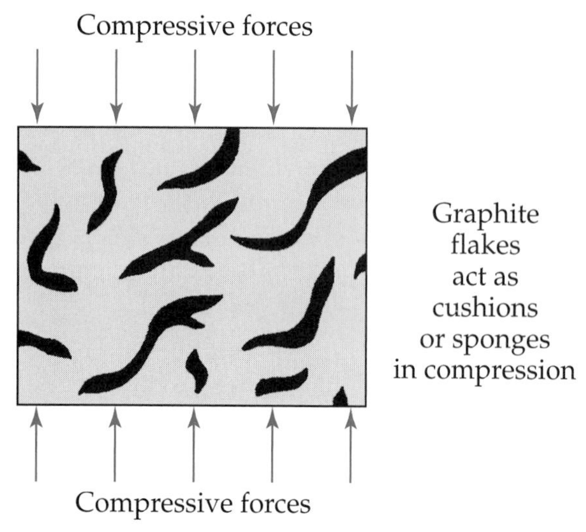

Figure 5-27. *Carbon flakes serve as a cushion for iron and protect it during compression.*

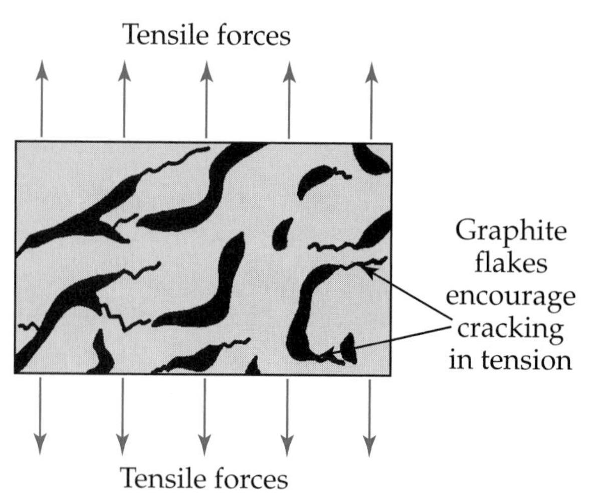

Figure 5-29. *The carbon flakes in gray cast iron make it weaker than pure iron in tension.*

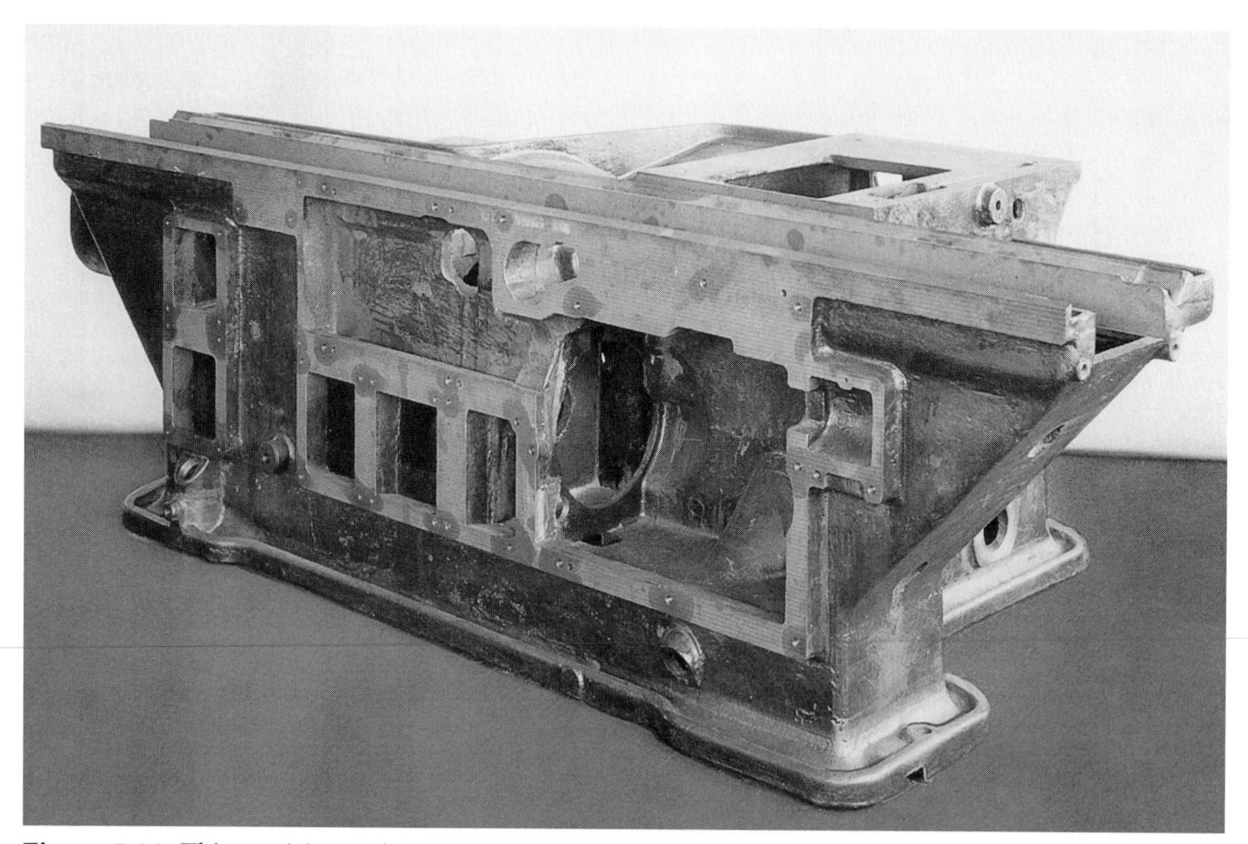

Figure 5-28. *This precision tool grinder base absorbs vibration. Making the base from cast iron gives it a long service life. (Iron Castings Society)*

Cast iron is very easy to machine. It is also easy to cast, due to the significantly low melting point relative to steel. This is a direct result of the higher carbon content of cast iron. Cast iron casting is also easier to control than steel casting.

Applications of Cast Iron

Cast iron is used extensively for the frames of large equipment and machine tools, Figure 5-30. Its damping capacity and compressive strength make it very suitable for these applications. Because cast iron has very good resistance to wear, it is used for engine blocks, piston rings, brake drums, rolls, and crushers.

In architecture, cast iron is used for stair treads and lampposts. Cast iron is also used for manhole covers, furnace grates, and any other applications where its castability, machinability, damping characteristics, and compressive strength are of benefit.

Types of Cast Iron

The following are the five basic types of cast iron (see Figure 5-31):

- Gray.
- White.
- Malleable.
- Ductile.
- Special alloy.

Gray cast iron

When people refer to a material as *cast iron*, but do not specify the type, chances are

Types of Cast Iron
• Gray (most common)
• White (most brittle)
• Malleable (higher quality)
• Ductile (higher quality)
• Special alloy (special properties)

Figure 5-31. *This chart lists the five basic types of cast iron.*

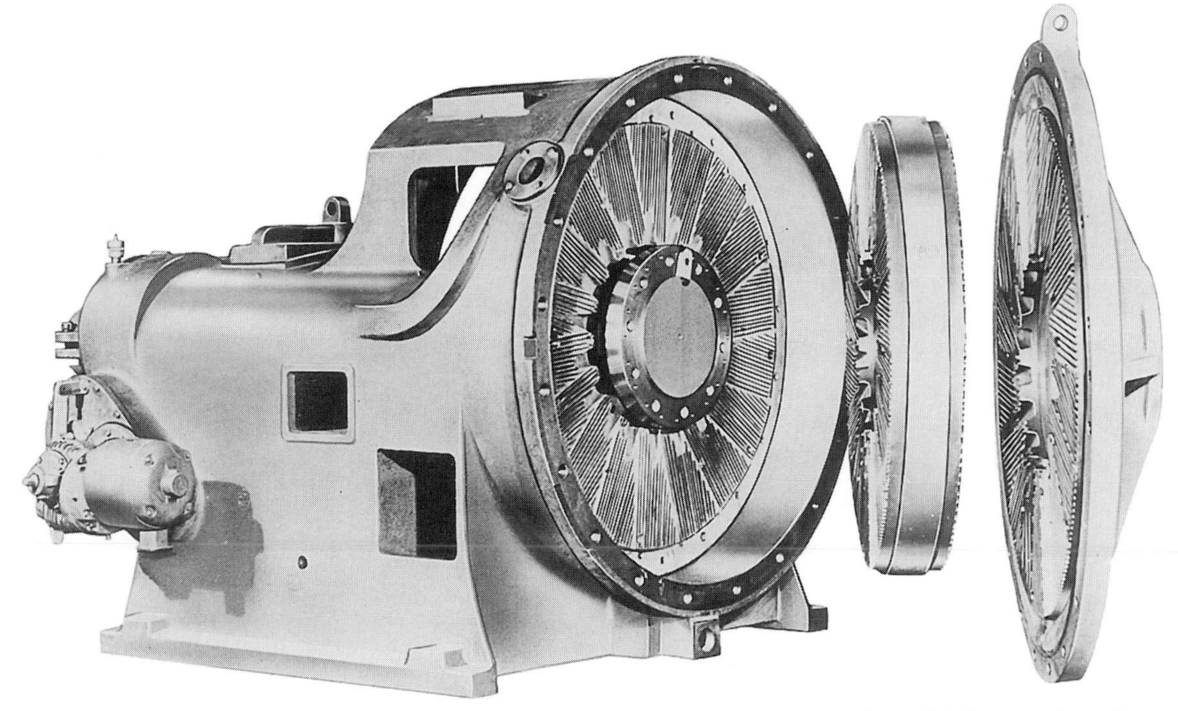

Figure 5-30. *Iron castings are used extensively in paper mill equipment, such as this large pulp refiner. Cast iron meets rugged service and corrosion resistance requirements. (Iron Castings Society)*

they are talking about gray cast iron. *Gray cast iron* is the most widely used type of cast iron. In fact, there is more gray cast iron used than the other four types combined.

Although gray cast iron does not have some of the good qualities of ductile, malleable, or special alloy cast iron, it is considerably less expensive. The differences in cost and quality must be considered when a type of cast iron is chosen for a given application.

Gray cast iron is very hard and brittle, and it has relatively poor tensile strength due to the graphite flakes in its structure. However, it has excellent compressive strength and damping capacity, and it can be easily cast.

White cast iron

White cast iron is not used as extensively as gray cast iron because it is harder, more brittle, and more difficult to machine. It also has less impact strength.

The extreme hardness of white cast iron, however, makes it a valuable material for some applications. For example, white cast iron rolls, used in mills, need great hardness to break up stone and other materials. See Figure 5-32. White cast iron is also produced as an intermediate step in the production of malleable cast iron.

When white cast iron is observed under a microscope, it looks much different from gray cast iron. See Figure 5-33.

Malleable cast iron

Malleable cast iron has several special properties that make it superior to both gray cast

Figure 5-32. *White cast iron is used in these muller tires because of its superior hardness and abrasion resistance. (Iron Casting Society)*

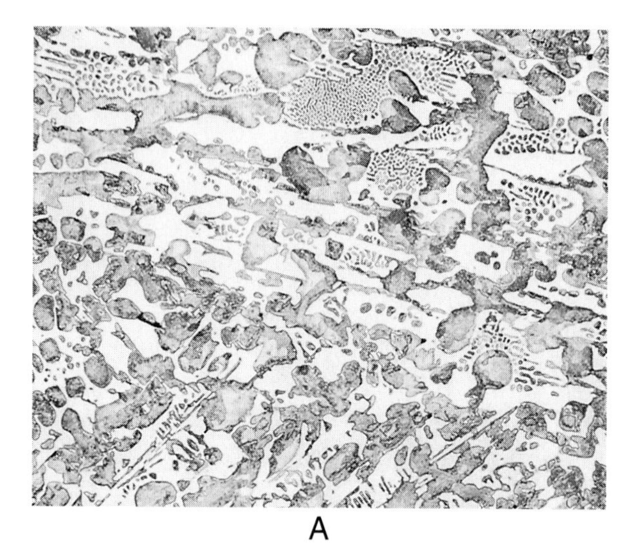

A

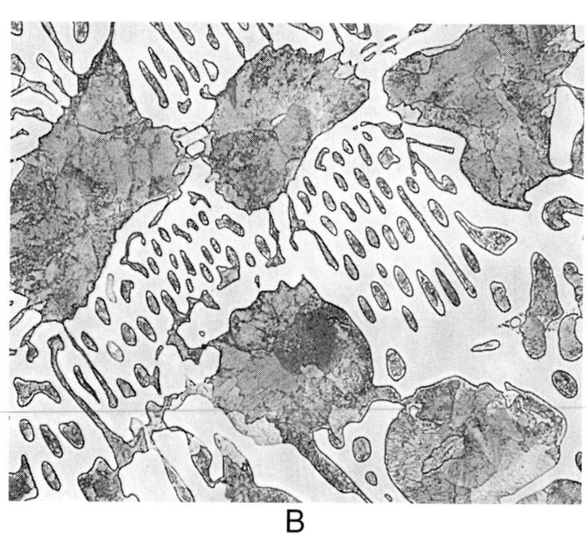

B

Figure 5-33. *Microscopic views of white cast iron. A—Magnification is 100X. B—Magnification is 400X. (LECO Corporation)*

iron and white cast iron. Malleable cast iron has more tensile strength, ductility, and impact strength. It also costs more than gray cast iron or white cast iron. The table in Figure 5-34 compares the properties of the various types of cast iron.

Malleable cast iron is made from white cast iron. The white cast iron is heated extensively at high temperatures to refine it. Eventually, carbides break down into carbon and free iron. The carbon collects in small, roughly spherical particles known as *temper carbon particles.* The formation of these particles causes malleable cast iron to become more ductile and workable, without sacrificing high compressive strength. Common applications for malleable cast iron are shown in Figure 5-35.

Ductile cast iron

Ductile cast iron is sometimes referred to as *nodular cast iron* because its graphite particles are shaped like tiny spheres (or nodules). Unlike the temper carbon particles in malleable cast iron, the nodules in ductile cast iron are precipitated directly from the liquid iron by treatment with magnesium.

Ductile cast iron, as its name implies, has great ductility. Its tensile strength is comparable to that of malleable cast iron, Figure 5-34. The price of ductile cast iron has become more competitive, and it is replacing gray cast iron in many applications. Some typical uses of ductile cast iron are shown in Figure 5-36.

Special alloy cast iron

Most cast iron has a very basic alloy content. It primarily contains just carbon and silicon as its alloys. A number of types of *special alloy cast iron* have been developed for specific applications.

A few grades of special alloy cast iron contain high percentages of nickel, copper, chromium, and other alloys. Nickel, copper, and chromium produce good corrosion resistance and good chemical resistance to acids.

Properties of Cast Iron

	Gray	White	Malleable	Ductile
Weight lbs/in^3	.25-.27	.27-.28	.26-.27	.25-.27
Tensile Strength psi × 10^3	20-70	20-50	60-120	60-120
Compressive Strength psi × 10^3	100-170	100-150E	200-290	120-300
Impact Strength V-Notched Charpy ft-lb	Low	3-10	14-17	2-30
Hardness BHN	140-290	300-580	110-270	140-330
Modulus of Elasticity psi × 10^6	12-20	Low	25-28	18-25
Coefficient of Thermal Expansion in./in. °F × 10^{-6}	6	5	6-8	6-10

Figure 5-34. *A comparison of the properties of different types of cast iron.*

Figure 5-35. *Malleable cast iron is used for applications requiring high ductility and compressive strength. (General Motors Powertrain Group)*

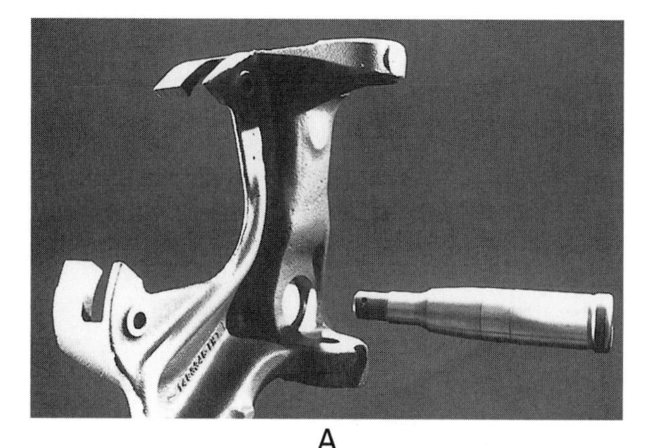
A

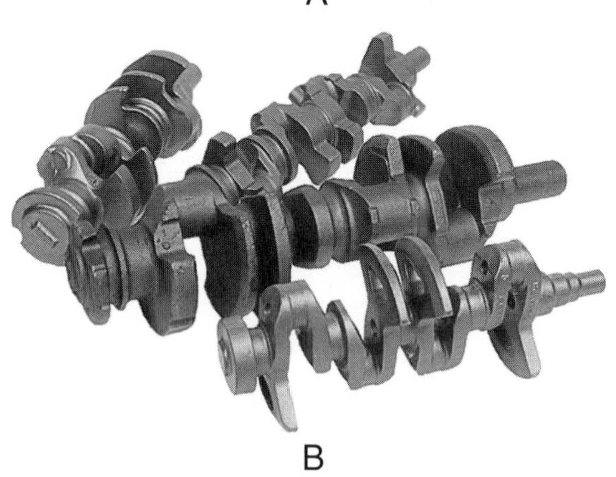
B

Figure 5-36. *Ductile cast iron applications. A—The wheel spindle and support shown are ductile cast iron parts. B—Automotive crankshafts are typically ductile cast iron parts. (General Motors Powertrain Group)*

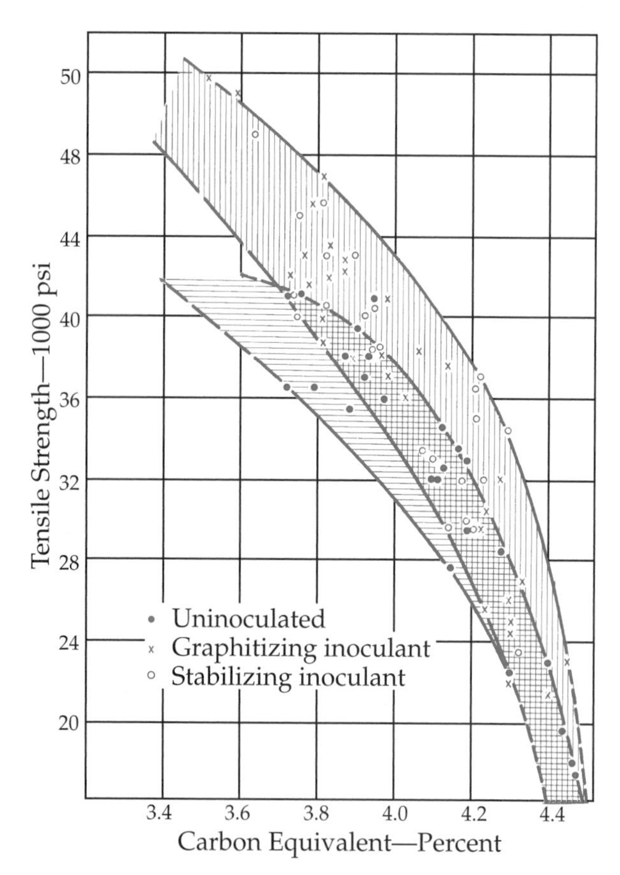

Figure 5-37. *The addition of small quantities of alloys to cast iron (inoculation) improves its properties. This graph shows the influence of alloys on the tensile strength of cast iron. (Iron Castings Society)*

Other special alloy cast irons have been developed that have greater strength and better high-temperature properties, Figure 5-37. Special alloy cast iron is used in cylinders, pistons, piston rings, and turbine stator vanes, Figure 5-38.

Wrought Iron

Wrought iron is very different from cast iron. *Wrought iron* is almost pure iron; it has very little carbon in its composition, Figure 5-5.

Wrought iron has very low strength and hardness, since it contains little carbon. However, it is very ductile and resistant to corrosion. Before 1860, wrought iron was the most important structural metal available. However, due to advances brought about by metallurgical research, wrought iron has been widely replaced as a structural material.

The key asset for wrought iron today is good corrosion resistance. Many fibrous stringers of slag are distributed throughout wrought iron. If corrosion attacks the iron, the deterioration proceeds only until it meets a slag stringer. The corrosion stops there, and forms a protective coating. The slag then becomes a barrier against further corrosion.

A microscopic view of slag stringers is shown in Figure 5-39. The slag contains considerable amounts of silicon. A typical chemical makeup of wrought iron is shown in Figure 5-40.

Figure 5-38. *The vanes of this turbine stator are integrally cast of a high-alloy cast iron. They are not machined. (Iron Castings Society)*

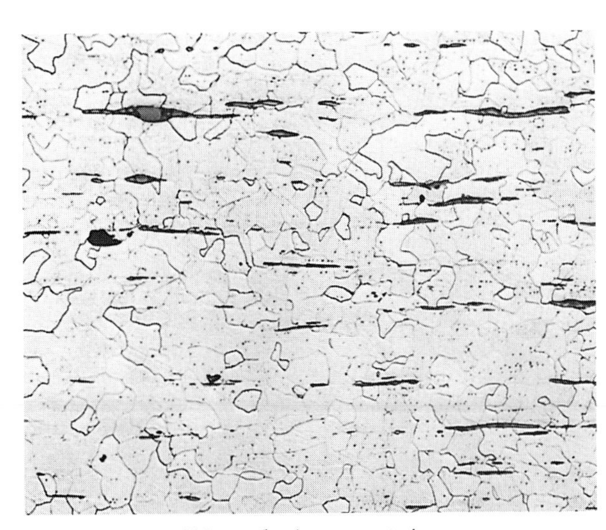

Figure 5-39. *Wrought iron contains many fibrous, elongated stringers of slag. The stringers are composed largely of ferrous oxide and silicon dioxide.*

Typical Composition of Wrought Iron	
Element	**Percentage**
Iron	Over 99.6%
Carbon	0.06%–0.08%
Silicon	0.10%–0.16%
Manganese	0.02%–0.05%
Sulfur	0.01%
Phosphorus	0.06%–0.07%

Figure 5-40. *This chart shows the typical chemical makeup of wrought iron.*

The amount of wrought iron used today is small, as compared to the amount of cast iron and steel in commercial use. Applications for wrought iron include processing tanks, wastewater pipes, sheeting on ships, and gratings.

Test Your Knowledge

Write your answers on a separate sheet of paper. Do not write in this book.

1. Name the two elements found in every type of steel.
2. What percentage of carbon is in steel?
3. Most steel contains more than _____ percent iron.
4. As the carbon content in steel increases, does the steel become more brittle or more ductile?
5. As the carbon content in steel increases, does steel become harder or softer?
6. As the carbon content in steel increases, does steel become stronger or weaker?
7. If the numerical name of a type of steel contains four digits, what do the first two digits tell you about the steel?
8. If the numerical name of a type of steel contains four digits, what do the last two digits tell you about the steel?
9. Why do some steels have five digit names?
10. What alloy(s) are found in 4024 steel?

11. What alloy(s) are found in unusual amounts in 4147 steel?
12. What alloy(s) are found in unusual amounts in 52100 steel?
13. What alloy(s) are found in unusual amounts in 8630 steel?
14. Compare the carbon contents of steel, cast iron, and wrought iron.
15. Name two alloying elements that improve the corrosion resistance of steel.
16. What is the percentage of nickel in 3310 steel?
17. What is the percentage of chromium in 6150 steel?
18. Name four ways in which alloying elements can improve steel.
19. Most steel is classified as either carbon steel or alloy steel. Which of the two is generally less expensive?
20. Which has the greatest use today, carbon steel or alloy steel?
21. What are the ranges of carbon content in the three types of carbon steel?
22. What is *maraging steel?*
23. What category of steel has members with names such as W1 and A6?
24. What does the *S* stand for in a steel identified as S1?
25. In addition to carbon, what two alloys are common in stainless steel?
26. List at least six uses for spring steel.
27. What is the difference between steel and cast iron?
28. Name the five basic types of cast iron.
29. Which type of cast iron is the most commonly used?
30. What is the biggest disadvantage of white cast iron?
31. What is the advantage of gray and white cast iron over malleable and ductile cast iron?
32. What type of cast iron is produced when carbon is broken down into temper carbon particles?
33. What are the main advantages of malleable and ductile cast iron over gray and white cast iron?
34. Wrought iron offers very good resistance against _____.
35. An engineer is designing a high-speed dental drill. The materials used for the drill must have high strength, good heat resistance, and thermal stability. What types of steel should the engineer consider for this application?
36. An engineer is designing a head for a nine-iron golf club. The part requires high impact strength and resistance to distortion. What types of steel should the engineer consider for this application?
37. An engineer is designing a sheet metal frame for a small business machine. What mechanical properties would be important for this material? What materials should the engineer consider for this application?
38. An engineer is designing a gear train for use in a large rock crusher. What properties would be important for the materials to be used for the gears? What materials should the engineer consider for this application? What are the advantages and disadvantages of using cast iron, compared to steel?

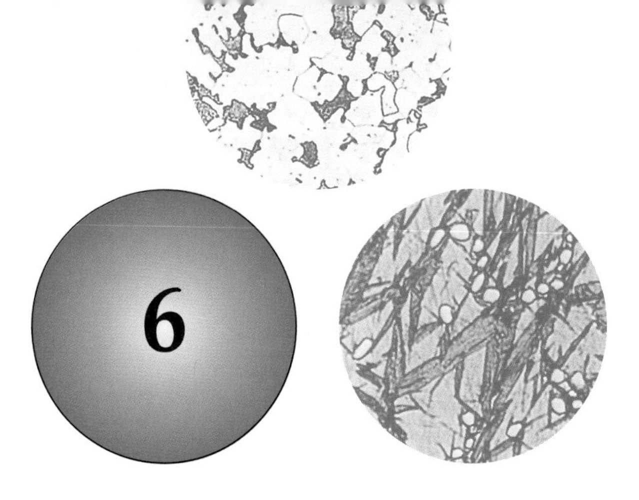

Manufacture of Iron and Steel

After studying this chapter, you will be able to:

- Identify the basic steps in the production of steel.
- Describe how cast iron is made.
- Identify the properties and uses of iron ore and pig iron.
- Differentiate between common steel-making and cast iron-making processes.
- Discuss how rolling mills change steel ingots into different shapes.

The manufacture of steel is done in two basic steps, Figure 6-1. Two furnaces—or metallurgical processes—are required. First, *iron ore* is converted to *pig iron* in the *blast furnace*, Figure 6-2. Then, *pig iron* is made into *steel* in a *steel-making furnace*. See Figure 6-3.

The Steel-Making Process

The entire process for making steel is diagrammed in Figure 6-4. In the first step, *iron ore* is mined from the ground and shipped to a steel-making center. There, the iron ore is mixed with coke, limestone, and hot gases inside a *blast furnace.* The products coming out of the blast furnace are pig iron, slag, and hot gases. The *pig iron* is used to make steel or cast iron.

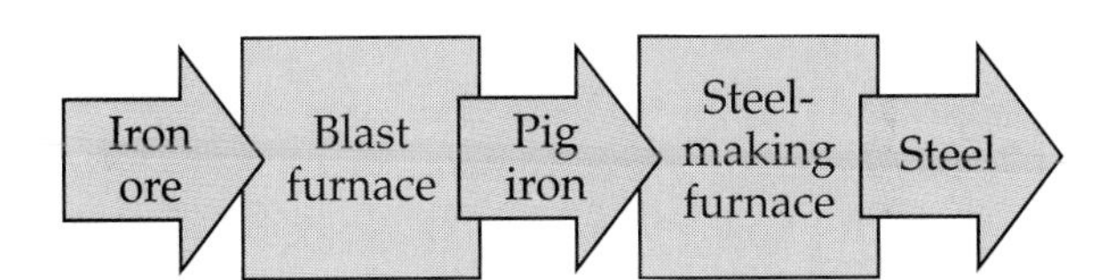

Figure 6-1. *Producing steel is a two-step process. First, iron ore is converted to pig iron in a blast furnace. Then, the pig iron is changed to steel in a steel-making furnace.*

Figure 6-2. *This huge blast furnace located at the Armco Inc. Ashland Works stands 234′ high with a hearth diameter of 33 1/2′. A computer selects the materials for the charge.*

Figure 6-3. *In the steel-making process, an overhead crane pours a ladle of molten iron into a basic oxygen furnace. The furnace can manufacture 300 tons of steel in less than an hour.*

Steel-Making Furnaces

There are two main types of steel-making furnaces:

- Basic oxygen furnace.
- Electric arc furnace.

In the steel-making process, the pig iron enters one of these furnaces along with fuel, alloys, steel scrap, limestone, and small amounts of iron ore. Then the pig iron is transformed into steel.

Cast Iron-Making Furnaces

There are two types of cast iron-making furnaces:

- Cupola.
- Electric induction furnace.

In this process, the pig iron goes into the furnace and is joined by fuel, limestone, and steel scrap. The fresh cast iron normally is poured directly into molds for castings.

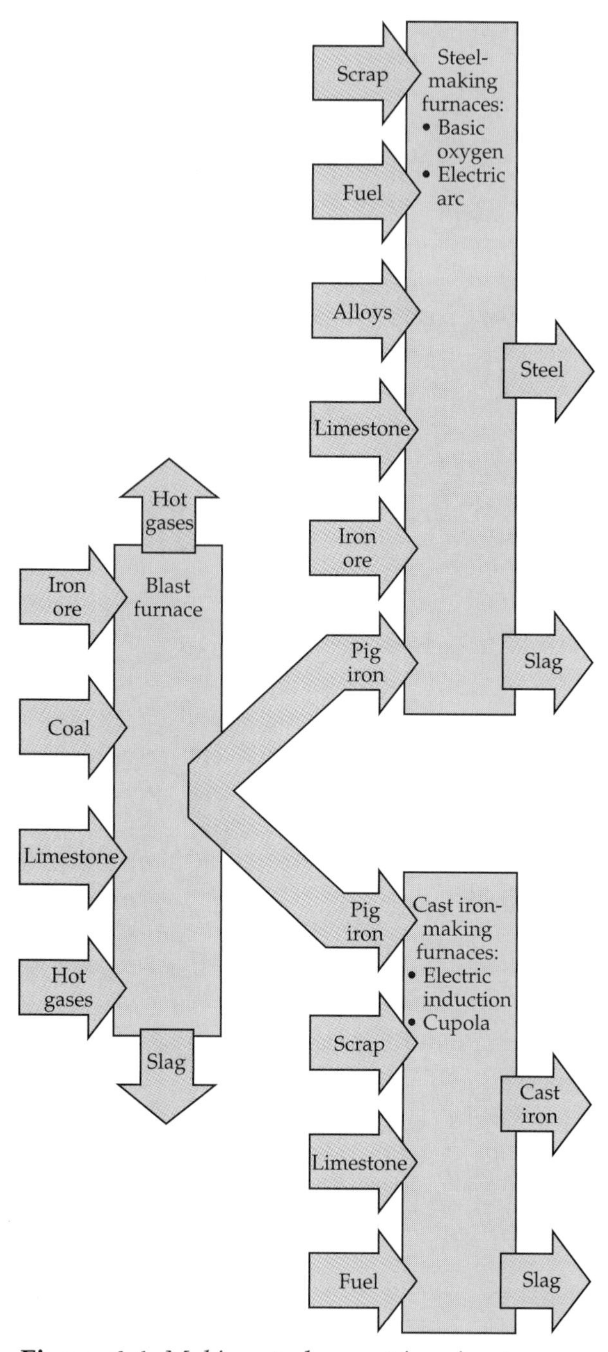

Figure 6-4. *Making steel or cast iron is a two-step process.*

Figure 6-5. *Iron ore has a high iron content. (Erie Mining Company, managed by Pickands Mather & Co., Cleveland, Ohio)*

Thousands of companies throughout the United States have cast iron furnaces. They pour final castings in their foundries.

Steel Foundries

While cast iron foundries are common, there are very few small steel foundries. Most of the steel-making furnaces are located near Chicago, Pittsburgh, Bethlehem (Pennsylvania), or one of the other great steel-making centers.

The steel that is poured from a steel-making furnace is just at the beginning of its long journey. Next it travels to the ingot building, then to the rolling mill. There the steel is made into its final commercial form.

Steel making and cast iron making are exciting businesses. Each of the steps shown in Figure 6-4 will be discussed in more detail later in this chapter.

Iron Ore

Iron ore is mined from the ground. It is an unimpressive looking rock, Figure 6-5, that may contain only 30% iron. The iron in the rock is in chemical compounds (ores) in which the iron is combined with oxygen or sulfur. This ore is found intermixed in rock, gravel, clay, sand, or even mud. Getting the pure iron out of this form is very costly, but necessary.

Iron Ore Deposits

Iron ore deposits are found in several locations in the United States. The largest concentration is in the Lake Superior region, Figure 6-6. Most of the iron ore is mined from open pit mines. Here the rock is broken up by explosive charges, then loaded with power shovels onto trucks, Figure 6-7, or railroad cars. It is then taken to a nearby processing plant.

Iron Ore Processing

At the processing plant, the ore goes through many operations. First, the large chunks of stone are broken up by giant *crushers,* Figure 6-8. These smaller pieces are then fed into rotating *roll mills* and *ball mills,* Figure 6-9. Hardened steel rollers or balls rotate with the ore and steadily pound it into finer particles.

After being crushed and milled, the rock particles become so fine that they feel like sugar or dry sand. In this condition, valuable iron ore particles can be separated from the worthless particles. This is done with *magnetic separators,* Figure 6-10. Strong magnets pull the iron ore particles away from the rest of the powder as the separator rotates.

Next, the powder goes to *flotation cells,* where the particles are immersed in a liquid. See Figure 6-11. Impurities bubble to the top and are skimmed away, leaving an iron-rich mixture. After magnetic separation and flotation, the ore may contain as much as 70% iron. It is then ready to be shipped to the steel mill.

Types of Iron Ore

There are many different types of iron ore, Figure 6-12. One of the most valuable types is *hematite,* which is reddish in color. This is a valuable ore because it contains a high percentage of iron.

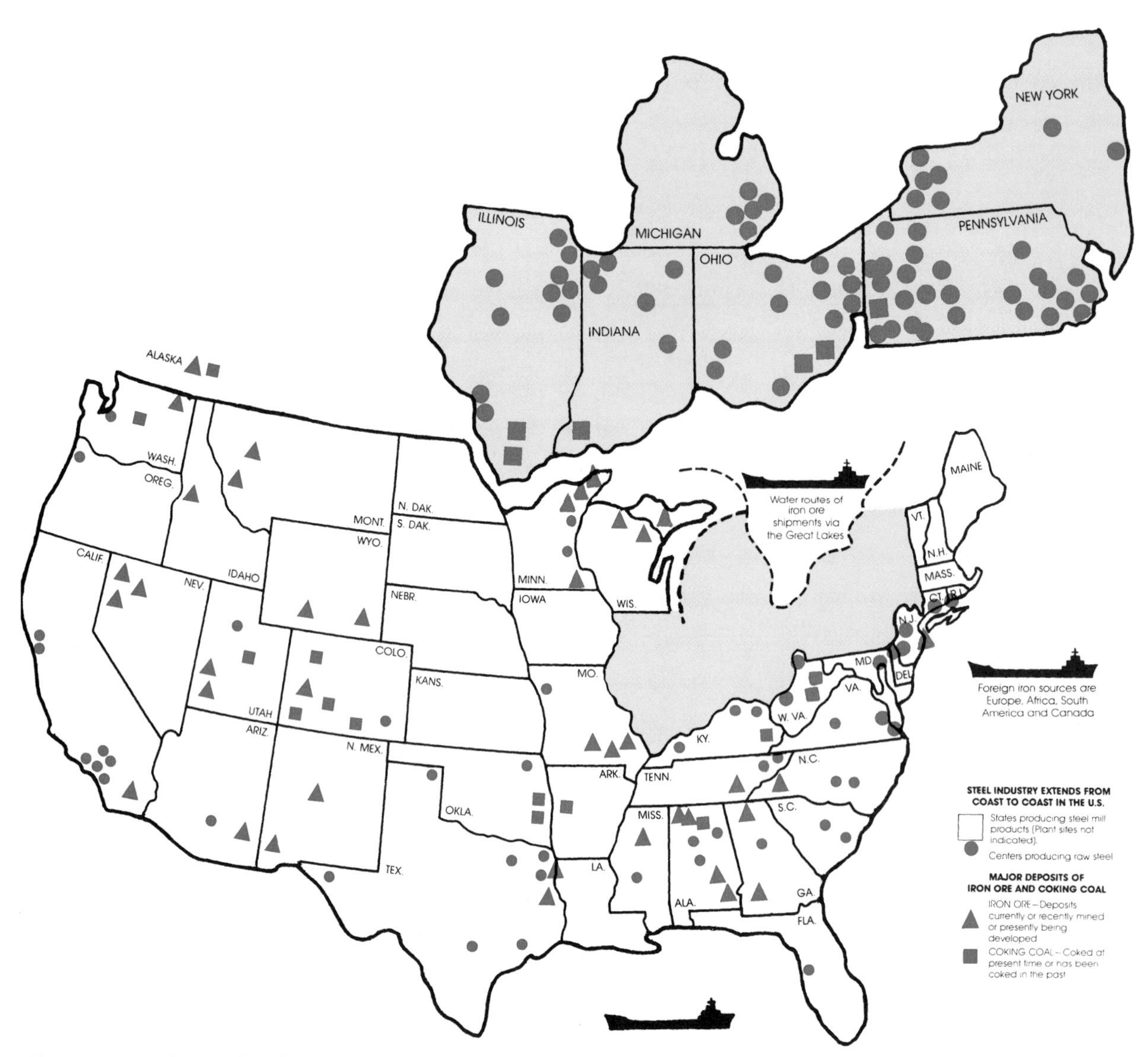

Figure 6-6. *The steel industry is widespread in the United States. Lake deposits of iron ore are concentrated in the Great Lakes region. (American Iron and Steel Institute)*

Taconite is one of the most interesting types of iron ore. At one time, taconite was considered worthless because it contained only 20% to 30% iron. Getting iron out of the ore required processing that was too expensive and time consuming. Today, due to improvements in processing methods, taconite is one of the most popular ores. Due to modern mining technology, the refining of taconite is now more efficient.

After the taconite ore has been milled to a fine powder and separated from the waste powder, it is poured into a *balling drum.* See Figure 6-13. Here, the taconite is mixed with a binder, or "glue," and rolled into small, round pellets about 1/2" in diameter. See Figure 6-14. The pellets are baked and hardened. The hardened pellets are then shipped to steel centers, where they are stored for use. See Figure 6-15. It is much more convenient to

Figure 6-7. *A power shovel loads large chunks of iron ore onto trucks at the iron ore mine. (Erie Mining Company, managed by Pickands Mather & Co., Cleveland, Ohio)*

Figure 6-8. *Chunks of iron ore are broken into smaller pieces by large crushers. (Erie Mining Company, managed by Pickands Mather & Co., Cleveland, Ohio)*

Figure 6-9. *Ball mills break iron ore into smaller pieces during processing. (Erie Mining Company, managed by Pickands Mather & Co., Cleveland, Ohio)*

Figure 6-10. *Valuable iron ore particles are recovered by magnetic separators. (The Cleveland-Cliffs Iron Company)*

handle the ore in pellet form than to handle it in powder form.

Blast Furnace

As discussed earlier, the blast furnace converts iron ore to pig iron. Iron ore, coke, and limestone are carried in skip cars, Figure 6-16, to the top of the blast furnace. The skip cars resemble little railroad cars riding on a very steep, inclined track.

A blast furnace may be more than 250' high, taller than a 12-story building. See Figure 6-17. The inside is about 30' wide. The outside surface of the blast furnace is a thick steel shell. The inside wall is lined with fire-resistant brick.

Figure 6-11. *Flotation cells generate air bubbles to remove impurities from iron ore particles. Impurities adhere to the bubbles and float over the top of the cell while the iron particles are drawn off the bottom. (The Cleveland-Cliffs Iron Company)*

Iron Ore	Iron Compounds in the Ore
Hematite	Fe_2O_3
Magnetite	Fe_3O_4
Siderite	$FeCO_3$
Pyrite	FeS
Limonite	Fe_3O_4
Taconite	Various
Jasper	Various

Figure 6-12. *This chart lists the more common types of iron ore.*

Up to 350 tons of molten iron are produced and tapped from the furnace every three to five hours. Over 80 million tons of molten iron is produced annually.

A blast furnace in operation is spectacular to see, especially at night. The tremendous heat needed to melt the coke, limestone, and iron ore is attained by a continuous blast of hot air introduced at the bottom of the furnace. The coke burns first. It, in turn, melts the iron ore and limestone,

Figure 6-13. *Taconite pellets are formed in a balling drum. (The Cleveland-Cliffs Iron Company)*

Figure 6-14. *Taconite pellets stockpiled at the Erie Mining Company processing plant. These pellets, with an iron content of about 65%, are comparable to high-grade ore. (Bethlehem Steel Corporation)*

Figure 6-15. *An enormous pile of iron ore feeds a pair of high-capacity blast furnaces. (Armco Inc.)*

Figure 6-16. *Skip cars are used to carry the coke, limestone, and iron ore to the top of the blast furnace. (American Iron and Steel Institute)*

and a chemical reaction takes place that produces free iron.

As everything melts, it trickles down through the tall stack and collects in a molten pool at the bottom of the blast furnace.

Since the iron is the heaviest, it sinks to the bottom of the furnace. The limestone reacts with impurities, forming *slag,* a relatively useless byproduct. The slag floats on top of the molten pool and looks like a scum.

As the hot gases work their way to the top of the blast furnace, they are filtered by dust catchers, cleaned by special equipment, and collected. These hot gases are then routed through one or more *stoves.* The stoves are tall, thin, and cylindrical in shape. They may be 120' tall and 28' in diameter. They look like junior-sized blast furnaces.

Inside the stoves, there is a complex arrangement of bricks. These bricks heat up as the hot gases pass through them. Then, a short time later, the incoming air (to be used for the hot air blast in the bottom of the furnace) is reheated by the hot bricks.

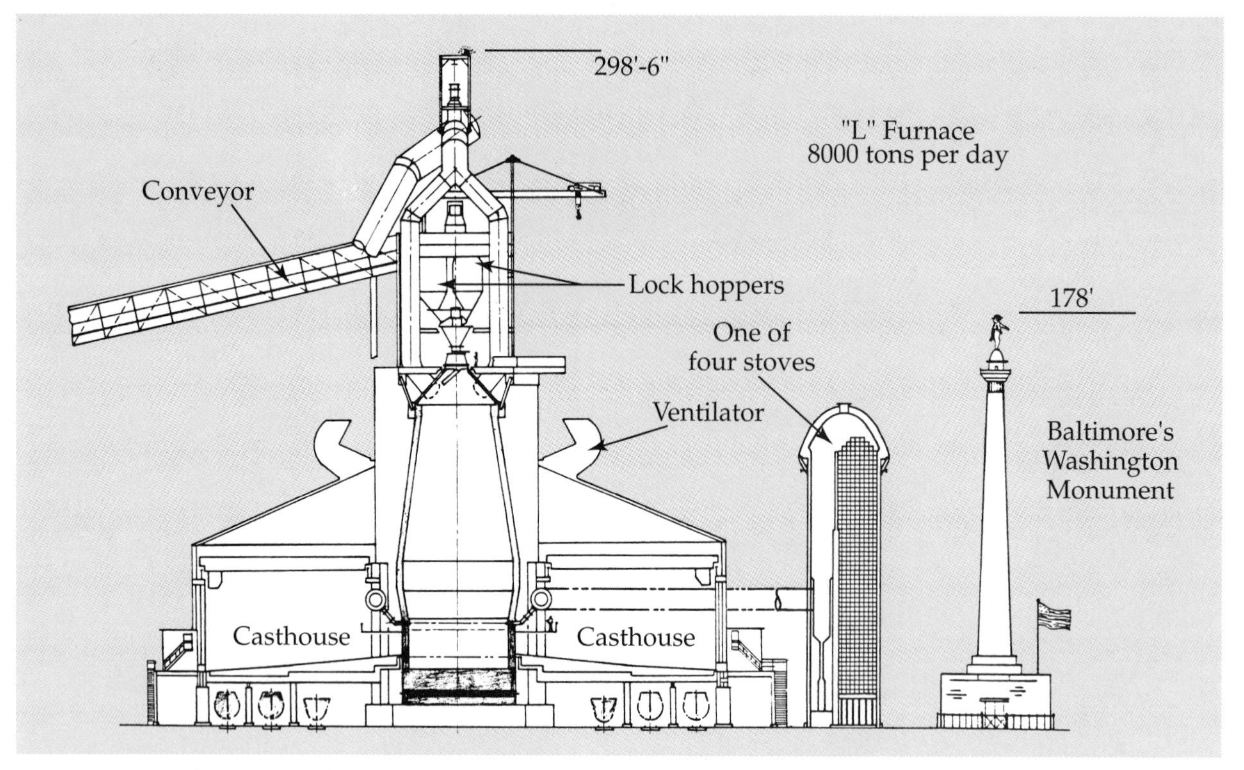

Figure 6-17. *This sketch of a modern blast furnace illustrates its mammoth size. This blast furnace is located at the Bethlehem Steel Corporation Sparrows Point Maryland plant.*

Modern technology in this field has come a long way. Today, technicians watch the overall blast furnace operation from a large air-conditioned control room. This control room looks like a computer center with switch panels, readouts, and colored lights throughout the room. See Figure 6-18.

Tapping the Blast Furnace

Every three to five hours, some iron (not all) is removed from the blast furnace. See Figure 6-19. A taphole is opened at the bottom of the furnace, and the molten iron flows along a long trough into a *bottle car,* Figure 6-20. The bottle car looks like a submarine with railroad car wheels. It is a gigantic drum. It is lined on the inside with refractory (heat-resistant) brick so that the iron retains its heat and stays molten. Iron enters the drum at about 2600°F (1430°C). Then, the bottle car takes the molten iron directly to a steel-making furnace.

The slag also is tapped off and collected in a large ladle. Slag has some commercial applications. It is most notably used as insulation. It is also used in brick making, as an asphalt filler, and as a spread on ice to prevent falling.

Pig Iron and Hot Metal

The most important product tapped from the blast furnace is the molten iron. *Molten iron* is commonly referred to as *hot metal* or *pig iron.*

The molten iron that comes out of the blast furnace will take one of two routes. One route leads to the production of castings as the end product, particularly cast iron castings. The other route leads to the steel-making building, where some type of steel shape becomes the end product.

In the first route, the molten iron is poured from the bottle car into long *molds,* which are cooled and hardened. Each mold

Figure 6-18. *This control center at the Bethlehem Steel Corporation Sparrows Point Maryland plant is computerized. Although the blast furnace is normally computer-controlled, it can also be manually controlled from these consoles. The 20,000 wire terminals in this control area illustrate the complexity of the system.*

Figure 6-19. *Every three to five hours, the blast furnace is tapped by burning out a plug at the bottom of the furnace. (American Iron and Steel Institute)*

Figure 6-20. *Molten iron flows down a trough into the bottle car, where it is carried to a steel-making furnace. (American Iron and Steel Institute)*

is referred to as a *pig* and weighs at least 40 lb. Pigs are refined in a casting machine and then sold to foundries. Molten iron that is cooled and solidified in this manner is called *pig iron.*

The term "pig iron" is certainly strange. Many years ago, molten iron was poured into a long trough with many gates or openings along its sides. The liquid metal flowed along the trough, through one of the gates, and into a sand mold shaped somewhat like a baby pig. After all the metal had cooled, the casting resembled an entire family of suckling pigs, all lined up along the sow. Thus, the name "pig iron."

In its initial state, pig iron is a rather useless material with essentially no product value. It is hard, brittle, and not very strong. However, pig iron is a vitally important ingredient in the manufacture of steel and cast iron.

Cast iron foundries (and some steel foundries) purchase pig iron and melt it in induction furnaces or cupolas, as shown in Figure 6-21. This metal is then used to make cast iron products, Figure 6-22.

Most of the molten iron travels the second route, from the blast furnace to the steel-making building. It is carried in bottle cars and poured into *ladles* for charging into a steel-making furnace, Figure 6-23. The hot metal remains in the molten state and never solidifies during its journey from the spout of the blast furnace to the mouth of the steel-making furnace.

Figure 6-21. *This cupola furnace is used to melt and refine pig iron into cast iron. (General Motors Powertrain Group)*

Steel-Making Furnaces

As discussed earlier, the two most modern types of furnaces used to convert molten

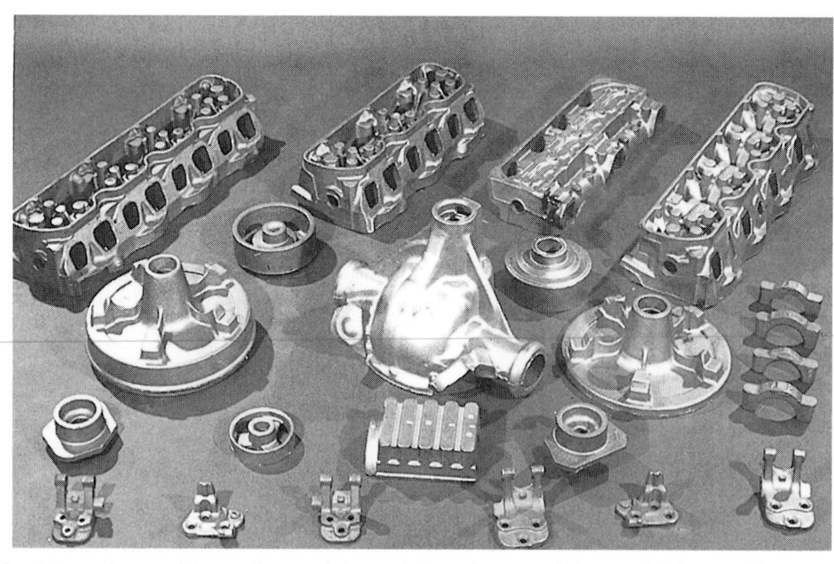

Figure 6-22. *Cast iron is used to make a wide variety of parts. (General Motors Powertrain Group)*

Figure 6-23. *Molten iron from the blast furnace is charged into a basic oxygen furnace. (Bethlehem Steel Corporation)*

iron into steel are the basic oxygen furnace and the electric arc furnace. Two other types—the open-hearth furnace and the Bessemer converter—were used as the main steel-making furnaces for many years. The open-hearth furnace is still in use to a limited degree today, but the Bessemer converter is practically obsolete.

Basic Oxygen Furnace

The *basic oxygen furnace (BOF)* replaced the open-hearth process as the number one method of steel making several years ago. Capital equipment for a basic oxygen furnace costs less, and the process is much faster.

A basic oxygen furnace can manufacture 300 tons of steel in less than an hour, which is five to ten times faster than the open-hearth process. What makes the process so productive is the oxygen blast. Massive amounts of oxygen enter the furnace at supersonic speeds and intensify the heat. The quality of the resulting steel is about the same as that of steel made by the open-hearth process.

Figure 6-24. *Molten iron is charged into a basic oxygen furnace for refining into steel. (Bethlehem Steel Corporation)*

The basic oxygen process

The entire basic oxygen process is rapid and colorful. It is described in the following eight steps:

1. The furnace uses a huge, pear-shaped steel barrel lined with refractory material. The barrel pivots on a shaft and tips to one side so that scrap (about 90 tons) can be poured into its mouth, Figure 6-24.
2. Immediately following the scrap charge, an overhead crane pours a ladle of molten iron (about 200 tons) from the blast furnace into the mouth of the basic oxygen furnace. See Figure 6-24.
3. The *oxygen lance* (pipe) is lowered to approximately 6' above the surface of the metal. The lance is locked in place. The mixture is ignited, and a high-pressure stream of oxygen hits the mixture at supersonic speeds. See Figure 6-25.
4. Shortly after ignition takes place, lime and fluorspar are added. These materials combine with carbon and other impurities to

Figure 6-25. *A high-pressure stream of oxygen is blown onto the molten iron and scrap at supersonic speeds through a "lance," which is lowered into the basic oxygen furnace. (American Iron and Steel Institute)*

form slag. Temperatures of about 3000°F (1650°C) are reached.

5. An intense chemical reaction follows. It keeps the molten mass churning as if it is in agony, Figure 6-26. This continues uninterrupted for about 20 minutes.
6. When the chemical reaction ends, the oxygen lance is removed. The molten steel is poured into a large ladle. The slag is poured into another ladle.
7. Alloying elements are added after the steel is poured into the ladle.
8. The ladle of liquid steel is then taken to the ingot center, where the steel is poured into ingot molds.

The entire process is computer-controlled. A computer determines the precise amounts of each charge of material to be added. It also determines cycle time for the operation. Basic oxygen furnaces are generally installed in pairs, so one can be filled with raw material while the other is manufacturing hot steel.

Electric Arc Furnace

The *electric arc furnace,* Figure 6-27, differs from the basic oxygen furnace in the following three ways:

- It uses electricity rather than gaseous fuel to produce the heat.
- The quality of the steel can be controlled more accurately than it can when other furnaces are used.
- It is used primarily for special types of quality steel, such as stainless steel, tool steel, and high-alloy steel.

The furnace itself looks like a giant teakettle with a spout on one end. It has a round, tightly closed chamber. The roof pivots and swings aside so that raw materials can be loaded into the furnace. The steel outer shell of the furnace is lined with heat-resistant refractory brick. Three retractable electrodes extend up through the roof of the furnace. They may be as large as 2' in diameter and 24' long. These electrodes are used to ignite the metal charge, Figure 6-28.

The electric arc furnace presents several advantages over the basic oxygen furnace. The controls of an electric arc furnace can regulate the temperature more precisely because electrical flow is easier to regulate than gaseous fuel. Also, no air is introduced, so there is more control over the oxygen content.

In general, the electric arc process is more costly than the basic oxygen or the open-hearth processes. Therefore, it is normally used only for high-quality steel. However, in recent years, the size of new electric arc furnaces has increased, and the electric arc process is becoming competitive in the low-carbon and medium-carbon steel market.

The electric arc process

The main ingredient in an electric arc furnace is steel scrap. This is another way in

Figure 6-26. *A heat shield protects members of the basic oxygen furnace crew taking steel samples and temperature measurements from the bath of molten metal. Samples are sent by pneumatic tube to the basic oxygen furnace spectrometer laboratories for analysis. (Bethlehem Steel Corporation)*

which electric arc furnaces differ from other furnaces. The scrap is the first to go in. It is analyzed and examined much more carefully than scrap that goes into any other type of furnace. It is separated, graded, and sorted into as many as 65 different classes of steel scrap. As a result, the composition of the finished steel can be predicted more accurately. This again reflects the outstanding control of the electric arc process.

After the scrap has been introduced, the electrodes are lowered through the retractable roof into the furnace near the metal charge. Electricity jumps from one electrode to the metal charge, through the metal itself, and then back to another electrode. Heat is developed by both the resistance of the metal to the massive flow of electricity, and by the heat of the electrical arc itself.

The operation of an electric arc furnace is spectacular. As soon as the electricity is

Figure 6-27. *High-quality steel is tapped from an electric arc furnace after it is made to a very exact composition. (Armco Inc.)*

Figure 6-28. *Heat for an electric arc furnace is made available by three large retractable electrodes. The electrodes are lowered into the furnace near the charge of selected scrap iron and steel. (American Iron and Steel Institute)*

Figure 6-29. *A maze of cables carry the electrical charge to the electrodes of an electric arc furnace. (American Iron and Steel Institute)*

turned on, a large roar occurs, and electric arcs flash like lightning in a thunderstorm. The arcs strike the metal charge in the bottom of the furnace and produce turmoil in the furnace ingredients. The electrodes turn white hot as more than 800,000 volt-amperes of electricity flow through them. The electrical charge is so great that the cables carrying the electrical charge literally sway when the electricity is turned on, Figure 6-29.

After the scrap has melted, a variety of other ingredients are added. Some iron ore may be used to remove carbon, or carbon may be added if the carbon content is low. Limestone is added as the fluxing agent. Alloying elements are added at the very end of the operation. Oxygen gas is sometimes fired into the melt to speed up the *heat* (the cycle that converts molten iron to steel). Mill scale and small amounts of molten iron may also be added.

As much as 300 tons of steel may be manufactured per heat, although most electric arc furnaces are designed for smaller amounts. A period of about three to seven hours is required for a complete cycle. When the steel is ready, the slag is poured off. Then the steel itself is poured into a ladle, Figure 6-30. Eventually, the steel travels to the ingot center and is poured into molds.

Open-Hearth Furnace

Prior to the emergence of the basic oxygen process, the *open-hearth process* was the primary method of manufacturing steel. Open-hearth furnaces were used for over fifty years. Today, the basic oxygen and electric arc are the standard steel-making processes. The

Figure 6-30. *Molten steel pours out of an electric arc furnace during a tap at Bethlehem Steel Corporation's Steelton, Pennsylvania, plant. This furnace has a capacity of 160 tons and can produce a heat of steel in three hours. (Bethlehem Steel Corporation)*

chart in Figure 6-31 illustrates the differences between the three methods.

The open-hearth furnace looks like a giant washbasin (hearth) that is exposed (open) to a powerful flame of fire that shoots against the contents of the basin, Figure 6-32. Steel scrap, limestone, pig iron, and iron ore are loaded into this giant hearth and heated to about 3000°F (1650°C). This produces a violent, spectacular boiling action. Despite its appearance, this bubbling, boiling fluid is actually steel.

An open hearth measures about 30' by 90'. As much as 600 tons of steel may be manufactured in a single heat. At most steel mills, several open hearths are located side by side for convenience in loading, tapping, and overall layout. Two floors are required for operation. The hearth itself takes up most of the second floor, where the loading is also done. The first floor houses the heating system.

The open-hearth process

Natural gas mixed with air is the fuel most often used to heat an open-hearth furnace. Sometimes, liquid fuel oil, tar, or gases from the coke ovens are also used. In this process, a single heating cycle takes up to ten hours. In recent years, the speed of the reaction has been increased by introducing an oxygen blast from overhead. The oxygen gas flows directly onto the burning ingredients in the open-hearth fire.

The open-hearth process is illustrated in Figure 6-32. After the burning gases pass over the molten mass in the center of the hearth, they are collected on the other side of the furnace and routed through a complex brick arrangement known as a *checker chamber.* As these hot gases pass through the checker chamber, the bricks are heated. After 15 to 20 minutes, the burning flame reverses direction and enters the hearth from the opposite side. As incoming air is introduced, it passes over these same hot bricks. This heats the air and increases the efficiency of the furnace. It also reduces the amount of pollution that escapes from the furnace. Hot gases then heat the bricks on the other side as they pass over a second checker chamber. This operation is reversed every 15 to 20 minutes.

Molten iron is added after the other ingredients have melted into a liquid. The molten iron pours out of a ladle spout into a large accepting beak at the mouth of the furnace, Figure 6-33. Alloying elements are the last ingredients added.

Steel-Making Furnace	Steel Cost	Application	Rating (Based on Quantity Produced)
Basic Oxygen Furnace	Lowest Cost and Most Efficient	General Steel	1
Electric Arc Furnace	Slightly More Expensive	Special Steel and High Alloy Steel	2
Open-Hearth Furnace	Low Cost	General Steel	3

Figure 6-31. *A comparison of the three main steel-making processes.*

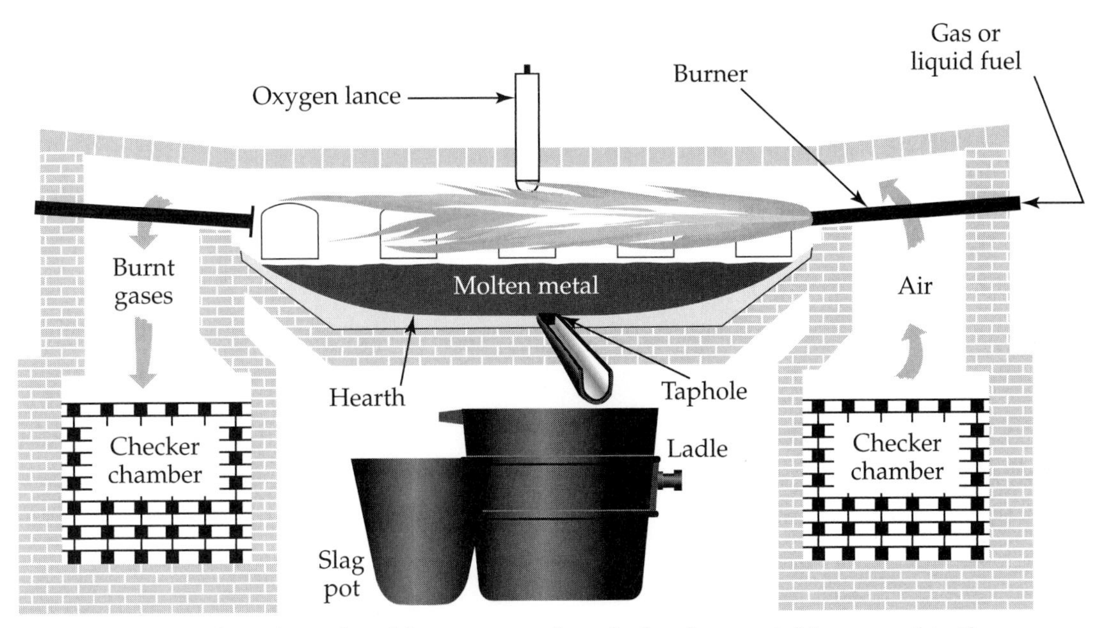

Figure 6-32. *The open-hearth steel-making process. A pool of molten metal is exposed to the sweep of flames flowing alternately from either side of the hearth. The entire cycle takes up to ten hours. (Bethlehem Steel Corporation)*

Figure 6-33. *Hot iron from the blast furnace is poured into an open-hearth furnace. This molten iron is subjected to 3000°F (1650°C) temperatures before being purified into steel. (Bethlehem Steel Corporation)*

At the end of the heat, a plug is removed from the taphole. An explosive charge may be used to remove the plug. As the plug is removed, fresh steel flows out in a trough and into another ladle. As the steel enters the trough, sparks fly in a spectacular display resembling fireworks. The slag floats to the top of the ladle, where most of it is skimmed off into a *slag pot.* See Figure 6-34. Next, molten steel is poured from the ladle into *ingot molds,* which give shape to the steel when it cools and solidifies.

Processing the Ingots

An *ingot* is a large steel casting that has cooled and solidified into a workable shape. Ingots are processed by means of three separate operations:

- Teeming. Pouring molten steel into the ingot molds.
- Stripping. Removing ingots from the ingot molds.
- Soaking. Heating ingots to obtain more uniform properties in the metal.

Figure 6-34. *As molten steel fills the ladle at the rear of an open-hearth furnace, some of the lighter slag rises to the top and flows into an adjacent slag pot. The remaining slag at the top of the ladle provides a "protective blanket" for the steel to retain its heat until it is ready to be poured into molds. (Bethlehem Steel Corporation)*

Teeming

When steel arrives from the furnace, it is poured into ingot molds from a large ladle. See Figure 6-35. The molten metal flows through a hole in the bottom of the ladle. A long row of ingot molds are lined up and the ingots are poured at one time. See Figure 6-36.

The cross-sectional shape of the ingot is normally square or rectangular. It is tapered, with rounded corners and corrugated sides. Ingots may be anywhere from 3' to 8' high and 6" to 3' wide.

Stripping

As soon as the ingots are partially cooled, they are separated from their molds as quickly as possible. A stripper crane grips lugs located at the top of each mold. The crane then lifts up the mold shell to separate it from the hot ingot.

Figure 6-35. *In pouring or "teeming," molten steel is transferred from the ladle into ingot molds. (Bethlehem Steel Corporation)*

Soaking

After stripping, the hot ingot is carried to a soaking pit, Figure 6-37. The soaking pit is actually a furnace. It is usually below floor level, thus resembling a pit. The ingot is heated in the soaking pit for six to eight hours at about 2200°F (1200°C) so that a uniform temperature is reached throughout the entire ingot.

The purpose of soaking is to prevent the outer surfaces of the ingot from solidifying before the inside. If ingots were not soaked, carbon, phosphorous, and sulfur (which tend to solidify last) would congregate around the top center portion of the steel.

After soaking is completed, the hot ingot is removed from the soaking pit, Figure 6-38, and carried to the rolling mill.

Figure 6-36. *A row of ingot molds. The molds are used to shape the steel after it is poured from the ladle. (American Iron and Steel Institute)*

Figure 6-37. *Ingots are removed from the ingot mold and transferred to a soaking pit. (American Iron and Steel Institute)*

Rolling Mill

A *rolling mill* is the part of the steelmaking complex wherein a series of large, hard rollers that compress steel ingots into different shapes. For example, an ingot that is 2' thick will be "squeezed" until it becomes considerably thinner. Eventually, this same steel may be pressed to 1/32" sheets or drawn into 1/16" diameter wire. Some stock is compressed to even smaller sizes.

The "squeezing" is done by using a pair of rollers, as shown in Figure 6-39. The ingot passes between the two rollers, which are rigidly supported in strong bearings. The space between the rollers is slightly less than the thickness of the ingot. Since the ingot is hot and pliable, it is reduced to the thickness of the space between the rollers.

After passing between these rollers, the steel moves to another pair of rollers where the gap is slightly smaller. Here, the steel is compressed again. Then, the steel moves to a third pair of rollers with a smaller gap than the previous pair. This process continues until the steel has reached the desired thickness.

A *two-high reversing mill,* Figure 6-40, uses only one pair of rollers. The ingot passes back and forth between this pair. After each pass, the gap between the rollers is narrowed until the desired thickness is attained.

Figure 6-38. *This ingot has been heated to an even temperature throughout and is being removed from the soaking pit. (Bethlehem Steel Corporation)*

Figure 6-39. *A glowing ingot takes on a new shape as it is worked by the rolls of a blooming mill. Rolling not only shapes the steel, it also improves its mechanical properties. (Bethlehem Steel Corporation)*

Three-high and four-high mills are also used to roll steel, Figure 6-41. In both cases, the ingot first passes between the two lower rollers and is compressed. Next, it is raised and passes between the two uppermost rollers. The gap between the upper two rollers is slightly less than the gap between the lower pair, so it is compressed further. Then, both gaps are decreased slightly and the steel repeats its journey, forward and back. After many passes, the steel attains the proper thickness.

As the steel becomes thinner, it also becomes longer. Steel from an ingot that starts out 9' long may be several miles long before it leaves the last rolling mill, Figure 6-42. The steel may make as many as 50 passes before it is ready for commercial sale.

Slabs, Blooms, and Billets

In a rolling mill, rolling is a two-stage process. The first stage occurs in the *primary rolling mill*. The second stage occurs in the *secondary rolling mill*.

In the primary rolling mill, the ingot is converted into a *slab* (in a slabbing mill), a *bloom* (in a blooming mill), or a *billet* (in a blooming and billet mill). Slabs receive dimensional reduction primarily in thickness. The typical dimensions of a slab are 10" thick, 6' wide, and 32' long. See Figure 6-43. The size of a bloom, Figure 6-44, is reduced equally in two different directions. A billet is made from a finished bloom. It is essentially a small bloom.

From slabs, blooms, and billets, all other commercial shapes are made in a secondary rolling mill. These shapes include sheet, strip, bar, structural shapes, plate, pipe, tube, rod, and wire.

The entire rolling operation is monitored from a control center. The control center uses computers, dials, gauges, and switches to monitor the operation. See Figure 6-45.

Figure 6-40. *This two-high reversing slab mill is driven by a pair of 6000 horsepower motors to convert 35-ton ingots into steel slabs. (Armco Inc.)*

Figure 6-41. *This four-high finishing mill produces plates from 3/16" to 15" in thickness, up to 150" in width, and up to 120' in length at Bethlehem Steel Corporation's Burns Harbor, Indiana, plant. (Bethlehem Steel Corporation)*

Figure 6-42. *This hardened steel roll is shown in operation at a Bethlehem Steel Corporation plant. Long rolls of this type are used for cold-rolling steel sheet.*

Figure 6-43. *Slabs are reduced in thickness. They may be further rolled into plates, which are used in ships, bridges, and machinery. (American Iron and Steel Institute)*

Figure 6-44. *Blooms are rolled into beams and other structural shapes typically used in the construction industry. (American Iron and Steel Institute)*

Figure 6-45. *The entire rolling operation at a bar mill is monitored from the mill's principal control center. (Bethlehem Steel Corporation)*

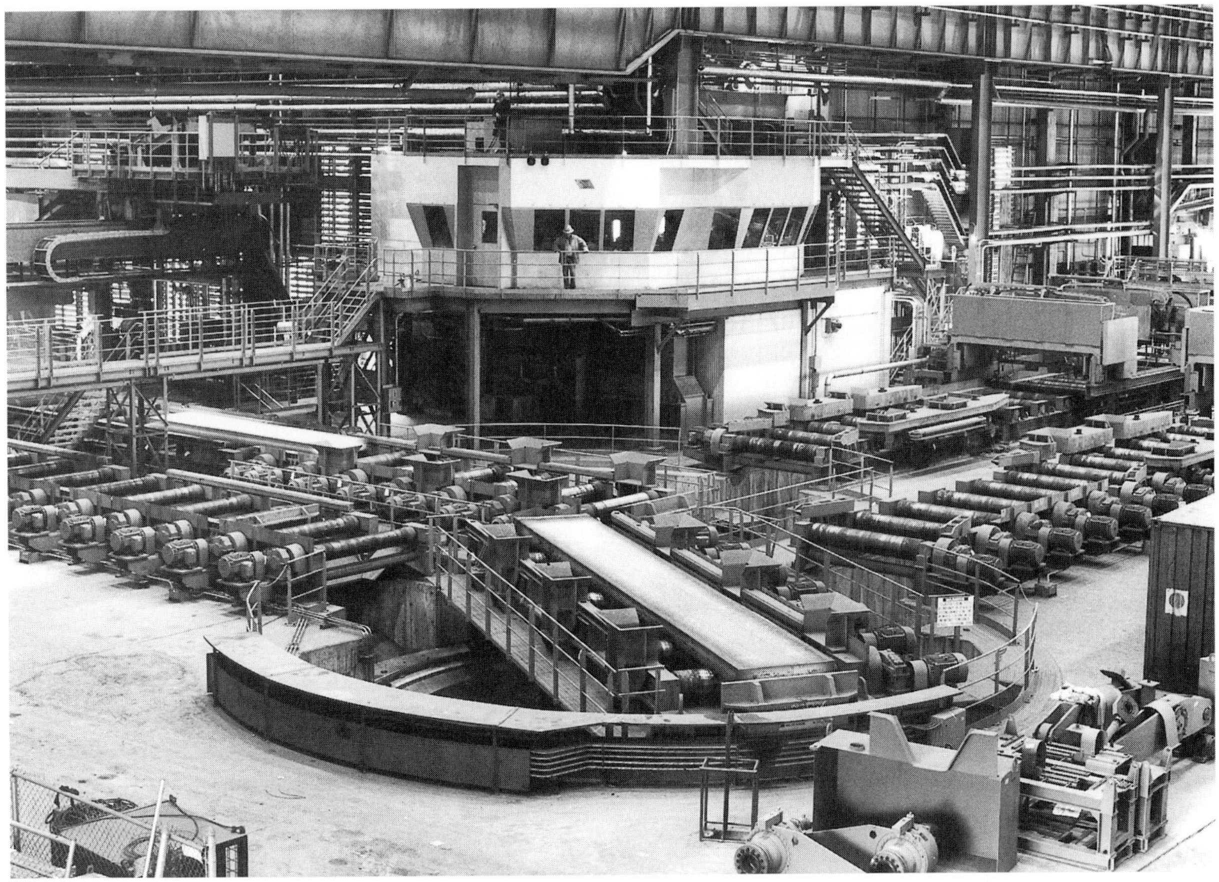

Figure 6-46. *Continuous slabs of steel are produced in this continuous casting mill. (Bethlehem Steel Corporation)*

Continuous Casting

Slabs, blooms, and billets can be made in one continuous and uninterrupted operation in a process known as *continuous casting*, or *strand casting*, Figure 6-46. A strand casting machine bypasses the ingot operation and the primary rolling operation (where ingots are turned into slabs, blooms, or billets). The continuous casting process produces a continuous length of steel.

There are three common methods of continuous casting:

- *Curved mold method.* In this method, the steel is roller-straightened. See Figure 6-47.
- *Vertical cutoff method.* This method uses a straight mold.
- Combination of the curved mold and vertical cutoff methods. The types of rolling mills used in this method vary.

The *curved mold method* is the most widely used. It is done in seven steps:

1. The molten steel is brought from the steel-making furnace in a ladle. A stopper at the bottom of the ladle is removed, and the metal flows into a reservoir called a *tundish*. See Figure 6-48.
2. The tundish is filled with the entire load of steel from the ladle. As it is being filled, it releases the molten steel in a continuous stream through a hole in its base. The steel flows out at a steady rate and is distributed into a series of moving molds.
3. The hollow interior of each mold has inside dimensions corresponding to the

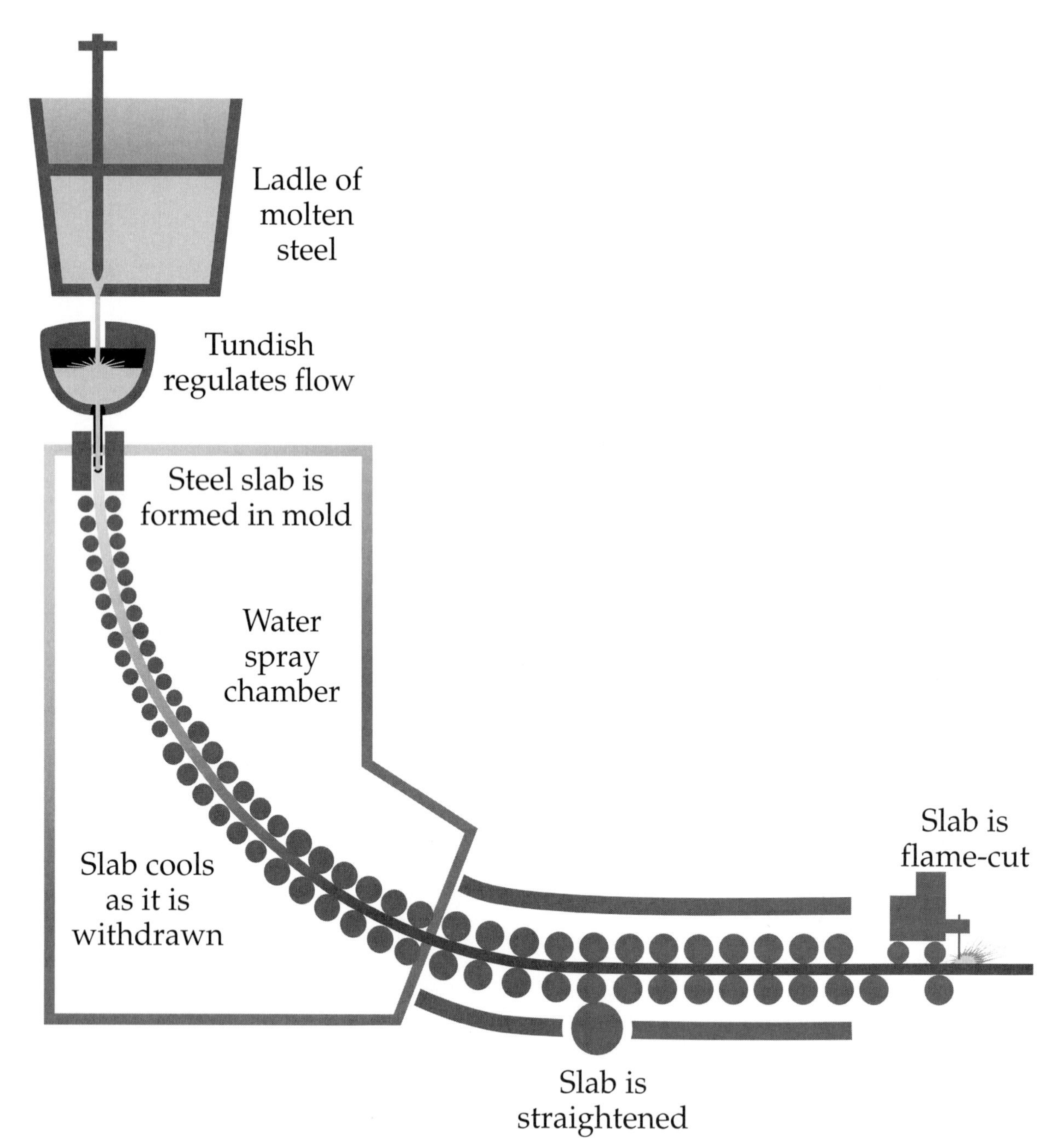

Figure 6-47. *The curved mold method is used in the continuous slab casting process. Molten steel in a ladle is teemed into a tundish, which regulates the flow of the liquid metal into a water-cooled mold. In the mold, a thin shell forms as the molten steel begins to solidify. The strand that is formed is further solidified below the mold by a direct spray of water as it is guided by a curved roller apron. The apron is divided into seven segments. Below the roller apron is the straightener-withdrawal section, which supports, straightens, and withdraws the cast strand as solidification is completed. The strand then moves into the slab processing area. It is cut into slabs of predetermined lengths, weighed, and stamped with identification numbers. (Bethlehem Steel Corporation)*

Figure 6-48. *An overhead crane positions the ladle of molten steel in preparation for continuous casting. The tundish is supported on an independently driven car below the ladle arms. (Bethlehem Steel Corporation)*

width and thickness of the slab, bloom, or billet being formed. Lining the walls of the molds are pipes that carry water to cool the outside surface of the metal. As the metal surface begins to "freeze," a thin skin is formed. During this operation, the metal continuously moves downward, and the molds oscillate up and down in order to keep the metal from sticking.
4. The metal then moves into the roller apron. This area is curved and contains bending rollers and a spray apron. The secondary cooling process takes place here. The metal solidifies from the outside skin toward the center as it moves through the roller apron. By the time the metal reaches the bottom of the apron, it is solid throughout.
5. Next, the metal enters the *straightener.* The straightener contains rollers that reshape the slightly curved metal into a flat slab, bloom, or billet. It carries the long ribbon of steel onto a level table for cutting. See Figure 6-49.
6. Flame-cutting torches cut the metal to a predetermined length. In order to make a straight cut, the torches move forward with the metal as they cut. A mechanism on the roller table controls this movement. After the cut is completed, the torches return to the starting position and begin a new cut. This operation is shown in Figure 6-50.
7. After the slab, bloom, or billet is cut to size, it is usually reheated, and then taken to the secondary rolling mill for finishing.

The speed of a strand casting machine may be as fast as 15' per minute. The entire journey from the ladle to the cutting machine takes less than thirty seconds.

When continuous casting was first introduced, vertical continuous casting machines were primarily used. Today, virtually all continuous casting operations in the United States are performed by curved mold machines.

Manufacture of Cast Iron

Cast iron-making furnaces convert pig iron to cast iron, Figure 6-4. The ingredients used in this process are pig iron, iron or steel scrap, limestone, and fuel.

There are thousands of cast iron manufacturers, in contrast to the relatively few numbers of major steel-making centers. There are several reasons for this:

- Cast iron is easier to cast into complicated shapes because of its high fluidity. Most steel, by contrast, is used in the wrought condition.
- Cast iron is slightly less expensive to manufacture than steel, so more manufacturers can compete.
- Cast iron has a superior damping ability (the ability to absorb vibration). Thousands of machine frames and other large support parts for equipment are made of cast iron.

Figure 6-49. *A continuous casting machine moves long ribbons of steel to a level table for cutting operations. Slabs are cut to exact lengths. Continuous casting can produce the longer slabs needed for the huge, weld-free coils required in high-speed stamping and forming operations. (Armco Inc.)*

Figure 6-50. *Slabs are cut to required lengths by the automatic flame torches on a continuous slab casting machine. This two-strand caster can convert 300 tons of molten steel into heavy slabs in 45 minutes. (Bethlehem Steel Corporation)*

Therefore, most parts made of cast iron are large and thick, as compared to parts that are machined from steel. See Figure 6-51.

Cupola

For many years, the *cupola* served as the workhorse of the cast iron industry. Almost all cast iron was produced by the cupola prior to 1970.

The cupola is round and tall, Figure 6-52. It resembles a small blast furnace. A typical cupola is two or three stories tall and measures about 5' in diameter on the inside. The outer shell is made of steel. The inside is lined with refractory brick. All of the ingredients are charged into the top of the stack, Figure 6-53.

There are two discharging spouts near the bottom of the cupola. One spout discharges cast iron. The other spout, located a little higher on the furnace, provides the exit for the slag.

The ingredients used to make cast iron in a cupola are pig iron, iron or steel scrap, limestone, and coke. The coke supplies the heat, and is added first. Added on top of the coke are layers of scrap, pig iron, limestone, more coke, more scrap, more pig iron, etc. The close proximity of the fuel (coke) to the metallic ingredients makes the cupola a very efficient melting unit. Usually, the furnace is continuously loaded and regularly tapped. This is a different process than the one used by a steel-making furnace, which manufactures steel in batches. To accelerate the heat in a cupola, air is charged through a *wind box* and through *tuyeres.* See Figure 6-54.

The furnace normally has a sand bottom. The temperature at the bottom reaches 3700°F (2040°C). The original charge of coke that forms the *coke bed* in the bottom of the furnace is typically about 3000 lb. (1360 kg).

The limestone helps flux out impurities (such as coke ashes, sand, and any foreign matter) that enter the cupola with the scrap. Alloys are added to the cast iron after it is tapped from the tapping spout.

Figure 6-51. *This diesel engine block is typical of the many large, thick parts made of cast iron. (General Motors Powertrain Group)*

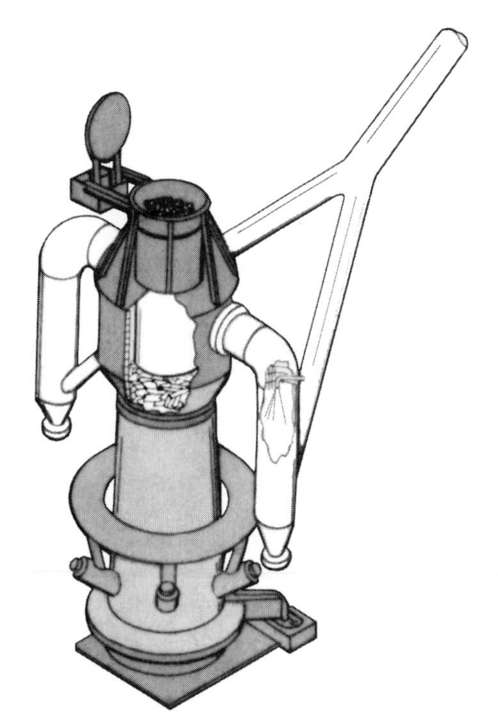

Figure 6-52. *This diagram shows a pictorial representation of a gas-recovery cupola. (General Motors Powertrain Group)*

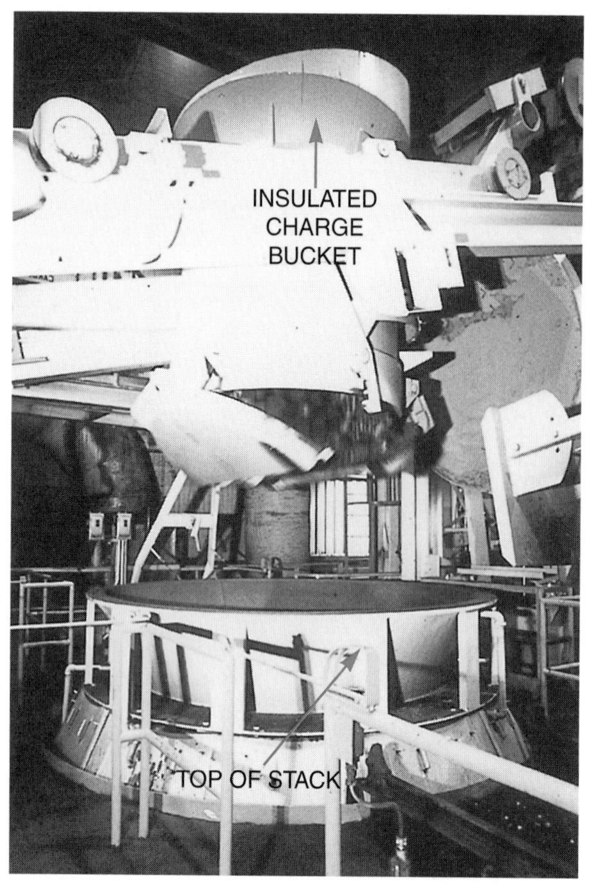

Figure 6-53. *The ingredients for a cupola are charged into the top of the stack. (General Motors Powertrain Group)*

Figure 6-54. *Air is charged through tuyeres into a cupola to accelerate the heating process. (General Motors Powertrain Group)*

Electric Induction Furnace

In the 1970s, many cupolas were removed by cast iron manufacturers and replaced with electric induction furnaces due to the threat of air pollution. The burning coke used by cupolas presented a hazard to the environment. Manufacturers whose business depended on the cupola were forced to do one of three things:

- The companies could install very expensive cleaning equipment to reduce pollution emitted from furnace stacks.
- Another option was to replace cupolas with another type of furnace that did not burn coke, usually an electric induction furnace. This was also a costly route.
- The companies could opt to go out of business because of insufficient funds to install newer equipment.

The choice was a difficult one. Each of the three alternatives was taken by thousands of companies. It was a difficult time for owners of cast iron foundries, but air pollution was considered to be very dangerous to the environment. Large foundries, needing high tonnages of liquid iron, retained cupolas because of their high melting rate and efficiency. They invested in the necessary pollution control equipment.

A high percentage of the smaller cast iron foundries took the second alternative and converted to the *electric induction furnace.* Since the fuel used by this furnace is electricity, the pollution hazard could be nearly eliminated.

The electric induction furnace presents other advantages for cast iron manufacturers:

- It can be operated economically as a small furnace, with loads of less than one ton.
- Metal melts very rapidly in an electric induction furnace.
- Less oxidation takes place because of the high melting speed.
- There is less chance for alloys to boil off during the casting process.

Electric induction furnaces are generally round and small. The inside usually consists of a *crucible,* a vessel that resists high temperatures. It is made of magnesia. The outside of the crucible is surrounded by a layer of refractory material. A large coil of copper tubing is wound around the refractory material. The copper tubing acts as an electrical coil and serves as the heat source.

The electric induction furnace acts as a batch furnace (like a basic oxygen furnace) rather than a continuous furnace (like a cupola). After the furnace is loaded, Figure 6-55, a high-frequency electric current is passed through the large copper coil. The coil acts as the primary coil of a transformer. The metal charge on the inside (in the crucible) acts as both the core of the transformer and the secondary coil.

Figure 6-55. *The raw materials needed to produce molten iron are charged into this electric induction furnace (General Motors Powertrain Group)*

The metal charge resists the flow of electrical current. Thus, heat is developed. The induced current in the molten metal creates a stirring action, which serves to mix the metal. The melting is very rapid and very quiet.

To speed up the manufacturing process, two or more electric induction furnaces can be used to produce cast iron. One can be loaded while the other is melting and pouring cast iron. This provides a more continuous and steady supply of molten metal. The use of more than one furnace also allows the foundry to remain operating while a furnace is being cleaned or repaired.

Pollution Control

Many safeguards are employed by steel manufacturers to protect against air and water pollution. Air pollution is controlled by dust catchers, electrostatic precipitators, and wet scrubbing systems. Water is cleansed of oils and solids in settling basins, clarifiers, and treatment plants before being returned to its source or before it is reused in closed circuit water systems.

Test Your Knowledge

Write your answers on a separate sheet of paper. Do not write in this book.

1. What task is performed by magnetic separators?
2. Which type of iron ore contains a higher percentage of iron: hematite or taconite?
3. What three materials are melted in a blast furnace?
4. Which ingredient in a blast furnace combines with impurities to form slag?
5. When a blast furnace is tapped, what material is removed first?
6. What is a *pig*?
7. What are the two most popular types of steel-making furnaces in use today?

8. What type of furnace is used to convert iron ore to pig iron?
9. Describe the basic oxygen process.
10. Why are lime and fluorspar added to basic oxygen furnaces?
11. Electric arc furnaces are normally used to produce which types of steel?
12. Describe the electric arc process.
13. What is *teeming*?
14. What is *stripping*?
15. What is *soaking*?
16. Why are ingots soaked?
17. Where are ingots taken after they are removed from the soaking pit?
18. Explain the difference between a batch furnace and a continuous casting furnace.
19. Name the two types of furnaces used to produce cast iron.
20. What is the function of a tundish in the continuous casting process?
21. Steel is the material that has been chosen for a small link in the fuel injection mechanism in a space vehicle. Which steel-making process would probably be used to manufacture the steel for this application?
22. You are involved in the design and construction of a new steel-making factory. Which steel-making process would you incorporate into this new plant? How would the geographic location of the plant affect the selection of the process?

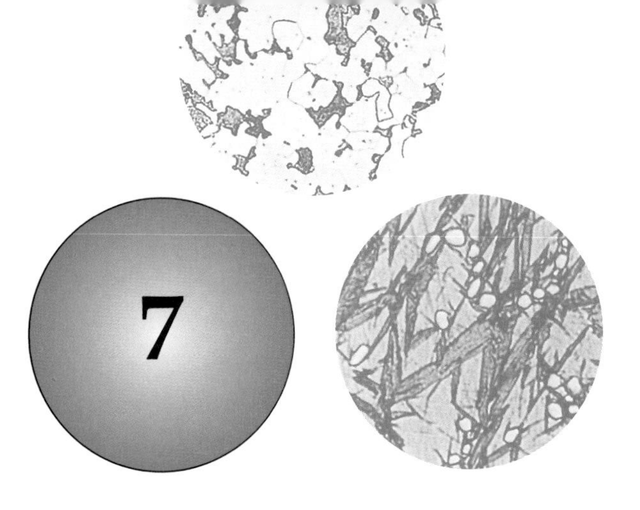

7

Crystal Structure

After studying this chapter, you will be able to:

- Explain how a crystal is formed in metal.
- Discuss the formation of space lattice structures and dendrites.
- Describe the appearance of atoms inside crystals.
- Explain how temperature affects the growth of a crystal.
- Tell what is meant by grain size.

Have you ever admired the crystalline pattern of snowflakes or frost on a window? Iron and steel form crystalline patterns the same way snow and ice form these structures. Snow crystals and frost patterns begin as water vapor in the air. Heat changes the liquid water vapor to solid crystals. Crystalline patterns of frost form when water vapor condenses on a window pane that is warmer than the outside air.

Iron and steel melt into liquid form at high temperatures. When this liquid gradually cools, *crystals* slowly begin to form and the steel solidifies. Tiny crystals form first. They keep growing until the crystals all stand tightly next to one another, "elbow to elbow." See Figure 7-1.

Would you think that after these crystals solidify, their atoms would appear in a regular, precise formation, or all mixed up? The answer may surprise you, Figure 7-2. The internal structure is very regular and precise; the atoms line up like a military band marching in a parade. Different crystals may not be

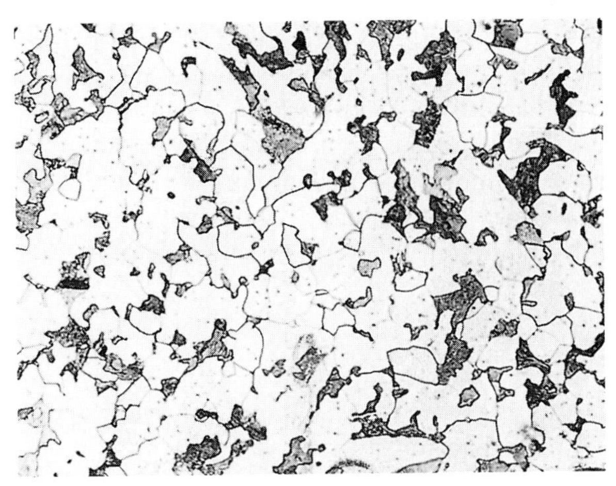

Figure 7-1. *This photomicrograph reveals the crystalline pattern of a 1020 steel ferrite structure with some pearlite. Magnification is 100X. (Buehler Ltd.)*

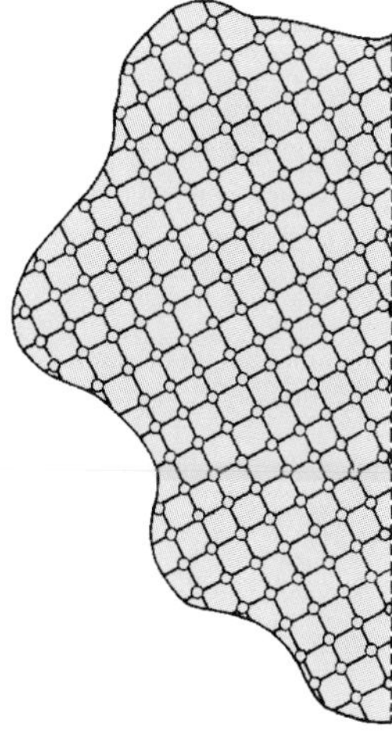

Figure 7-2. *Atoms in a crystal appear in a regular, precise formation.*

in the same formation with respect to each other. However, the tiny atoms form long, neat rows in all directions within each crystal.

Space Lattice

The organized arrangement of atoms in a crystal is known as a *space lattice.* Some examples are shown in Figures 7-5, 7-7, 7-9, and 7-11.

Atoms in a space lattice are so small that they are invisible. Not long ago, scientists were unable to see them with the most powerful microscopes. Today, electron microscopes are used to view particles as small as ten billionths of an inch. This permits scientists to faintly see the atom. Research is ongoing to develop more powerful microscopes so that someday we will be able to see atoms close-up. Not all metals have the same pattern of atoms (or space lattice).

Unit Cell

The most fundamental arrangement of atoms in a space lattice is the *unit cell.* A space lattice is simply a group of unit cells—perhaps billions—in which each unit cell is identical. Each type of metal has its own space lattice formation, so there are many different basic unit cells.

The following are four common types of unit cells for metals (see Figure 7-3):

- Body-centered cubic.
- Face-centered cubic.
- Close-packed hexagonal.
- Body-centered tetragonal.

Atoms assume many different crystal formations, but these four structures are the most common types in the study of metallurgy.

Body-centered cubic space lattice

The unit cell of a *body-centered cubic* space lattice is diagrammed in Figure 7-4. The cell has a cubic shape, with an atom located at each of the eight corners. In the center of these eight atoms is a ninth atom, which completes the formation. When a number of body-centered cubic unit cells are grouped, they produce the space lattice formation shown in Figure 7-5.

It should be pointed out that the black lines appearing between these atoms do not exist. They are merely shown to help you visualize the spatial relationship between the atoms. The atoms are also considerably closer to one another than Figure 7-4 suggests. The atoms depicted here as balls are packed so closely together that they nearly touch.

Metals that commonly have a body-centered cubic formation include chromium, molybdenum, tantalum, tungsten, vanadium, niobium, and the *ferrite form of iron.*

To help you visualize how unit cells become part of a space lattice, refer to Figure 7-5. Here, a series of body-centered cubic unit cells are connected in a small space lattice. Next, visualize billions of these unit cells side by side

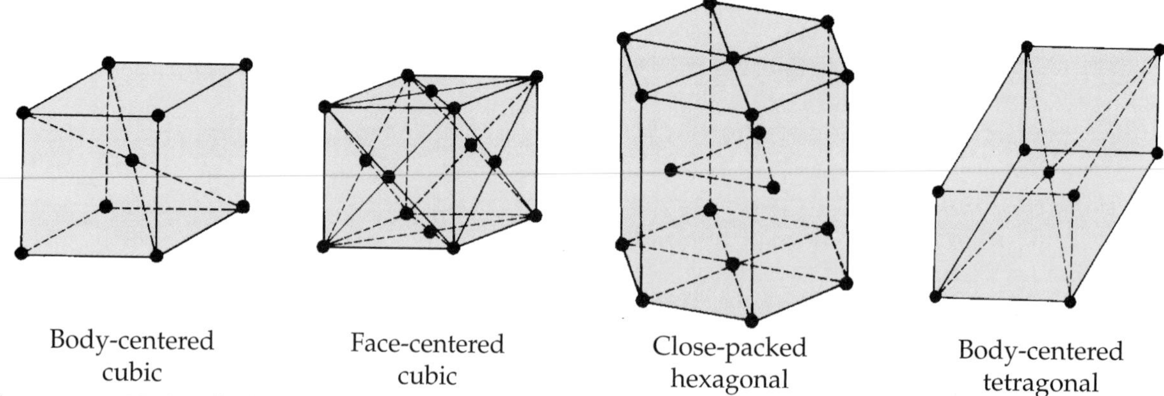

Figure 7-3. *Unit cells for common space lattice formations in metal.*

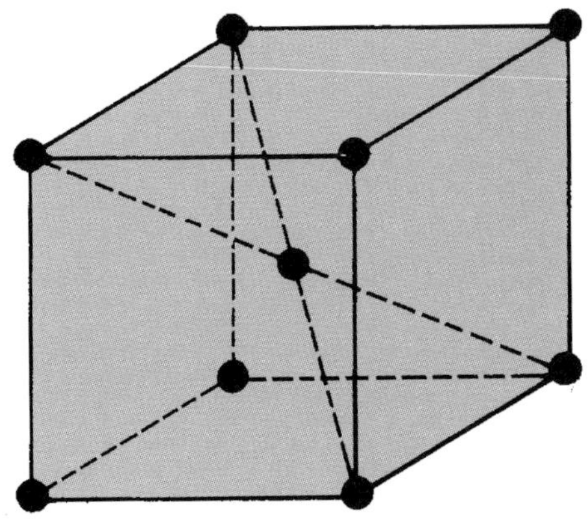

Figure 7-4. *A unit cell for a body-centered cubic space lattice.*

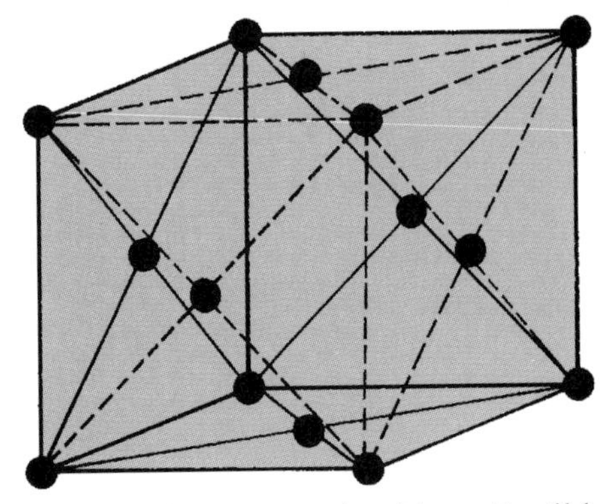

Figure 7-6. *A face-centered cubic unit cell has 14 total atoms.*

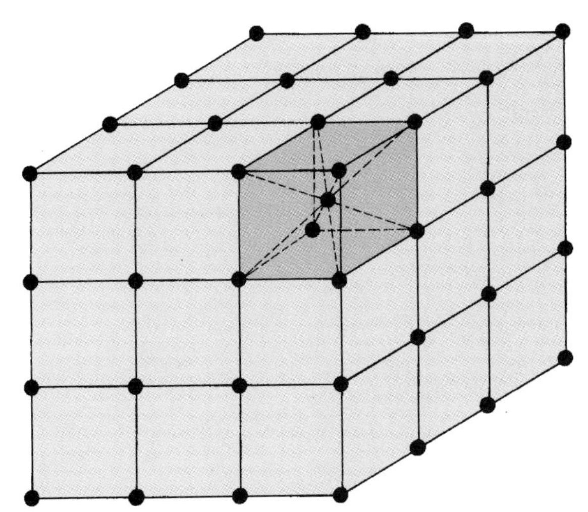

Figure 7-5. *A space lattice made up of body-centered cubic unit cells.*

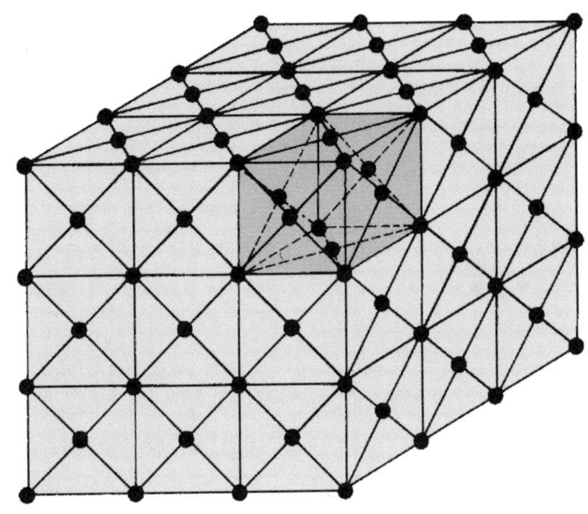

Figure 7-7. *A space lattice of face-centered cubic unit cells.*

instead of just the few that are shown. Billions of these cells are needed to form just one cubic inch of iron.

Face-centered cubic space lattice

The *face-centered cubic* space lattice is another very common unit cell found in metals. The unit cell is cubic in shape, with atoms at the eight corners and one atom located in the middle of each of the six faces. See Figure 7-6. Thus, the total number of atoms in the unit cell of a face-centered cubic space lattice is 14. A series of face-centered cubic unit cells forms the small space lattice shown in Figure 7-7.

Metals that commonly take this formation are aluminum, copper, gold, lead, nickel, platinum, silver, and austenitic iron.

Close-packed hexagonal space lattice

The *close-packed hexagonal* space lattice is a very brittle formation. It is found in metals that have little ductility. The structure of atoms is quite different from the body-centered and face-centered cubic structures.

A close-packed hexagonal unit cell consists of 17 total atoms—seven on each of two hexagonal ends and three more in the center of the structure. See Figure 7-8. A

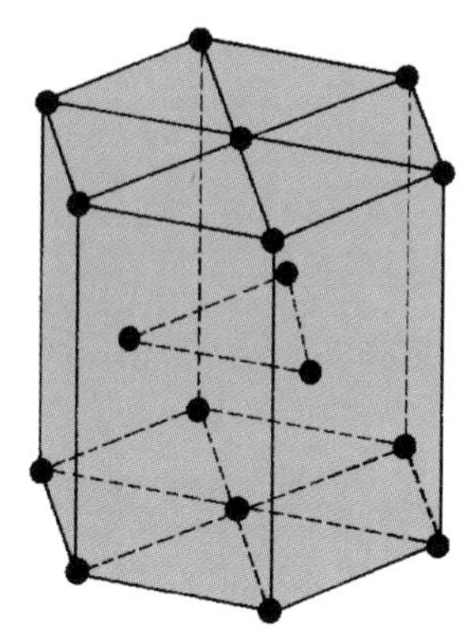

Figure 7-8. *A close-packed hexagonal unit cell has 17 total atoms.*

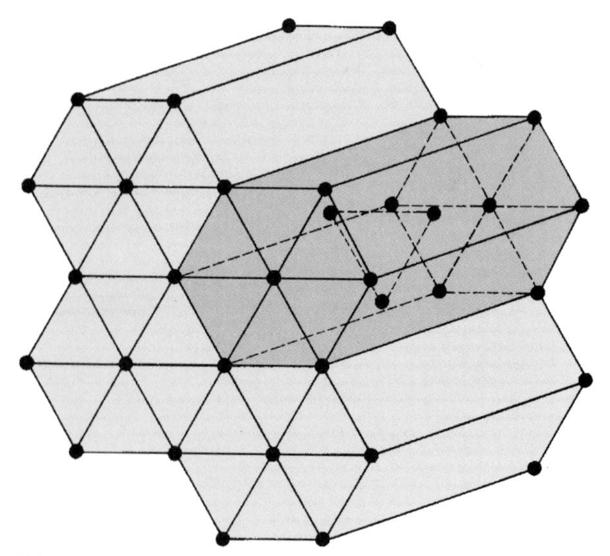

Figure 7-9. *A close-packed hexagonal space lattice.*

close-packed hexagonal space lattice is shown in Figure 7-9.

Metals that commonly take a close-packed hexagonal space lattice formation include cadmium, cobalt, magnesium, titanium, beryllium, and zinc.

Body-centered tetragonal space lattice

A *body-centered tetragonal* space lattice is similar to a body-centered cubic lattice. However, the faces of each unit cell are rectangular instead of square, Figure 7-10. The same number of atoms (nine) make up a basic unit cell. The space lattice for this structure is shown in Figure 7-11.

The body-centered tetragonal space lattice is the hardest, strongest, and most brittle of the four space lattice structures we have discussed. There is only one ferrous metal that takes this formation—martensitic iron.

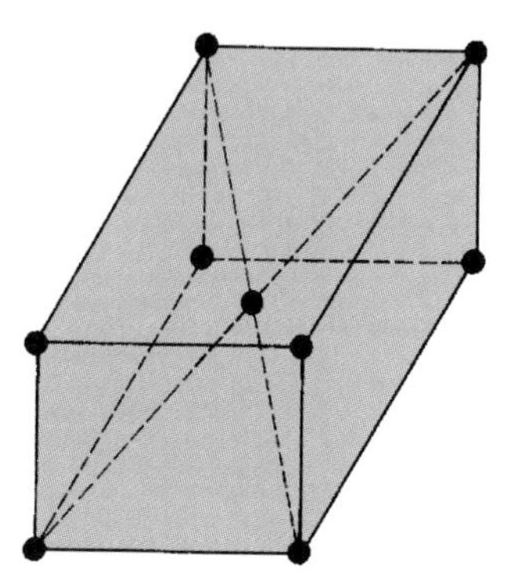

Figure 7-10. *A body-centered tetragonal unit cell resembles a body-centered cubic unit cell, but the faces of the structure are rectangular instead of square.*

Space Lattice Structures in Iron and Steel

It may have surprised you to see that iron was listed three times when we discussed the four different unit cell types. Iron is unusual in that it can take three different space lattice structures. As iron goes through a temperature change, its atoms realign into new geometric patterns. This has great effects on the strength, hardness, and ductility of iron.

Ferritic iron (ferrite) takes the body-centered cubic space lattice formation. Ferrite is basic iron at room temperature that has not been previously heat treated.

Austenitic iron (austenite) has a face-centered cubic space lattice structure. Austenite is the structure iron takes at elevated temperatures. In other words, ferrite becomes austenite when it is heated to a high temperature. During the transformation to austenite, the atoms are reshuffling within the crystal.

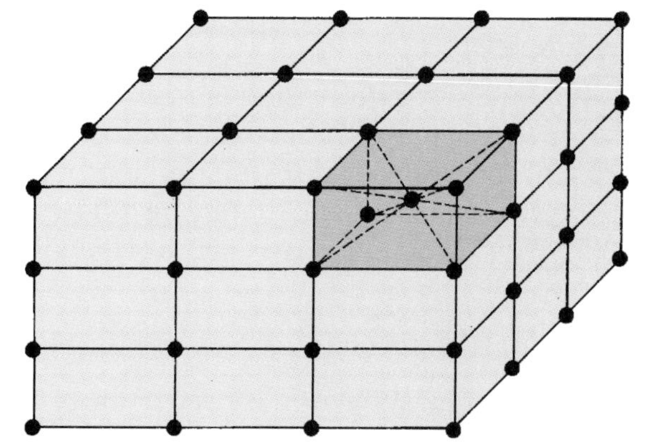

Figure 7-11. *A body-centered tetragonal space lattice.*

When the austenite is allowed to slowly cool, it returns to the ferrite structure.

Martensitic iron (martensite) has a body-centered tetragonal space lattice. Martensite is formed when steel is heated and then rapidly quenched. This process produces the body-centered tetragonal structure. Heating and sudden quenching tend to harden metal. Therefore, martensite is the strongest, hardest, and most brittle of the three iron structures.

The heat-treating processes that produce ferrite, austenite, and martensite are summarized in Figure 7-12.

While most metals tend to have one basic crystal lattice structure at all times, iron and steel are different. Iron assumes different properties when it is heated and its atoms realign into different space lattice structures. Iron can be transformed into ferrite, austenite, or martensite. In other words, "iron is not iron is not iron." The characteristics of the three basic forms of iron are summarized in Figure 7-13.

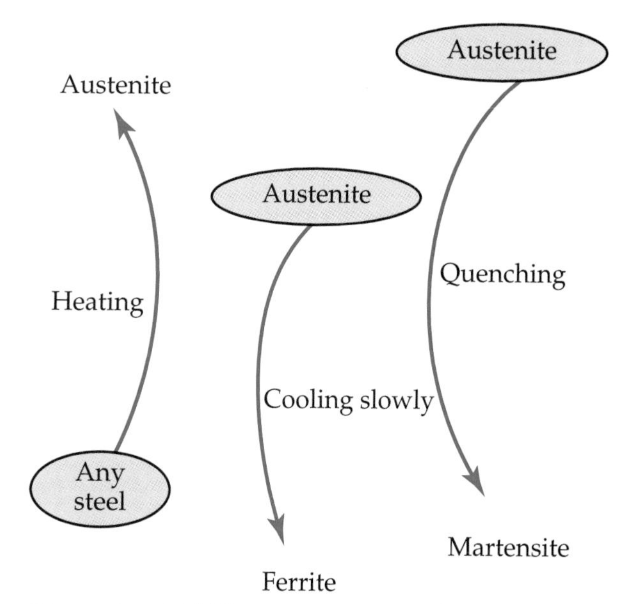

Figure 7-12. *Different structures of iron are produced by heating, slow cooling, and quenching.*

Ferrite	Austenite	Martensite
Body-centered cubic lattice formation	Face-centered cubic lattice formation	Body-centered tetragonal lattice formation
Exists at low temperature	Exists at high temperature	Exists at low temperature
Magnetic	Nonmagnetic	Magnetic
Less hardness than most steels	Essentially no hardness	More hardness than most steels
Less strength than most steels	Essentially no strength	More strength than most steels
Ductile	(Not applicable)	Brittle
Less internal stress than most steels	(Not applicable)	More internal stress than most steels

Figure 7-13. *This chart compares the characteristics of ferrite, austenite, and martensite.*

Transformation Temperatures

As ferritic iron changes to austenite, there are two important temperatures to understand:

- The *lower transformation temperature* is the temperature at which the body-centered cubic structure starts to change to the face-centered cubic structure. It is the temperature at which ferrite starts to change to austenite.

- The *upper transformation temperature* is the temperature at which the body-centered cubic structure has completely changed to face-centered cubic. At this temperature, no ferrite exists. The iron structure that occurs above the upper transformation temperature is 100% austenite.

As austenite is cooled back down to room temperature, it starts changing at the two transformation temperatures. When the upper transformation temperature is reached, austenite (face-centered cubic) starts to change back to ferrite (body-centered cubic). When the steel has cooled to the lower transformation temperature, the structure is 100% ferrite. The heating and cooling cycles are shown in Figure 7-14.

If the metal is quenched rapidly, the austenite changes to martensite (body-centered tetragonal), instead of returning to ferrite. See Figure 7-15.

The lower transformation temperature for all iron and steel is approximately 1330°F (720°C). The upper transformation temperature varies for each metal. It may be as low as 1330°F (720°C) or as high as 2000°F (1100°C).

Crystal Growth

Assume that some iron is heated above its upper transformation temperature. Assume, in fact, that it has been heated above its melting temperature. The iron has become liquid. If the iron is left at this elevated temperature, it will remain molten. If it is cooled, the structure will begin to slowly solidify, unit cell by unit cell, until the entire mass of metal has solidified and become ferrite.

The stages of solidifying (or crystal growth) are shown in Figure 7-16. The iron begins in a molten state. When the temperature is lowered very slowly, one spot in the

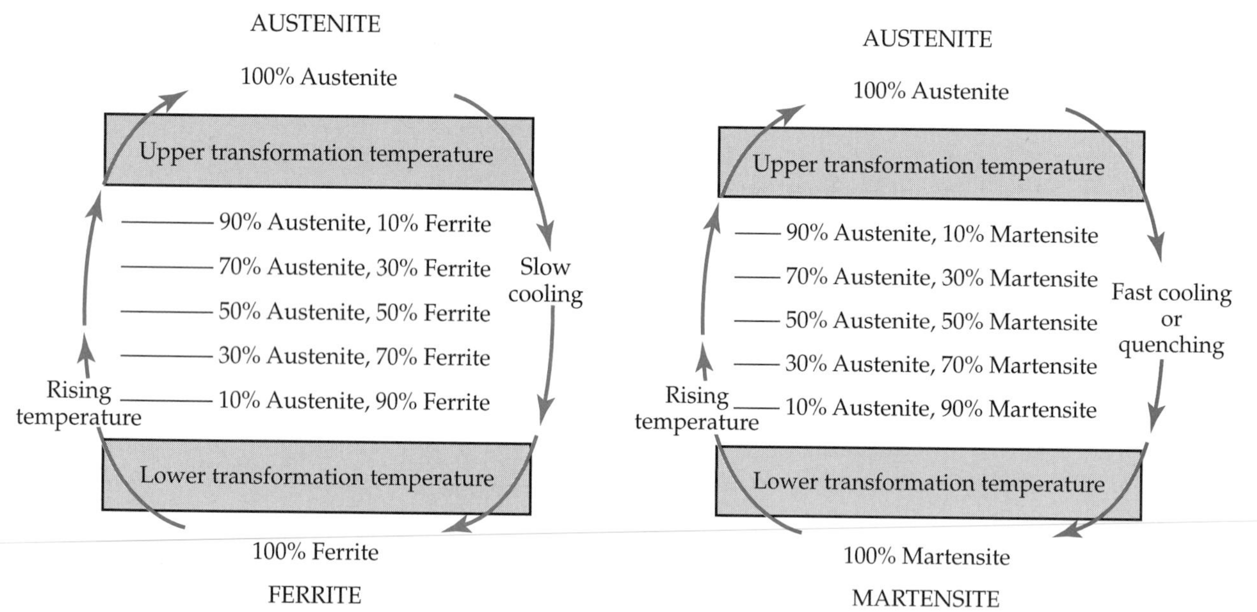

Figure 7-14. *Ferrite is transformed to austenite and back to ferrite in stages between the lower transformation temperature and the upper transformation temperature.*

Figure 7-15. *Martensite is transformed to austenite and back again in stages between the lower transformation temperature and the upper transformation temperature during fast cooling.*

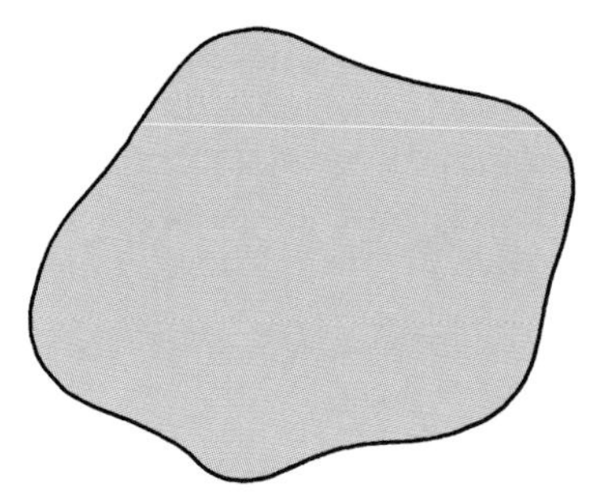

Step 1. Molten iron (far above upper transformation temperature).

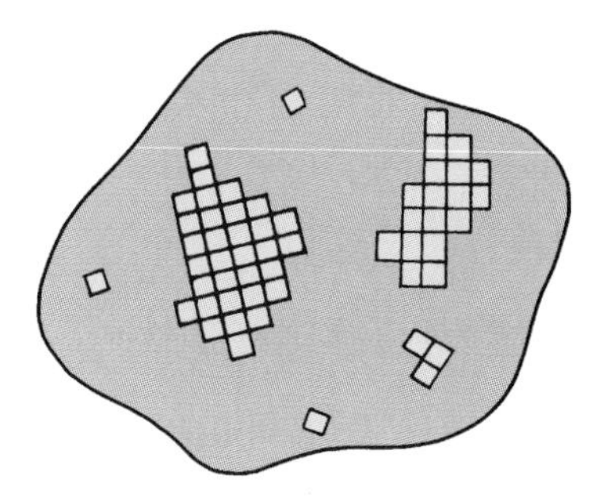

Step 4. As the metal continues to cool, several colonies (called dendrites) form.

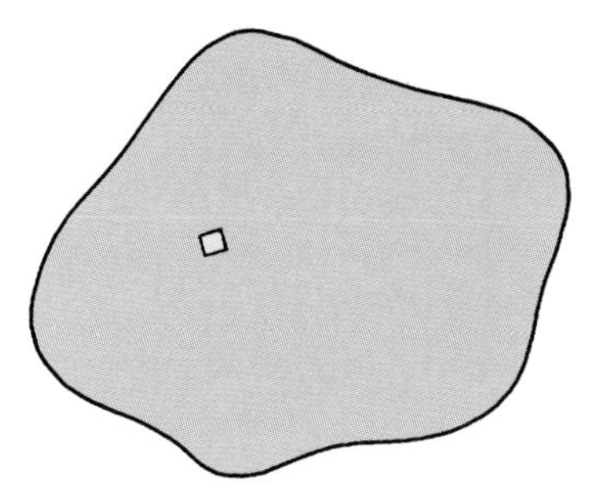

Step 2. As molten iron cools slowly, the first unit cell solidifies.

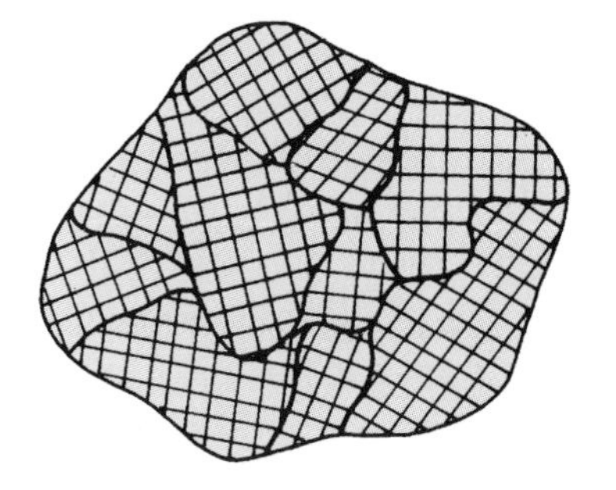

Step 5. As solidification nears completion, boundaries contact each other.

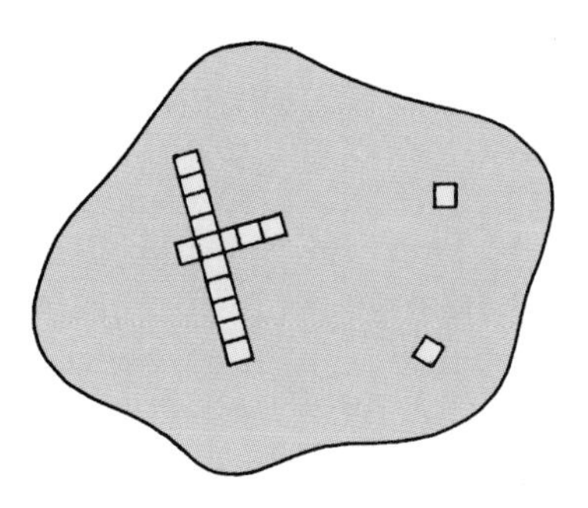

Step 3. Branches develop as the iron continues to cool.

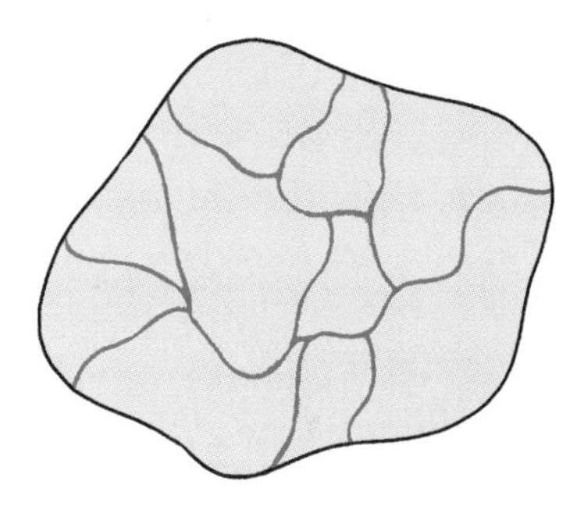

Step 6. The cooled iron has many grains.

Figure 7-16. *The crystal growth process.*

mass of iron eventually cools enough to solidify. A unit cell is formed at this point. As the temperature continues to drop, more unit cells are formed throughout the molten iron. Some of these newly solidified unit cells attach themselves onto the first unit cell, and branches develop.

As the unit cells solidify, they tend to collect in groups, or small colonies. If the temperature is lowered very slowly, many of these colonies grow quite large. As the colonies grow, they grow larger and longer branches and resemble a skeleton. These sprouts or growing colonies, which resemble an unfinished frost pattern, are known as *dendrites.*

The growth of the dendrites continues. As solidification nears completion, the boundaries of some of the colonies contact each other. Here a conflict exists. The unit cell axis of each colony is different; as a result, the colonies tend to compete for the membership of the last few unsolidified cells.

When the entire mass of iron finally becomes solid, there are boundaries between the colony groups. These boundaries are visible under a microscope. In fact, in many materials, the boundaries between colony groups are visible to the naked eye.

Each colony within a boundary is known as a grain. A *grain* is any portion of a solid that has external boundaries and an internal atomic lattice arrangement that is regular. A grain could also be called a "fully grown dendrite."

The term *crystal* has several meanings when it is used to describe structures. Most often, "crystal" means the exact same thing as grain. However, in a technical sense, a crystal can be anything from a unit cell to a fully grown grain. Thus, a crystal can be a grain, a unit cell, or a dendrite.

Grain Size Versus Cooling Time

The grain size of metal is affected by the rate of cooling after the metal is heated. If metal is cooled very slowly from a molten state, the colonies have more time to add on members. Therefore, in slow cooling, the colonies have time to grow larger and larger, and very large grain size results.

If, on the other hand, the material is cooled very rapidly, many more colonies are formed. The size of each colony is limited due to the large number of colonies. Therefore, rapid cooling produces a small grain size. The effects of slow and rapid cooling on metal are shown in Figure 7-17.

Effect of Grain Size

Grain size has a profound effect on the strength, hardness, brittleness, and ductility of metal. If you take a sheet of paper and start to tear it from one edge, the paper has little resistance to prevent the tear from moving across the entire surface. On the other hand, if a tear is started on a sheet formed by several bits of paper glued together, Figure 7-18, it is more difficult for the tear to move smoothly and rapidly across the sheet.

Metals behave in the same manner. Metals with large grain size are easier to tear, break, or fracture. Metals with small grain size have a higher resistance to fracture. A small crack in metal has more difficulty moving across a series of small grains than across one large open field, Figure 7-19.

The smaller the grain size, the greater the strength; the larger the grain size, the lesser the strength. Therefore, in metallurgy, efforts are made to keep the grain size as small as possible when strength is important.

Since strength, hardness, and brittleness are related properties, small grain size not only yields better strength characteristics, but also results in a harder and more brittle material. Therefore, if it is important for a material

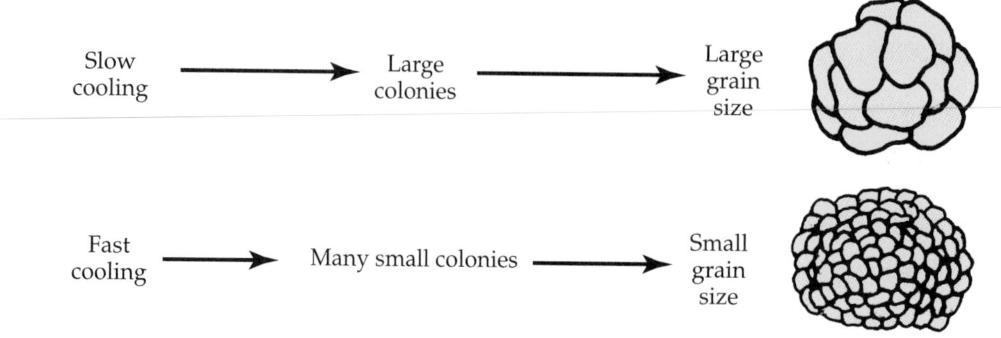

Figure 7-17. *Slow cooling produces a large grain size. Rapid cooling produces a small grain size.*

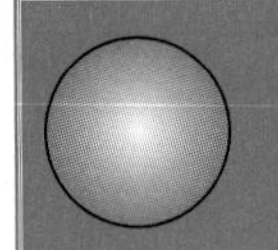
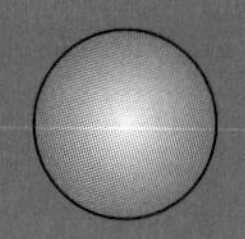

Self-Demonstration Formation of Crystals

Snowflakes and different structures of metal are examples of crystal formation. Crystals can also be grown from several household products. These crystals have the same behavior and appearances that are produced in metals and other structures.

Purchase a small jar of alum from your local drug store. Alum is a powdered material that is relatively inexpensive. It is commonly used in home pickling and contains ammonium aluminum sulfate. **Warning: Alum is harmful if swallowed.** If the drug store does not sell alum, ask the druggist to recommend another powdered material that will crystallize. There are many different materials that will work, but alum is one of the easiest to find.

Pour some of the alum into a glass of water and stir until the water is saturated or slightly oversaturated.

Next, dampen the end of a string and dip it into a small amount of powered alum until a small crystalline ball or cluster of alum forms on the end of the string.

Dip the cluster on the end of the string gently into the alum solution. If the string is left in the solution for a long period of time (such as several days), the crystal on the end of the string will grow. The crystal serves as a "seed" for crystalline growth.

If the string is left in the solution for a longer period of time (such as several weeks), a large crystal will eventually form.

After growing the crystal, examine its shape. Are the surfaces of the crystal flat and smooth?

to have strength, a wise metallurgist will attempt to produce a small grain size. If ductility is more important than strength, the metallurgist will want to attain a large grain size.

Figure 7-18. *Paper is more difficult to tear when it is made up of many small pieces glued together.*

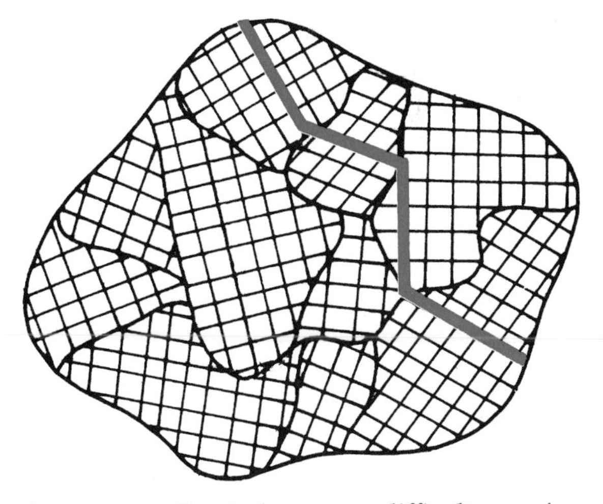

Figure 7-19. *Cracks have more difficulty moving across a series of small grains than across large grains.*

Ferrite and Martensite Properties

Some properties of ferrite and martensite are compared in Figure 7-20. Note that the ferrite lattice structure has the advantage of being more ductile. Therefore, ferrite is more easily machined and less prone to cracking.

The small crystal structure of martensite, on the other hand, has greater strength and hardness. Martensite is also more brittle and more prone to cracking and distortion than ferrite, due to the rapid cooling process required to attain the small grain size.

Comparison of Room-Temperature Iron		
	Ferrite	**Martensite**
Grain size	Large	Small
Cooling speed	Slow	Rapid
Strength	Lower	Higher
Hardness	Lower	Higher
Ductility	Ductile	Brittle
Distortion/ cracking	Little tendency	Greater tendency
Machinability/ formability	Good	Difficult

Figure 7-20. *This table compares two types of room-temperature iron structures, ferrite and martensite.*

Test Your Knowledge

Write your answers on a separate sheet of paper. Do not write in this book.

1. What is a *space lattice?*
2. What is the most fundamental arrangement that shows the basic structure of a space lattice?
3. How many atoms are in the unit cell of a body-centered cubic space lattice?
4. Which form of iron has a face-centered cubic space lattice? How many atoms are in a face-centered cubic unit cell?
5. Which common type of space lattice is not typical of any iron structure?
6. What is the difference between a body-centered tetragonal space lattice and a body-centered cubic space lattice?
7. What type of space lattice structure occurs in ferritic iron?
8. What type of space lattice structure occurs in martensitic iron?
9. Slow cooling will produce _____ iron. Rapid cooling will produce _____ iron.
10. Explain the different space lattice structures and forms of iron that occur at the lower transformation and upper transformation temperatures.
11. What are *dendrites?*
12. What is a *grain?*
13. What is another name for a grain, unit cell, or dendrite?
14. A large grain size is produced when metal is cooled _____. A small grain size is produced when metal is cooled _____.
15. What grain size is best suited for an application in which strength is more important that ductility?
16. In the manufacture of a motor home, which steel parts in the vehicle would most likely be made of ferrite and which parts would be made of martensite?
17. What grain size would be best suited for the steel required in each of the applications listed below?
    a. A long lever in an exercise machine.
    b. The linkage parts in a hardness tester.
    c. The framework for a seat in a space vehicle.
    d. The framework for the support of a chiropractic adjusting table.

8 Failure and Deformation of Metal

After studying this chapter, you will be able to:

- Explain what happens inside a piece of metal when it breaks.
- State what is meant by deformation of metal.
- Summarize the different types of metal failure or breakage.
- Describe work hardening and its applications.

Deformation

When a material is stretched, it deforms. *Deformation* is the amount that a material increases or decreases in length when it is *loaded* (when force is applied). Materials such as cast iron, concrete, and peanut brittle are very unwilling to stretch. They have little deformation. Other materials—such as aluminum, polyethylene, and rubber—stretch far more. They have a high degree of deformation.

Before a piece of metal fractures, it may stretch a lot or it may stretch very little. The material shown in Figure 8-1 breaks when the force on it reaches 500 lb. Before it breaks, the metal does not stretch at all.

The metal shown in Figure 8-2 also fails when the force reaches 500 lb. However, before it fails, it stretches considerably. Thus, the amount of deformation that a material goes through before failure does not indicate the total amount of force it is able to take before breaking.

Examples of fracture and deformation are shown in Figure 8-3. The materials are shown before and after testing. Aluminum, low-carbon steel, nylon, and polyethylene are *ductile* materials. They stretch considerably before failure. Cast iron is a *brittle* material because it does not stretch before it breaks.

Ductility and Brittleness

A ductile material stretches much before it fails (refer to Chapter 4). The magnitude of the load at failure may be small or great. A brittle material lacks ductility. It stretches very little before it fails.

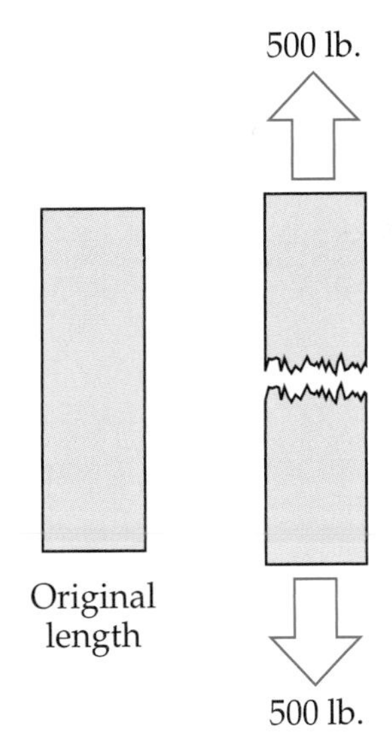

Figure 8-1. *This material does not stretch before breaking.*

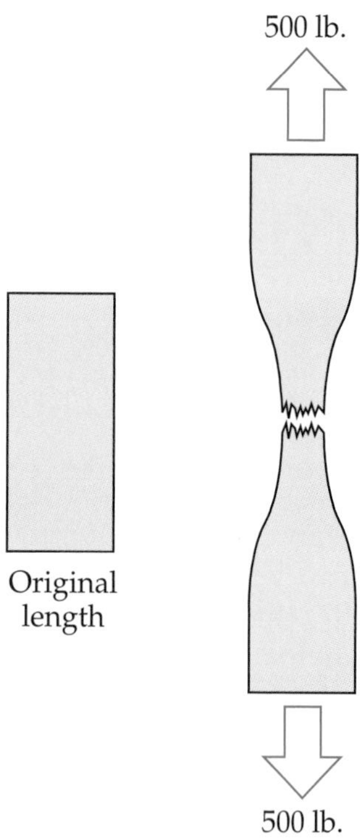

Figure 8-2. *This material stretches considerably before breaking.*

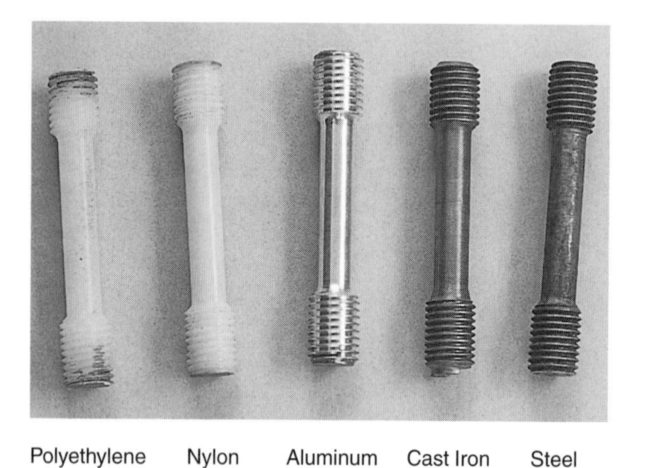

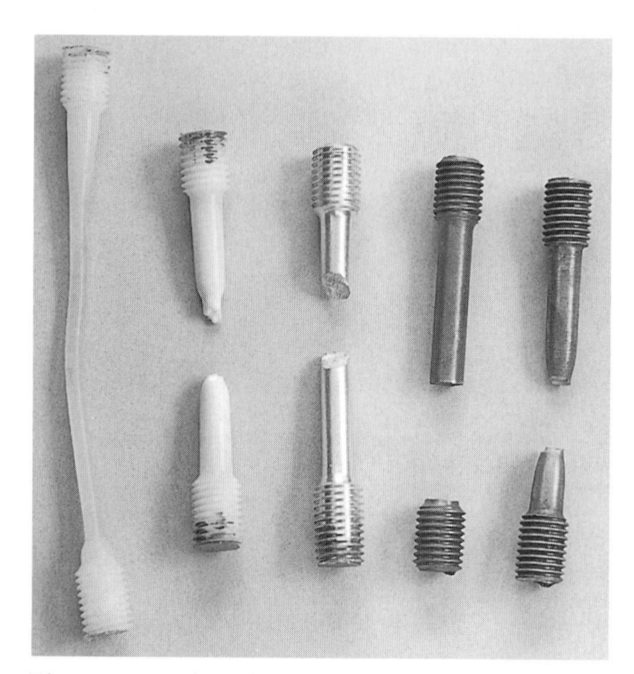

Figure 8-3. *These five samples have the same length, diameter, and general configuration before testing. After fracture, some materials show greater ductility than others. Polyethylene has the greatest ductility. Cast iron has the least ductility and most brittleness.*

Brittle materials fail in *cleavage.* Ductile materials fail in *shear.* The table in Figure 8-4 compares brittle and ductile failures.

Cleavage—Brittle Failure

A material that failed in cleavage is shown in Figure 8-5. Breakage in cast iron is a typical example of this type of failure. In a brittle failure, the atoms simply break apart from each other and separate. This type of failure appears bright, rough, and granular, Figure 8-6.

Occasionally, ductile materials may exhibit a "brittle" failure. This occurs when the material is subjected to a rapid shock load and does not have time to begin to stretch.

Shear—Ductile Failure

When a ductile material is loaded and fails, the atoms slide past each other within the crystals. This type of stress, called *shear,* causes the material to stretch, Figure 8-7. Shear will sometimes cause materials to "neck down" or become thinner and longer near the failure point. See Figure 8-8. Shear failures appear dull, smooth, velvety, and fibrous.

Shear failures take one of two forms, slip failures or twinning failures, depending on how atoms are displaced within the material.

Comparison of Brittle and Ductile Failures		
	Brittle Failure	**Ductile Failure**
Failure mode	Cleavage	Shear
Stretching before failure	Negligible	Much stretching
Material examples	Cast iron Concrete Wood	Aluminum Low-carbon steel Rubber

Figure 8-4. *This chart compares the characteristics related to brittle and ductile failures.*

Figure 8-5. *This brittle cast iron sample failed in cleavage.*

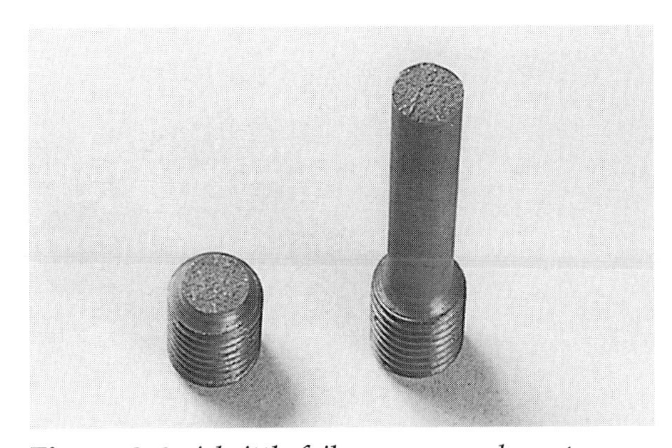

Figure 8-6. *A brittle failure occurs when atoms break apart and separate. This type of break has a rough, granular appearance.*

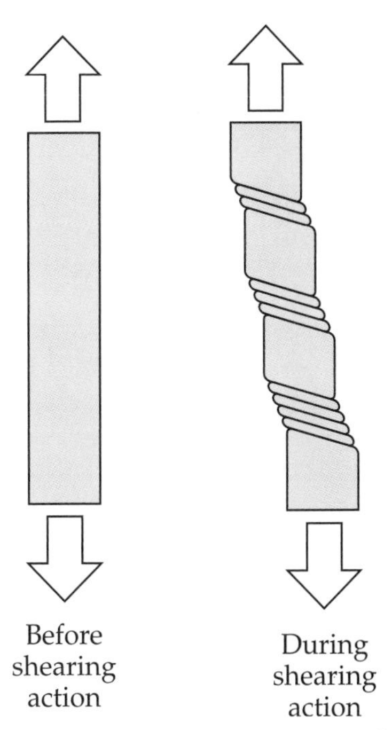

Figure 8-7. *Internal shearing causes a ductile material to stretch. The effect of shearing is exaggerated in this illustration.*

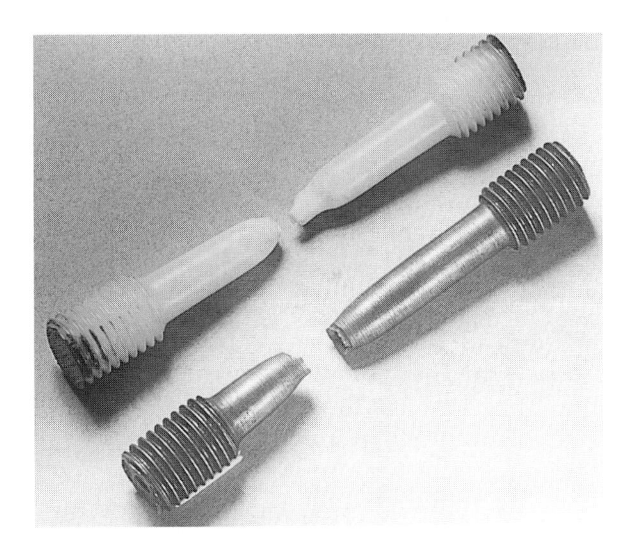

Figure 8-8. *Steel and nylon both have a tendency to "neck down" before failure.*

Slip failures

A shear failure for aluminum, a ductile material, is shown in Figure 8-9. This type of shear failure is known as *slip.* Instead of dividing in the middle of a crystal, the

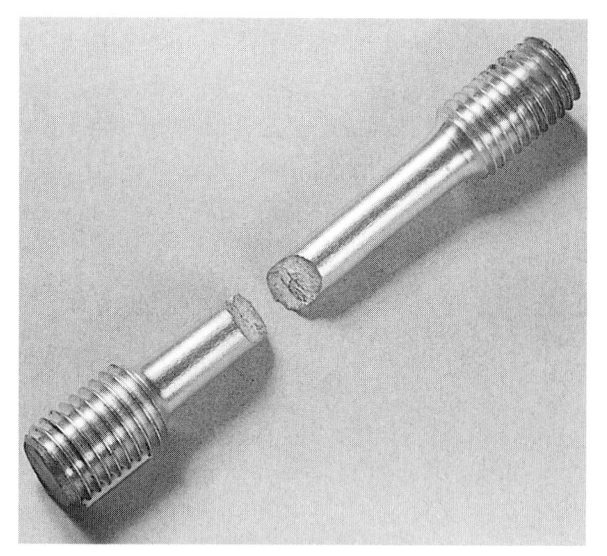

Figure 8-9. *Aluminum is a ductile material that fails by shear.*

atoms slide past each other and move down one row at a time. As they slide past each other, the part deforms, becoming longer and thinner. Eventually, a break may occur.

Slip takes place along certain crystal lines called *slip planes.* When failure occurs, an entire block of atoms moves over another block of atoms. This action takes place through several thousand atomic layers. The movement moves forward one step at a time, Figure 8-10. A body-centered cubic space lattice structure that has slipped two atomic layers is shown in Figure 8-11.

As a material stretches, several slip planes may be in action at the same time. A group of slip planes is called a *slip band,* Figure 8-12. When many slip planes are present, it is relatively easy for deformation to occur. When fewer slip planes are involved, it is more difficult for slip to occur. The metal is stronger but less ductile.

Twinning failures

Twinning is very similar to slip. Both are considered ductile failures. Both result from shear. Some metals are more susceptible to twinning deformation than slip deformation. Some metals deform as a result of both twinning and slip.

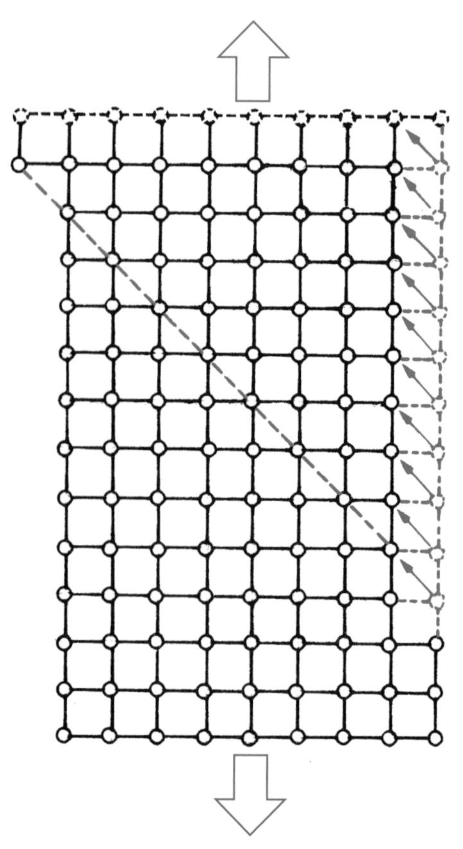

Figure 8-10. *A slip failure moves forward one step at a time.*

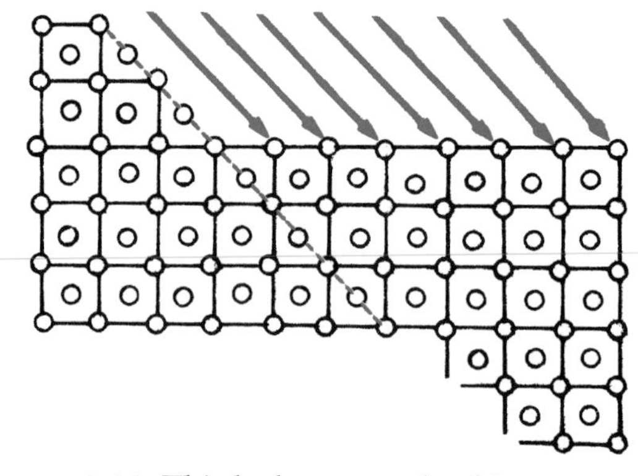

Figure 8-11. *This body-centered cubic space lattice structure has slipped two atomic layers.*

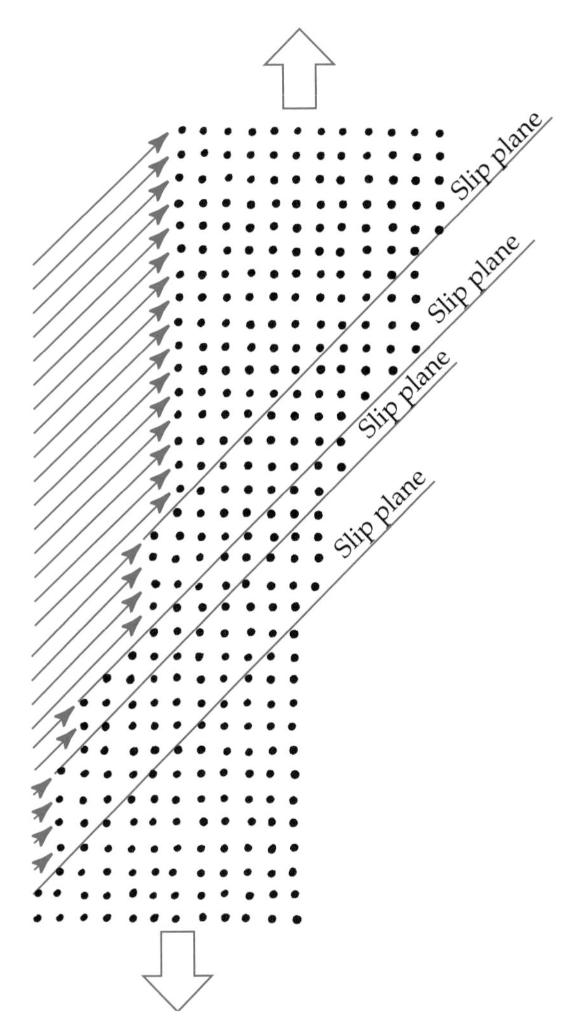

Figure 8-12. *A slip band is a group of slip planes that form during failure.*

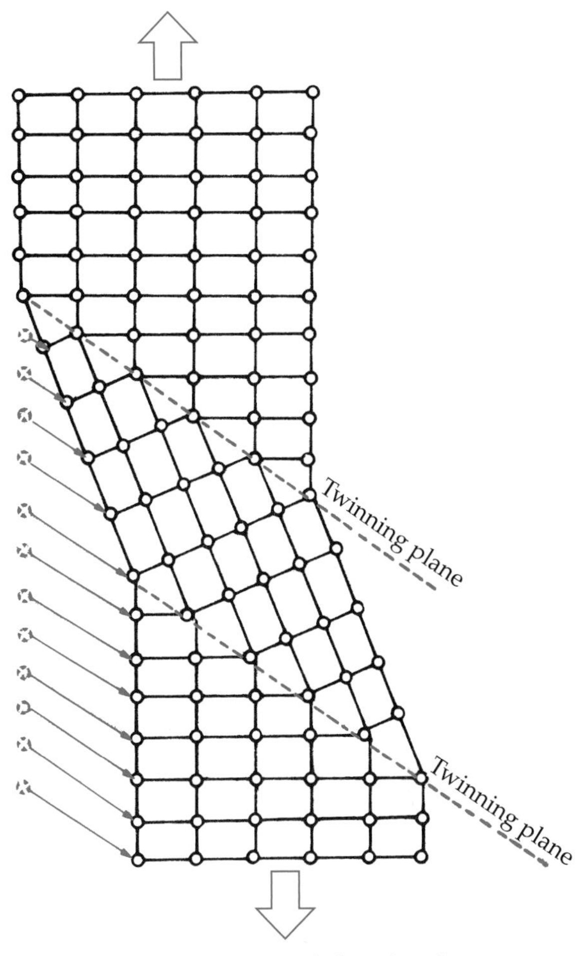

Figure 8-13. *A twinning failure involves two mirror lines (or twinning planes).*

In twinning, a zone within a crystal structure is deformed from its original space lattice formation. The two lines that separate the deformed zone from the original structure are known as *mirror lines* or *twinning planes.* See Figure 8-13.

The formation of atoms on either side of the twinning plane is the same. If a mirror were placed on the edge of the twinning plane, the image in the mirror would be identical to the orientation of the group of atoms behind the mirror, Figure 8-14.

Twinning causes a new space lattice structure in the twinning region. This region may consist of millions of atoms. Essentially, a long block or plane of atoms is displaced. The atoms move forward and slide past other atoms (in a pattern similar to the movement of atoms in slip). As twinning progresses, a separation failure may also take place.

Large and Small Crystals

The size of crystals (or grains) in metal has a direct effect on the strength and ductility of the material. This was covered in Chapter 7. A material with small crystals is stronger than a

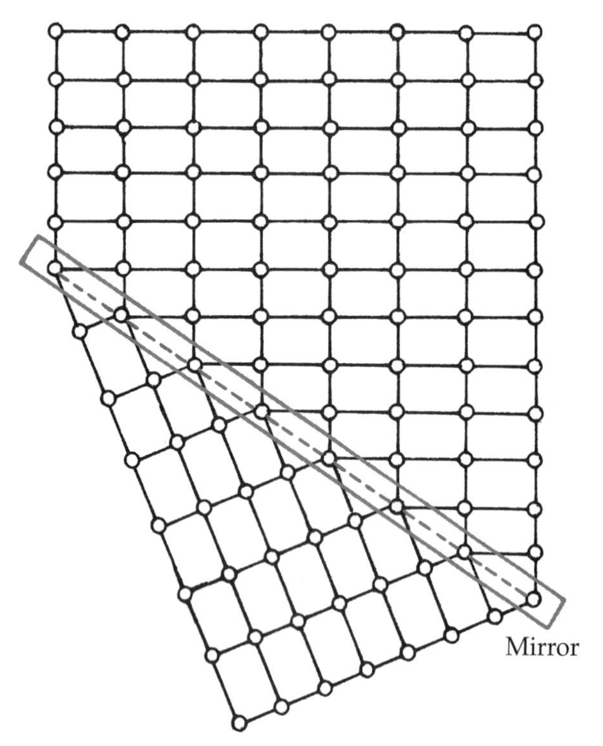

Figure 8-14. *In a twinning failure, the image on either side of the mirror line is identical.*

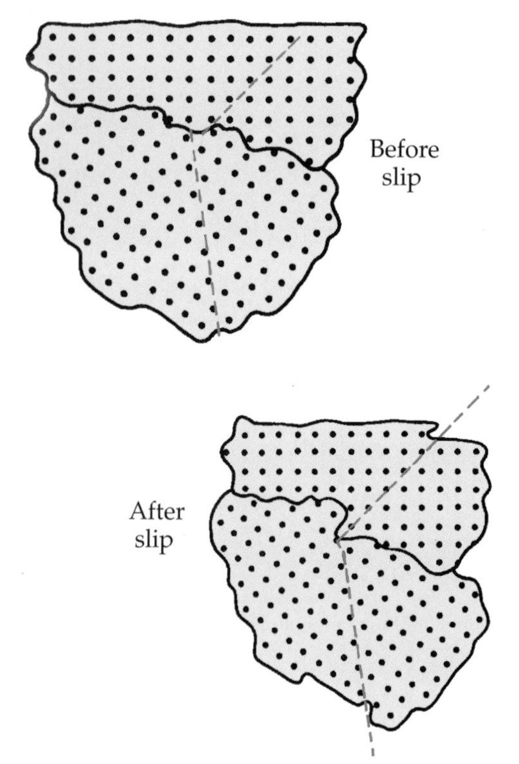

Figure 8-15. *Slip becomes more difficult when the slip plane reaches the end of a crystal.*

similar material with large crystals and is able to resist more force. Small crystals resist cracking better. Slip and twinning help clarify why this is true.

If slip or twinning occurs, a slip plane can move rapidly across a crystal. However, when the slip plane reaches the end of a crystal and must continue across a second crystal, fracture or additional slip becomes more difficult. See Figure 8-15. The second crystal will have a different direction of lattice orientation, so the crack must change direction. Failure becomes more difficult each time a crystal boundary is reached. Therefore, a material with many small crystals is generally more capable of resisting fracture than a material with large crystals. See Figure 8-16.

The effects of crystal size emphasize the importance of cooling metal rapidly. With rapid cooling, small crystals are formed. The small crystals formed in quenched metal are much more resistant to failure than the large crystals formed by slow cooling.

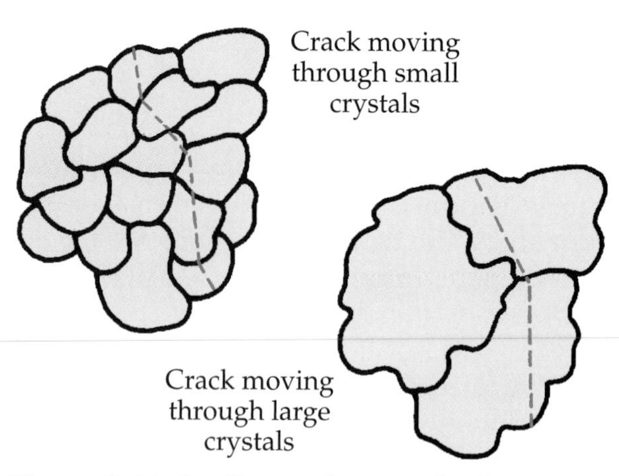

Figure 8-16. *Small crystals can resist fractures better than large crystals.*

Work Hardening

As materials are stretched out of their normal shape, a unique phenomenon called *work hardening* (or *strain hardening*) occurs. This process is discussed below.

Consider two machine parts. One part is subjected to a force. Before it breaks, the force is removed and the part relaxes. The force is again applied, then removed, and the part relaxes. The force is again applied, then removed before failure takes place. If the application of force is repeated over and over, do you think this material will eventually fail at a lower force level than a second machine part that is loaded until failure the first time? In other words, does the repeated application and removal of force weaken a material or strengthen it?

When a material receives a few preliminary applications of force, it *work hardens.* This makes the material stronger and harder (although the material also becomes more brittle). Thus, in Figure 8-17, the work-hardened machine part (A1) will break at a higher force value than the second machine part (A2), which is loaded until failure the first time.

The reason why work hardening occurs is not fully understood. However, it is assumed that as atoms are forced into a stretched position, they tend to lock into a formation that is stronger than the original formation. As stretching takes place, some atoms are torn from their original lattice structure and move to a new spot in between slip planes. Then, they become a roadblock and prevent one plane from sliding over another. Thus, there is an increase in strength and hardness, but some ductility and elasticity is lost.

In manufacturing, the phenomenon of work hardening is often a blessing. In processes such as forging, extruding, drawing, and rolling, "cold working" takes place as the metal is repeatedly compressed. The added strength that occurs as a result of work hardening can be considered a natural benefit.

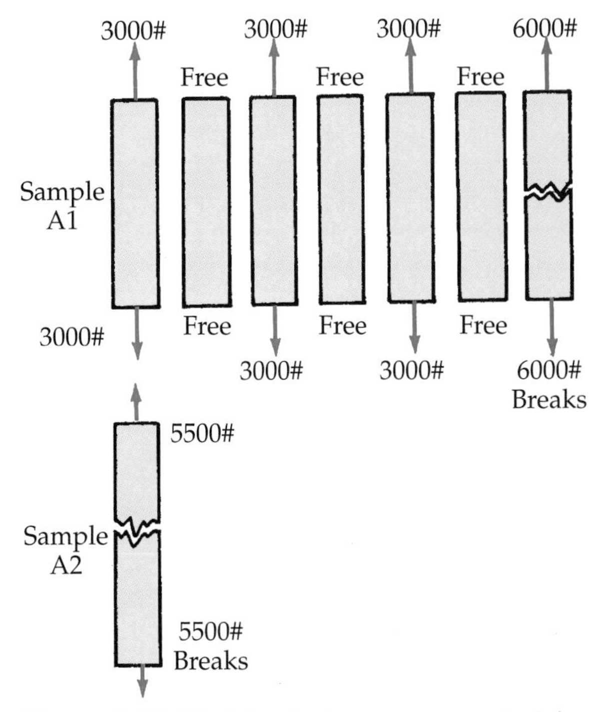

Figure 8-17. *Work hardening causes a material to become stronger, harder, and more brittle.*

Test Your Knowledge

Write your answers on a separate sheet of paper. Do not write in this book.

1. Define *deformation.*
2. Name two metals that have a high degree of ductility and resistance to fracture.
3. Name one metal that shows a very low degree of deformation.
4. If a metal stretches before it breaks, is it more likely to fail in shear or cleavage?
5. Under what circumstances will a ductile material fail in cleavage instead of shear?
6. What type of failure occurs when the atoms inside a crystal slide past each other, one row at a time?
7. What is a *slip band?*
8. What are *twinning planes?*
9. What is the difference between small crystal structures and large crystal structures in relation to deformation?

10. Which is generally stronger, a material with small, fine crystals or a material with large crystals?
11. Which is generally able to resist cracking better, a material with small, fine crystals or a material with large crystals?
12. What size crystal structure is formed by rapid cooling?
13. What is *work hardening?*
14. Discuss three medical applications for which deformation would be an important consideration in the design of equipment.
15. Discuss two applications where the deformation of materials would be considered an asset.
16. Discuss two types of manufacturing in which the deformation of materials would create a severe hazard.

Iron-Carbon Diagram

After studying this chapter, you will be able to:

- Describe five important structural forms of steel and iron.
- Identify the transformation regions and other major elements of an iron-carbon diagram.
- Use an iron-carbon diagram to determine the steel structures that occur at various temperatures.
- Explain how different cooling techniques are used to produce mechanical properties in steel.
- Use an iron-carbon diagram to determine the temperature to which steel must be heated to cause it to harden.

Structural Forms of Steel

Steel is iron with more than 0% carbon but less than approximately 2% carbon. Steel behaves very differently as the amount of carbon increases from 0% to 2%. Cast iron contains more than approximately 2% carbon. This was covered in Chapter 5.

Structural forms of steel are commonly classified by the amount of carbon in steel. Steel that has very little carbon is called *ferrite,* Figure 9-1. Steel that has approximately 0.8% carbon is called *pearlite.* Steel that has a carbon content above 0.8% contains some *cementite.* All three of these forms of steel—ferrite, cementite, and pearlite—represent the material at room temperature.

Steel that is heated to an elevated temperature is called *austenite.* This was briefly discussed in Chapter 7. When austenite is cooled from the upper transformation temperature back to room temperature, it is transformed into other structures (such as ferrite, cementite, and pearlite).

Ferrite

Ferrite is almost pure iron. It has little tendency to dissolve carbon when carbon is added. Therefore, a very small percentage of carbon is in ferrite. Since carbon gives steel the ability to become strong and hard, ferrite is a very weak steel. Ferrite exists at low temperatures only, and it is magnetic.

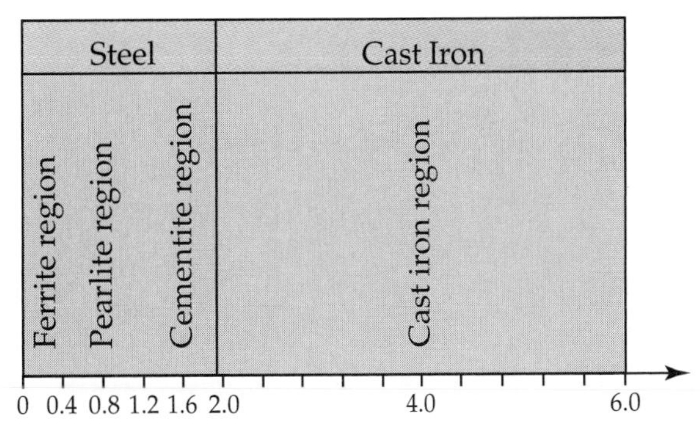

Figure 9-1. *Steel can be classified as ferrite, pearlite, or cementite, depending on its percentage of carbon. Cast iron contains more than approximately 2% carbon.*

Cementite

Cementite (also called *iron carbide*) is actually a compound of iron and carbon. Its chemical formula is Fe_3C. Pure cementite contains 6.67% carbon by weight. However, different amounts of cementite are present in steel that contains between 0.8% and 2.0% carbon.

As the percentage of carbon increases, more and more cementite is present. At 6.67% carbon, the entire mixture is cementite. Below approximately 2% carbon, the alloy is still considered to be steel; above that percentage, it becomes cast iron.

Cementite exists at room temperature, and it is magnetic. After heat treatment, cementite can become very strong and hard.

Pearlite

Pearlite is a mixture of ferrite and cementite. It exists at room temperature and is magnetic. Under a microscope, pearlite appears as a series of layers, resembling an aerial view of newly plowed fields. See Figure 9-2. The black ridges are cementite, and the white ridges are ferrite. Thus, pearlite is made up of alternating layers of ferrite and cementite.

In steel with approximately 0.8% carbon, the ferrite and cementite are sufficiently balanced so the entire microscopic view contains ridges. If there is less than 0.8% carbon, the steel will be a mixture of ferrite and pearlite, and only certain parts of the view will appear as ridges. If the steel has more than 0.8% carbon, it will be a mixture of cementite and pearlite. Figure 9-3 shows the relationship of the composition of steel at various percentages of carbon.

Figure 9-2. *This microscopic view reveals the pearlite structure of 1095 steel. Magnification is 500X. (Buehler Ltd.)*

Austenite

Austenite is the structural form of steel that occurs only at elevated temperatures. It is not magnetic.

When steel is heated to an elevated temperature and becomes austenite, its structure changes from body-centered cubic to face-

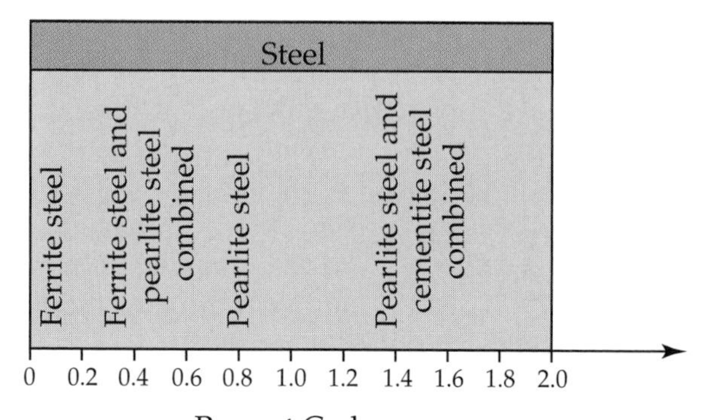

Figure 9-3. *This chart shows effect of percentage of carbon on presence of ferrite, pearlite, and cementite in steel.*

Ferrite	Austenite
• Low temperature • Body-centered cubic	• High temperature • Face-centered cubic

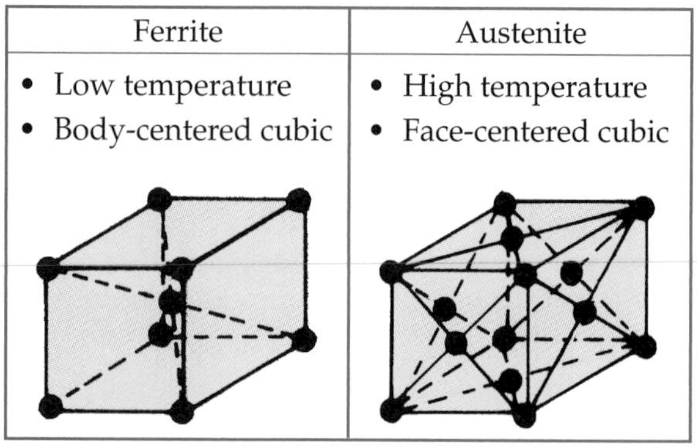

Figure 9-4. *This diagram compares the structural differences in low-temperature and high-temperature steel.*

centered cubic. See Figure 9-4. If it is slowly cooled back to room temperature, the face-centered cubic structure changes back to body-centered cubic and the steel becomes ferrite, pearlite, or cementite.

Iron-Carbon Phase Diagram

An *iron-carbon phase diagram* is used to identify the different structures of steel that occur at various temperatures for a given percentage of carbon. A simplified version of the diagram is shown in Figure 9-5. An industrial version is shown in Figure 9-6. From the iron-carbon phase diagram, you can tell what structure iron takes at any given temperature—if you know the percentage of carbon present. The diagram allows you to determine whether the steel is ferrite, pearlite, cementite, austenite, or any combination of these four structures. First, however, you must know the following information:

- The temperature of the steel.
- The percentage of carbon in the steel.
- The history of heat treatment for the steel.

There are two important lines shown on an iron-carbon phase diagram. The lower of the two lines is the *lower transformation temperature line.* The upper line is the *upper transformation temperature line.* Transformation temperatures were discussed in Chapter 7. At the lower transformation temperature, the transformation of iron to austenite begins. At the upper transformation temperature, the transition to austenite is complete.

Any steel structure that occurs above the upper transformation temperature line is pure austenite, Figure 9-7. Any steel structure that occurs below the lower transformation temperature line contains no austenite. Iron structures that occur below the lower

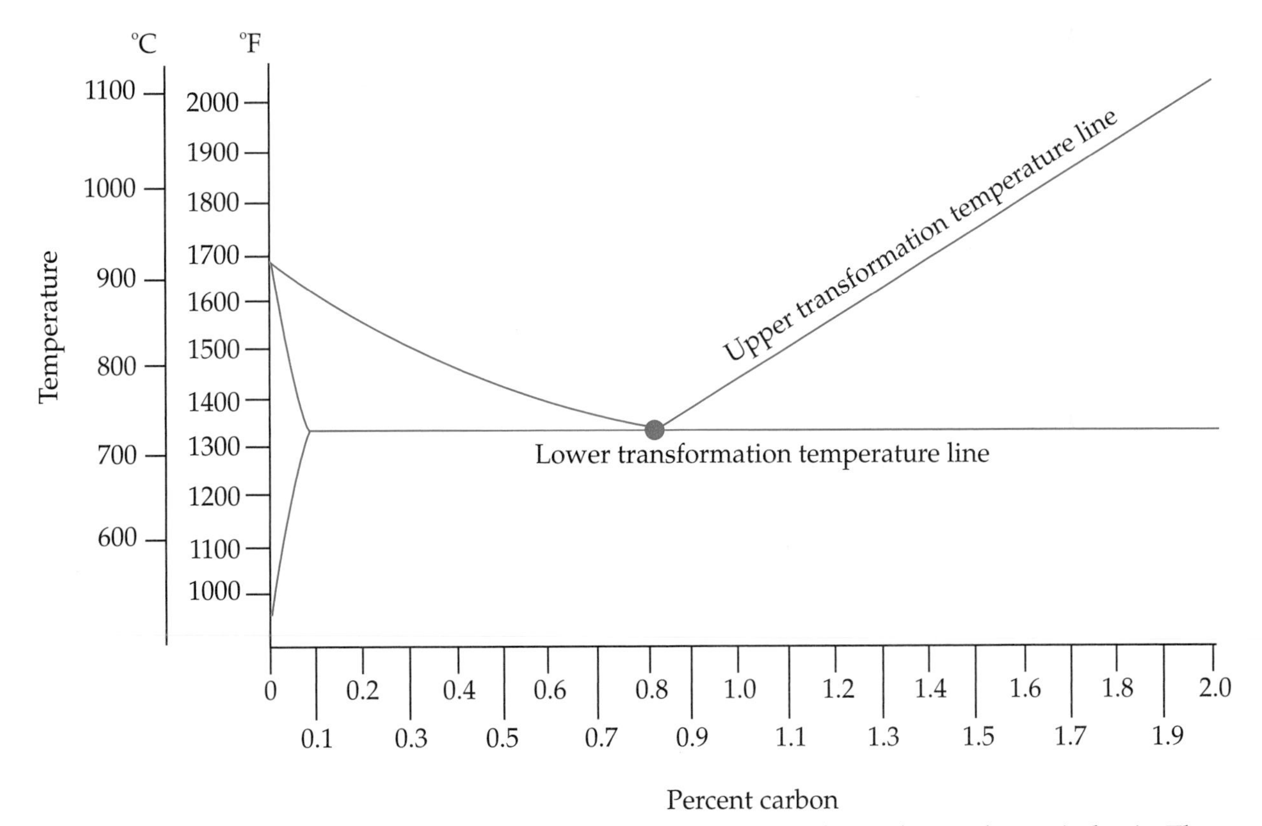

Figure 9-5. *A basic iron-carbon phase diagram. A temperature scale for steel is on the vertical axis. The percentage of carbon is on the horizontal axis.*

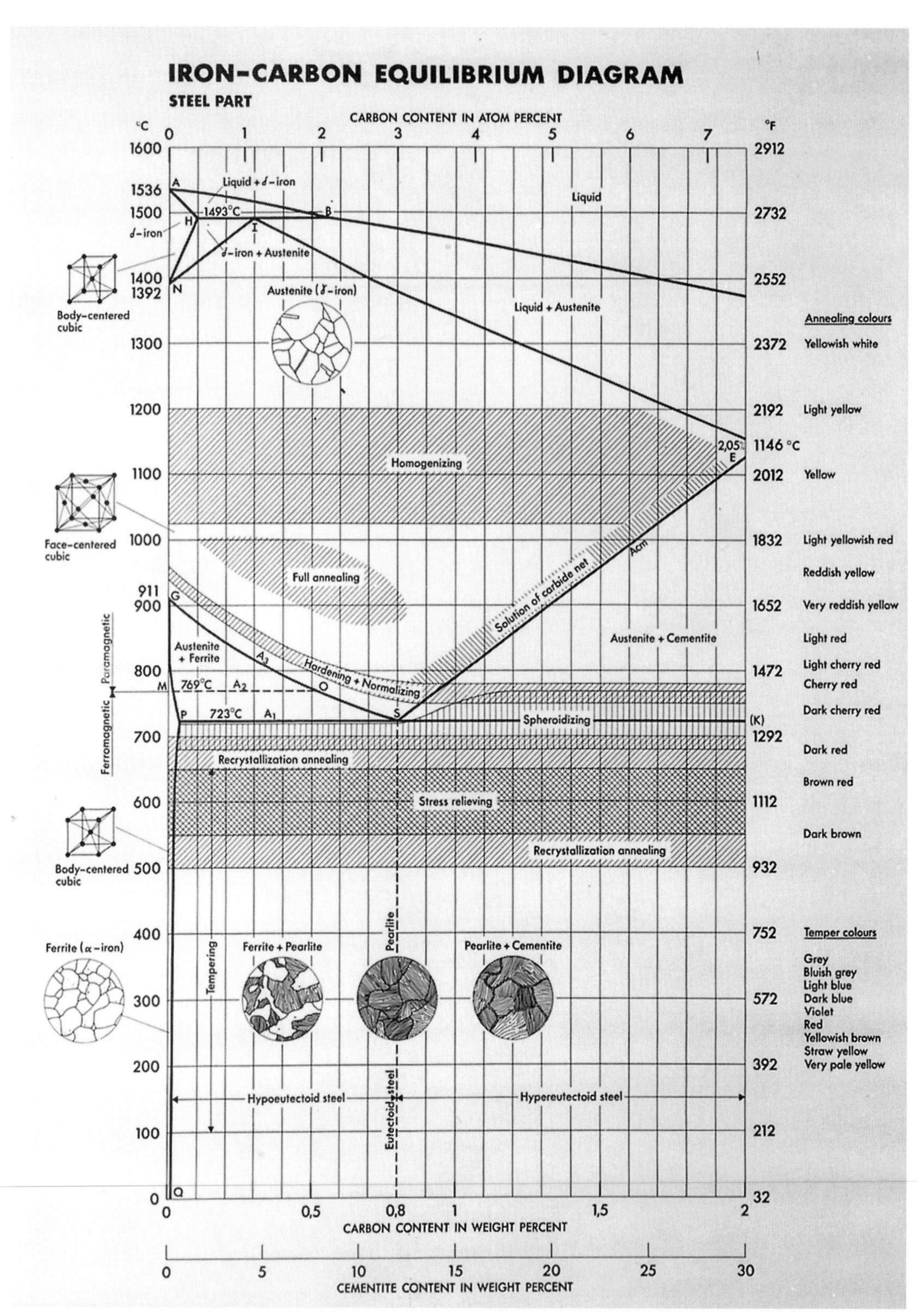

Figure 9-6. *An industrial version of the iron-carbon phase diagram. This diagram is expanded to include higher temperatures. Note the different types of crystal structures shown at left. (Struers Scientific Instruments)*

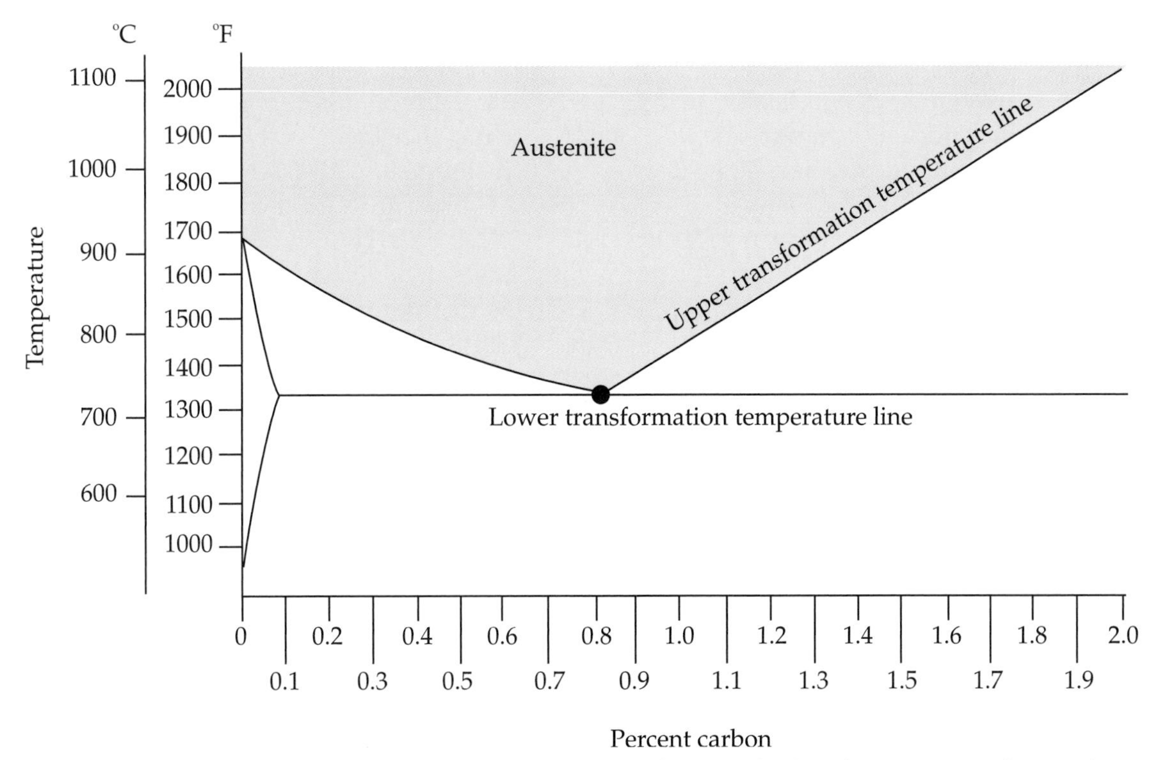

Figure 9-7. *Above the upper transformation temperature line, steel takes the structure of austenite.*

transformation temperature line are ferrite, pearlite, cementite, or combinations of these structures. See Figure 9-8.

In the two triangular areas between the upper and lower transformation temperature lines, a mixture of austenite with either ferrite or cementite occurs. See Figure 9-9. To the left of the ferrite and austenite transformation region is a smaller triangular area. This is the 100% ferrite region. In this area, all of the carbon added is dissolved in the iron. There is no pearlite or cementite.

Using an Iron-Carbon Phase Diagram

One of the best ways to learn to use an iron-carbon phase diagram is to take a sample of steel and follow its journey through the diagram. The following steps illustrate how to read a typical diagram and identify different structures of steel.

1. Assume that the steel is at room temperature and contains 0.4% carbon. This is represented by point A in Figure 9-10. Since the steel is below the lower transformation temperature, it contains no austenite. Point A is nearly midway between the pure ferrite limiting line and the pure pearlite line. The material is approximately half ferrite and half pearlite.
2. Next, assume the material is heated to about 1000°F (540°C) at point B. What structure is it now? Is it austenite, ferrite, pearlite, or cementite? Since the steel is still below the lower transformation temperature line, no austenite is present. Point B is slightly closer to the pure ferrite limiting line than it is to the pure pearlite line, since the pure ferrite line has moved slightly to the right. This structure contains 50%–51% ferrite and 49%–50% pearlite.
3. Now assume the material is heated to 1330°F (720°C) at point C. At 1330°F, the structure is 52% ferrite and 48% pearlite (point C falls just to the left of the 50% ferrite/pearlite line). However, at this

temperature, some key changes start to take place. Above 1330°F, the pearlite changes to austenite. Theoretically, all of the pearlite suddenly changes to austenite at a temperature near 1330°F. In reality, the change occurs over a small range of temperatures near 1330°F (perhaps 1310°F–1350°F).

4. Next, the material is heated to 1360°F (740°C) at point D. All the pearlite has changed to austenite. The structure is now approximately 52% ferrite and 48% austenite.
5. At 1400°F (760°C), the steel is at point E. It is closer to the pure austenite limiting line than it is to the pure ferrite limiting line. Therefore, the material now contains more austenite than ferrite. It is approximately 63% austenite and 37% ferrite.
6. Just before the steel reaches the upper transformation temperature line, at point F (at approximately 1450°F or 790°C), it has almost completely changed to austenite. Perhaps 10% ferrite is left, and the rest of the material has changed to austenite. Thus, the steel is now approximately 90% austenite and 10% ferrite.
7. At 1550°F (840°C), the material is at point G. It has been completely transformed into austenite. No ferrite or pearlite remains.
8. As the material is heated higher to 1700°F (930°C) at point H, there is no further structural change. It remains 100% austenite.

Now assume the steel is slowly cooled back to room temperature, Figure 9-11. If the material is slowly cooled from 1700°F to 1550°F to 1400°F and all the way down to 1000°F, it will change back to ferrite and pearlite. The structures produced at various temperatures during cooling are the same that occur when the steel is heated (in reverse order). Thus, when the steel is cooled and reaches 1550°F, the structure is still

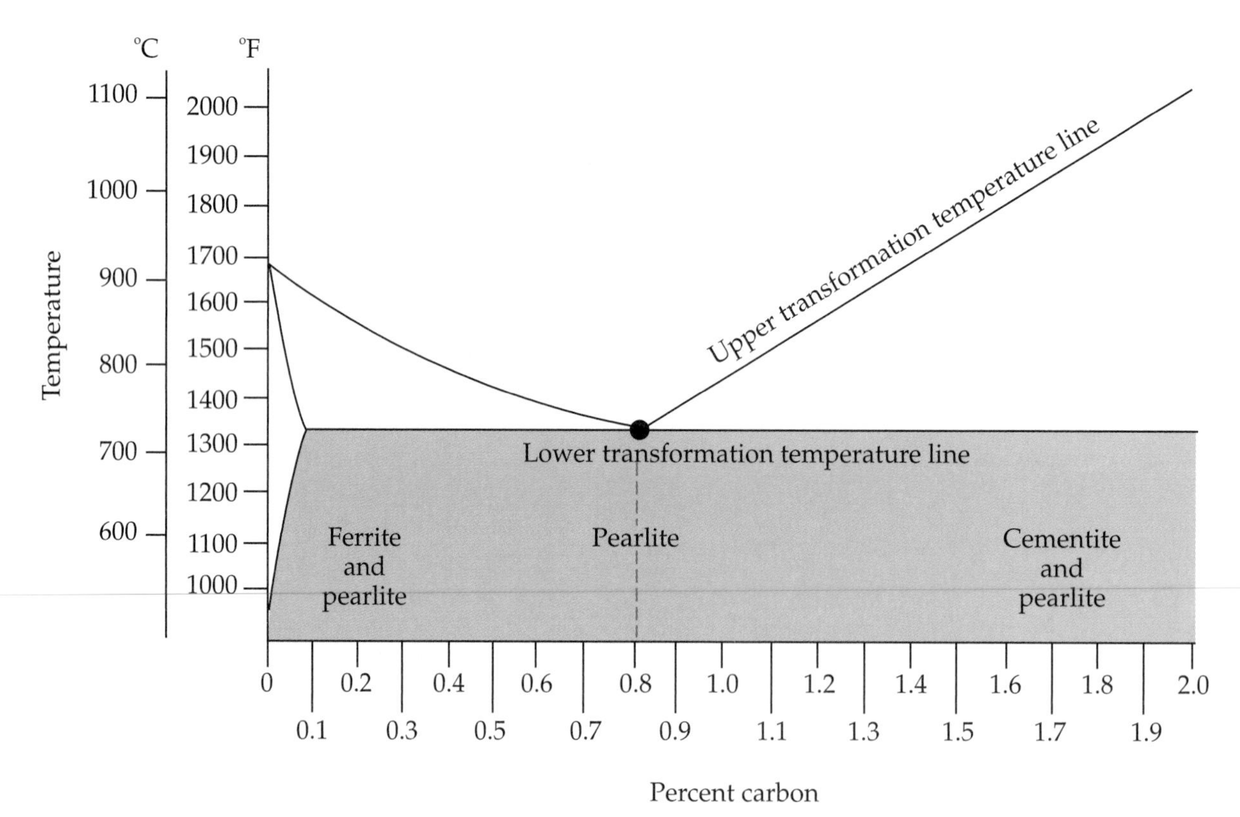

Figure 9-8. *Different structures of ferrite, pearlite, and cementite occur below the lower transformation line.*

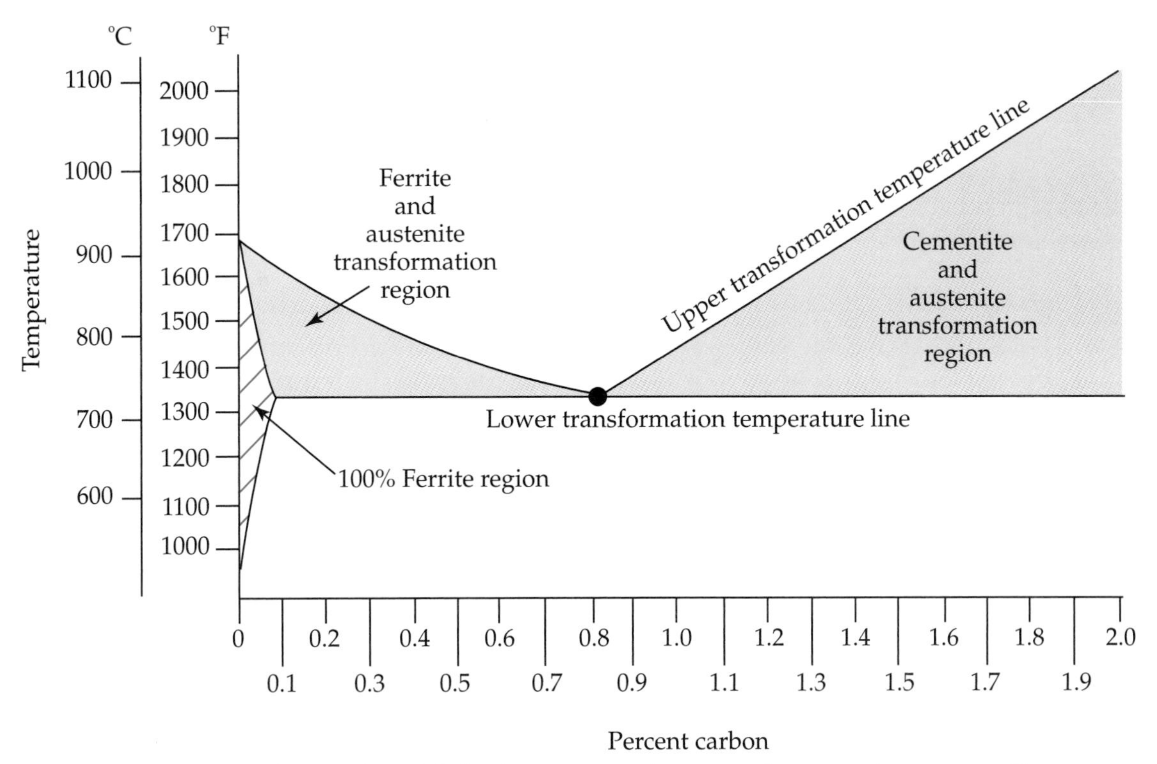

Figure 9-9. *A mixture of austenite with ferrite or cementite occurs between the upper and lower transformation temperature lines.*

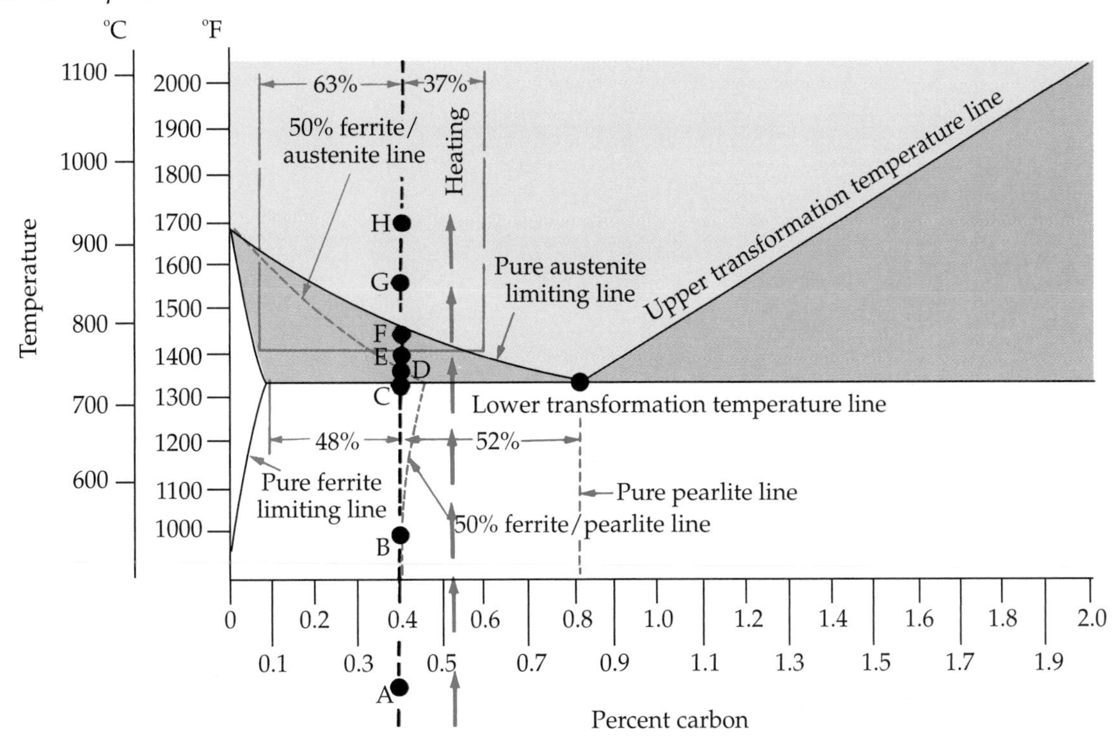

Figure 9-10. *This iron-carbon phase diagram shows the journey of 0.4% carbon steel when it is heated to an elevated temperature.*

100% austenite. At 1400°F (point E), the steel is 63% austenite and 37% ferrite. At 1360°F (point D), there is 48% austenite and 52% ferrite.

At 1000°F (point B), the entire steel structure has changed back to ferrite and pearlite and no austenite remains.

Transformation to Martensite

If steel is heated above the upper transformation temperature and then rapidly cooled, its internal structure changes to *martensite* instead of ferrite, pearlite, or cementite. Martensite was discussed in Chapter 7. It is a very hard, brittle form of steel. The process of cooling steel rapidly is known as *quenching*. This technique is covered in Chapter 11.

Rapid cooling produces structures of steel that are harder than the structures obtained by slow cooling. The effects of quenching steel that contains 0.4% carbon are shown in Figure 9-12. At point H (1700°F), the steel is 100% austenite. At point G (1550°F), the structure is still 100% austenite. At 1400°F, the steel has been transformed into 37% martensite and 63% austenite. At 1000°F, 100% martensite exists.

Identifying Steel Structures

The iron-carbon phase diagram in Figure 9-13 identifies locations of different structures of steel. The ten points shown are plotted at a variety of temperatures, and they indicate different types of steel with various percentages of carbon. Can you identify the structures that correspond to each of the points? Which samples are austenite? Which are ferrite? Which are pearlite? Which are cementite? Which samples are a combination of the four?

The samples at points A and B have been heated to temperatures above the upper

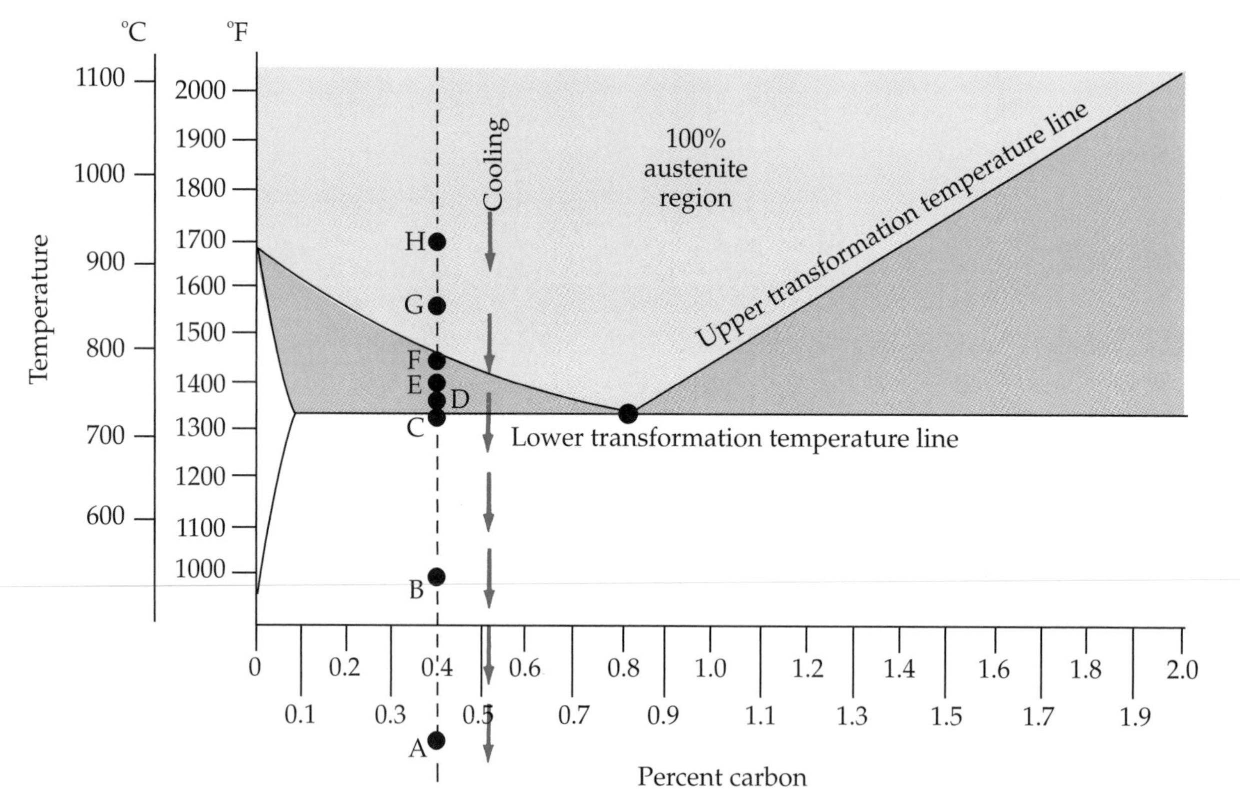

Figure 9-11. *This diagram follows the path of 0.4% carbon steel when it is cooled to room temperature.*

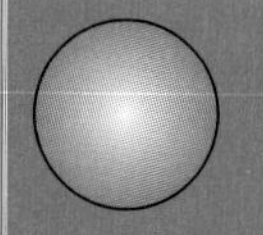

Self-Demonstration Effects of Heating and Cooling on Different Types of Steel

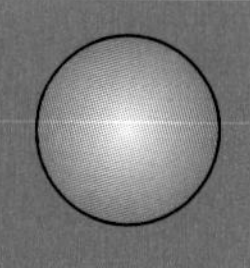

Obtain five small pieces of each of several different types of steel, including 1018, 1045, 1095, 4140, and 52100 steel. If other types of steel are substituted, try to obtain an assortment that includes low-carbon, medium-carbon, and high-carbon steel.

First, test the hardness of each of the samples on the Rockwell C scale and record the hardness values.

Load all the samples into a metallurgical furnace and heat it to 1300°F (700°C). After holding this temperature for one-half hour, remove one sample of each type of steel and immediately plunge the samples into cold water. After the samples have cooled, record each hardness value.

Increase the temperature of the furnace to 1400°F (760°C) and hold it for 15 minutes. Remove one sample of each type of steel and plunge the samples immediately into cold water. Record the hardness values when the samples have cooled.

Repeat this procedure at 1500°F, 1600°F, and 1700°F, recording the hardness values of all samples in chart form. Plot the hardness values on an iron-carbon phase diagram.

Did the hardness of each sample increase as the temperature at quenching was increased? Was the temperature of maximum hardening different for the various types of steel? According to the iron-carbon diagram, which type of steel should reach its maximum first?

Did the samples quenched at 1300°F show a minimal change in hardness? Normally, the hardness values for these samples should increase slightly. However, if the original material had been previously hardened, there may be a decrease in hardness. If the furnace temperature is not accurate, a major increase in hardness could take place for some of these samples.

Did your results agree with your expectations based on the different transformation temperatures on an iron-carbon phase diagram? Why can you expect to see some discrepancies?

transformation line. Thus, both structures are 100% austenite.

The samples at points C, D, E, F, and G contain no austenite because they are below the lower transformation temperature line. Since point C is far to the left and has very little carbon, it is almost 100% ferrite. A small percentage of pearlite is also present.

Point E is located on the 0.8% carbon line, which represents 100% pearlite. Point F is halfway between the pure ferrite and pure pearlite lines. Therefore, it is made up of approximately 50% ferrite and 50% pearlite. Points D and G are both located between the 100% pearlite line (at 0.8% carbon) and the 100% cementite line (at 6.67% carbon). Point G is closer to the 100% pearlite line than point D, but both points are closer to this line than they are to the 100% cementite line. Point D contains 81% pearlite and 19% cementite. Point G has an 89:11 ratio of pearlite to cementite.

Point H is in the ferrite and austenite transformation region. Theoretically, all the pearlite has changed to austenite. Since this sample is approximately midway between the pure ferrite line and the pure austenite line, its ratio of ferrite to austenite is 50:50.

Point K is also in the ferrite and austenite transformation region. However, point K is closer to the pure ferrite line than point H.

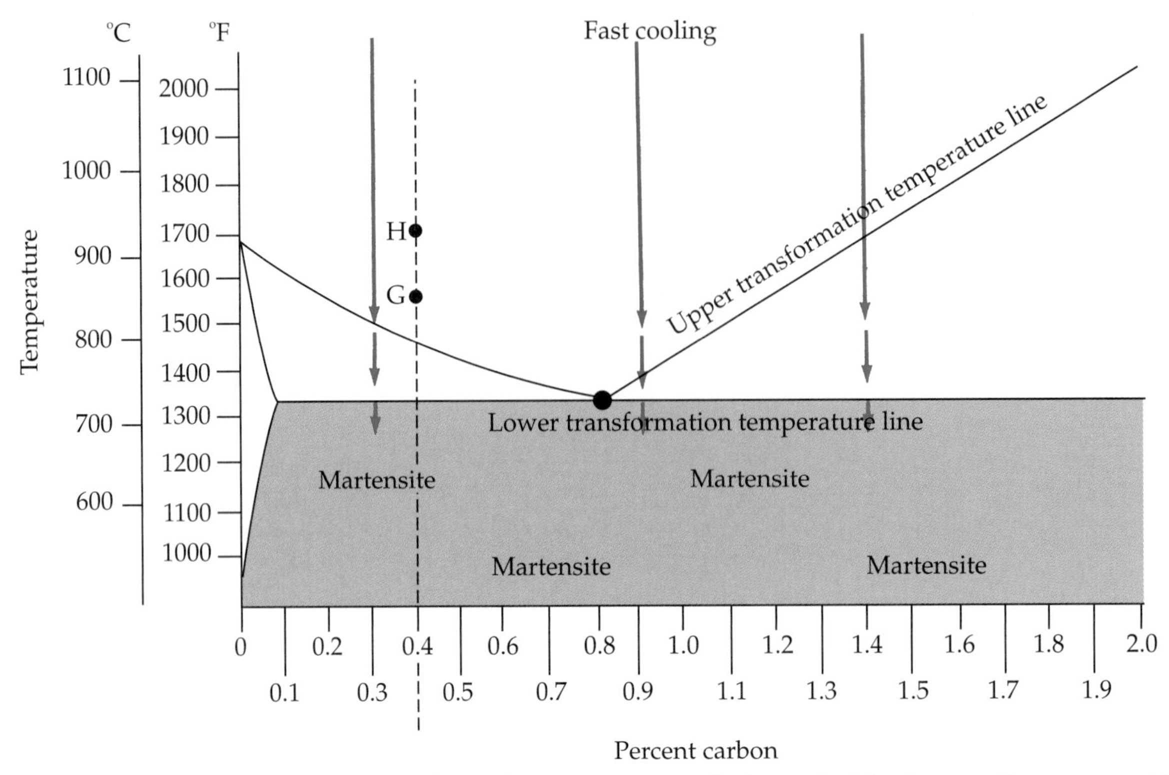

Figure 9-12. *If steel is quenched rapidly, it changes to martensite instead of ferrite, pearlite, or cementite.*

The sample at point K contains about 67% ferrite and 33% austenite.

Point J is in the cementite and austenite transformation region. There is no pearlite in this structure, since it lies above the lower transformation temperature line. Point J is closer to pure austenite than it is to pure cementite (6.67% carbon). Its structure can be calculated as 81% austenite and 19% cementite. This calculation would be essentially identical to the calculation that produced an 81:19 ratio for point D.

The samples of steel at points C, D, E, F, G, H, J, and K could have been transformed into martensite (instead of ferrite, pearlite, and cementite) if they had been rapidly quenched. Heating the samples above the upper transformation temperature line, followed by rapid quenching, would have given them a martensitic structure.

After rapid quenching, the samples at points C, D, E, F, and G would be 100% martensite. The sample at point K would contain 33% austenite and 67% martensite. The sample at point H would contain 50% austenite and 50% martensite. The sample at point J would contain 19% martensite and 81% austenite.

Transformation Regions

There are several important transformation regions on an iron-carbon phase diagram in addition to the regions previously discussed. These regions are used to classify basic structures of steel that are produced when material is heated and cooled. They are discussed below.

The *eutectoid point* refers to the point where the upper transformation temperature line, the lower transformation temperature line, and

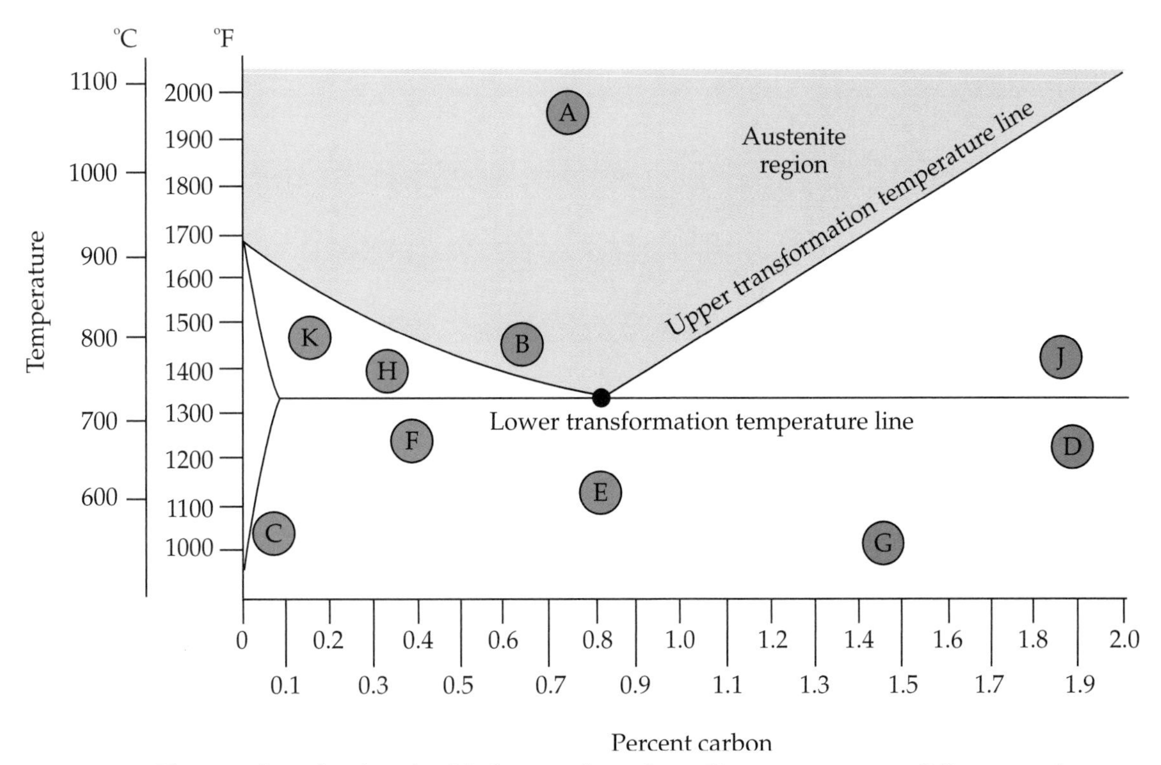

Figure 9-13. *The ten plotted points in this iron-carbon phase diagram represent different steel structures.*

the pure pearlite line (at 0.8% carbon) intersect. See Figure 9-14. This point divides the major regions of transformation on an iron-carbon phase diagram.

The *transfer temperature range* comprises the two regions between the upper and lower transformation temperature lines. Refer to Figure 9-14. Both of these triangular areas are transformation regions. Here, either austenite is changing to one of the low-temperature structures, or a ferrite, pearlite, cementite, or martensitic structure is changing to austenite.

The *hypoeutectoid region* is the region to the left of the pure pearlite (0.8% carbon) line, Figure 9-15. Any steel that falls in this region is known as hypoeutectoid steel. The *hypereutectoid region* is to the right of the 0.8% carbon line. Any steel that falls in this region is known as hypereutectoid steel.

As discussed earlier, different rates of cooling produce different types of transformation for steel. Quenching of austenite produces martensite, while slow cooling of austenite produces ferrite, pearlite, or cementite. The common effects of different quenching techniques on the transformation of steel are discussed next.

Temperature Change and Mechanical Properties

The hardness, strength, brittleness, ductility, and grain size of steel are greatly affected by different heating and cooling methods. If a sample is heated to the austenitic range and then quenched very rapidly, martensite forms. This makes the steel hard and strong, with a small grain size. However, the steel is also brittle.

If a sample is heated to the austenitic range and then cooled very slowly, it will

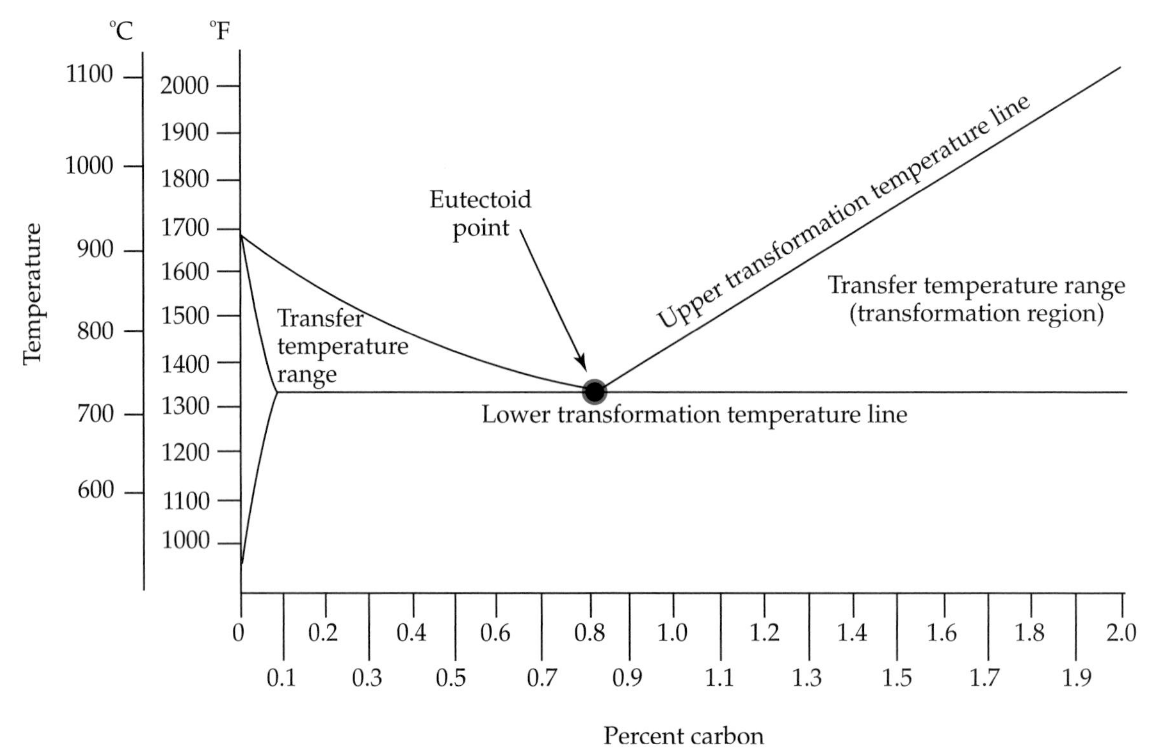

Figure 9-14. *This iron-carbon phase diagram shows the eutectoid point, the upper and lower transformation temperature lines, and the transfer temperature range.*

change to ferrite, pearlite, or cementite. These structures are comparatively softer, less strong, and more ductile than martensite. They have a large grain size.

Hardness, strength, ductility, and small grain sizes are generally considered to be assets in steel. Thus, it would be ideal to heat and quench a material in a manner that produces hardness and strength without a loss of ductility.

Metallurgists generally need to choose from the following three situations:

- If the material must be hard and strong, the steel is quenched. However, this material will also be brittle.
- If great ductility is required, the material is cooled slowly. Then, the steel will be very machinable and formable. However, it will not have good strength or hardness qualities.
- If both strength and ductility are required, special alloys can be added to steel. The addition of alloys will increase the cost of the material. Often, however, the higher cost is justified if strength and ductility are very important.

A summary of the relationship between quenching, slow cooling, and the characteristics resulting from each process is given in Figure 9-16.

Improving Hardness

Hardness is very important for many applications involving steel. Hardness can be improved by applying the following two general factors:

- Quenching speed. The faster a steel is quenched, the harder it will be after quenching.

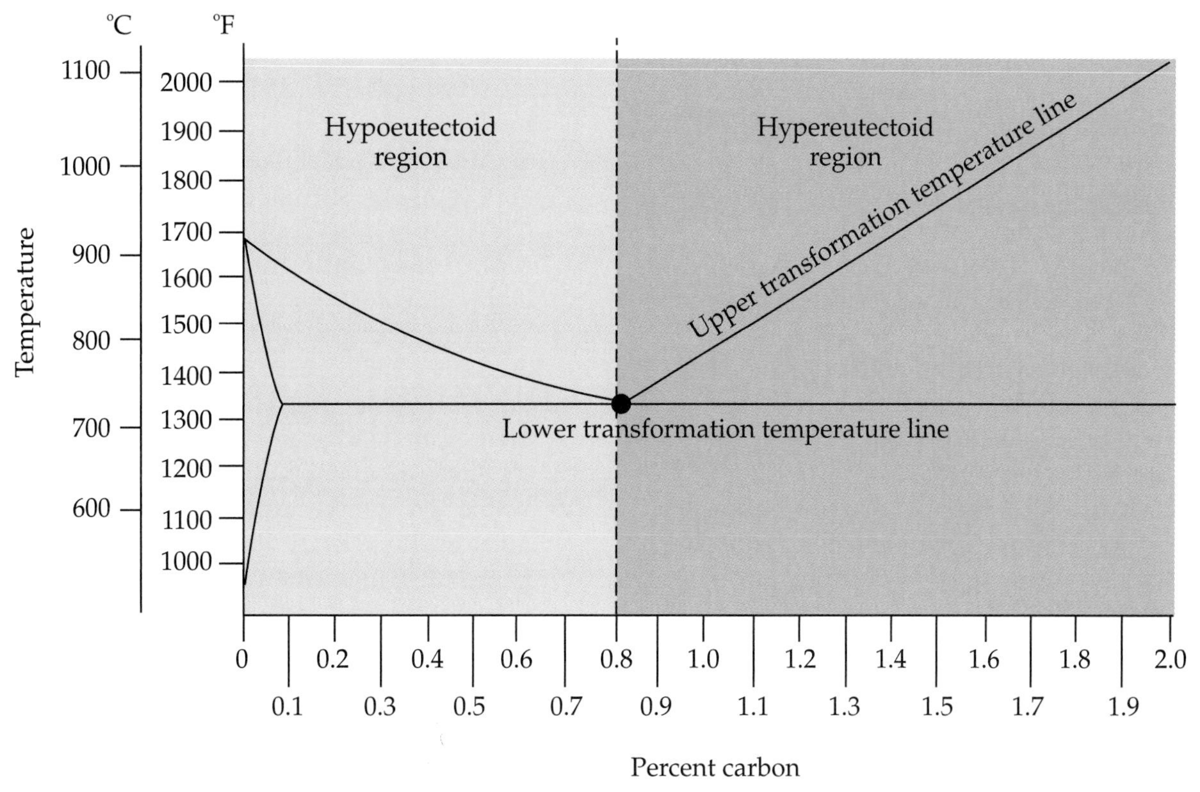

Figure 9-15. *Steel that falls to the left of the 0.8% carbon line is in the hypoeutectoid region. Steel that falls to the right of the line is in the hypereutectoid region.*

Comparison of Cooling Methods	
Quenching	**Slow Cooling**
Harder	Softer
Stronger	Less strong
Brittle	Ductile
Martensite	Ferrite-Pearlite-Cementite
Small grain size	Large grain size

Figure 9-16. *This table shows how cooling rates affect the hardness, strength, ductility, structure, and grain size of steel.*

- Percentage of carbon. The more carbon present in the steel, the harder it will be after quenching.

Thus, if an extremely hard steel is required, a high carbon content should be used and a rapid quenching method should be applied to cool the steel.

Test Your Knowledge

Write your answers on a separate sheet of paper. Do not write in this book.

1. Give the name of the structure of a solid solution of steel at 1200°F that contains a very small percentage of carbon (perhaps 0.02%). Assume that this steel has not been previously heat treated.
2. Name the structure of iron at room temperature that contains 6.67% carbon. Assume that this metal has never been heat treated.

3. Name the structure of a solid solution of steel at room temperature that contains 0.8% carbon. Assume that this steel has never been heat treated.
4. Approximately how much carbon is in a sample of pure pearlite?
5. What is *austenite?*
6. What structural change occurs when steel is transformed into austenite? What change occurs when the steel is slowly cooled to room temperature?
7. Of the following types of iron-carbon, which is the hardest: austenite, cementite, ferrite, or martensite?
8. Iron at 1800°F is rapidly quenched in water. What structural form of iron-carbon is the most predominant in the result if the alloy contains 0.9% carbon?
9. Assume a sample of steel is cooled from an elevated temperature to 1200°F and contains a very small percentage of carbon (perhaps 0.2%). The steel has not been previously heat treated. Name the structural form of the solid solution of this steel.
10. Assume a sample of steel has been rapidly quenched from an elevated temperature to room temperature. It contains 0.8% carbon. Name the structural form of the solid solution of this steel.
11. What point on an iron-carbon phase diagram marks the intersection of the upper transformation temperature line, the lower transformation temperature line, and the pure pearlite line?
12. Any type of steel that contains less than 0.8% carbon is called _____ steel. Any type of steel that contains more than 0.8% carbon is called _____ steel.
13. What structural form of steel is considered to be the hardest and most brittle?
14. What characteristics of steel are associated with small grain size? What characteristic is associated with large grain size?
15. Name two general rules that are used to improve the hardness of steel.
16. Sketch an iron-carbon phase diagram. Identify the low-carbon steel region, the medium-carbon region, and the high-carbon region.
17. Discuss three automotive applications that would require quenched steel for certain parts.
18. Discuss two applications for which quenched steel parts would be seldom necessary.
19. Name two parts for a medical application that would benefit from slow cooling. Name two parts for a medical application that would require rapid quenching.

20. There are eight points indicated by the letters on each of the iron-carbon phase diagrams shown in Figure 9-17. These 16 points correspond to different structural forms of steel. Using the letters in the left column for this question, match each point to one of the 22 structures listed in the right column. Some of these structures may be used more than once; some may not be used at all. Assume that all points on the diagram represent steel that has been slowly cooled, not quenched.

A. _____
B. _____
C. _____
D. _____
E. _____
F. _____
G. _____
H. _____
J. _____
K. _____
L. _____
M. _____
N. _____
P. _____
Q. _____
R. _____

1. All austenite.
2. All pearlite.
3. All martensite.
4. Almost all ferrite.
5. The maximum cementite possible in steel.
6. Half cementite and half austenite.
7. Half ferrite and half austenite.
8. Half pearlite and half cementite.
9. Half pearlite and half austenite.
10. Half ferrite and half cementite.
11. Half ferrite and half pearlite.
12. Over 60% austenite with some cementite.
13. Over 60% austenite with some ferrite.
14. Over 60% cementite with some austenite.
15. Over 60% ferrite with some austenite.
16. Over 60% pearlite with some austenite.
17. Over 60% pearlite with some cementite.
18. Over 60% pearlite with some ferrite.
19. Over 60% ferrite with some pearlite and some austenite.
20. A combination of austenite, pearlite, and cementite.

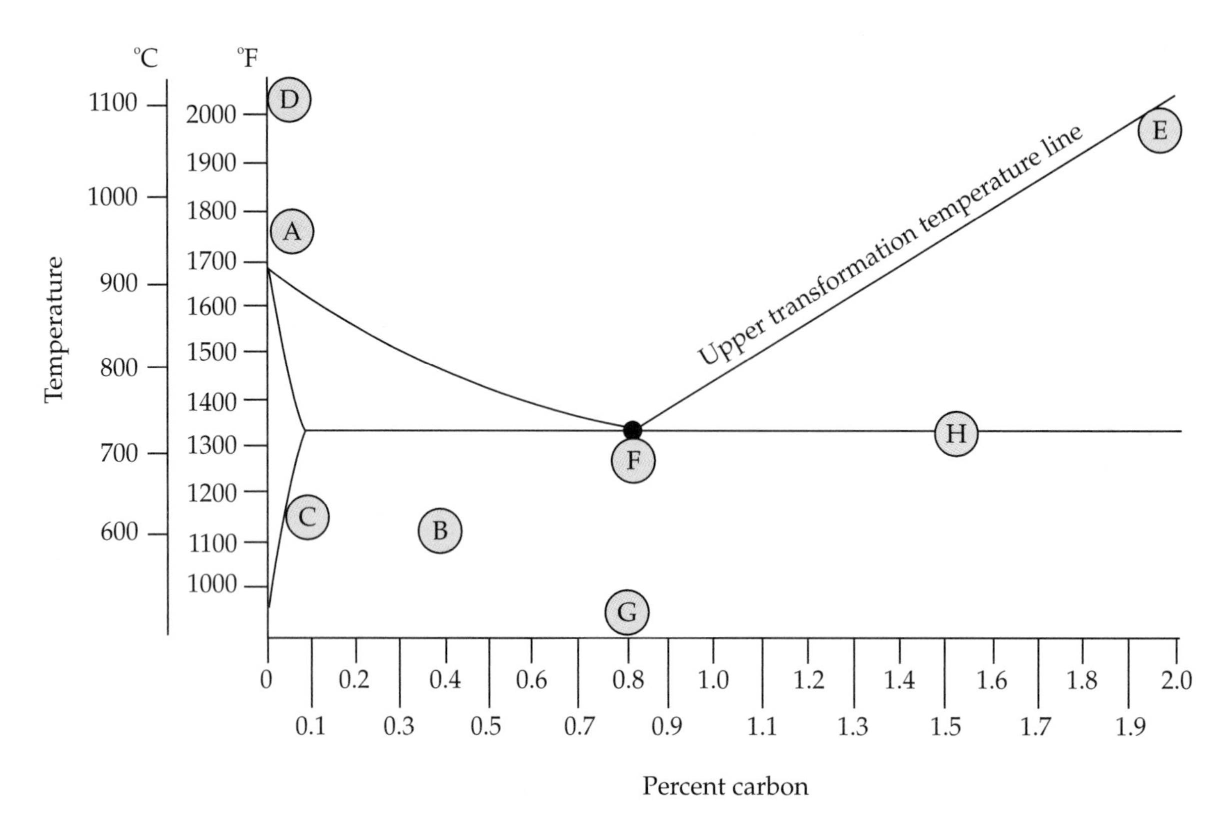

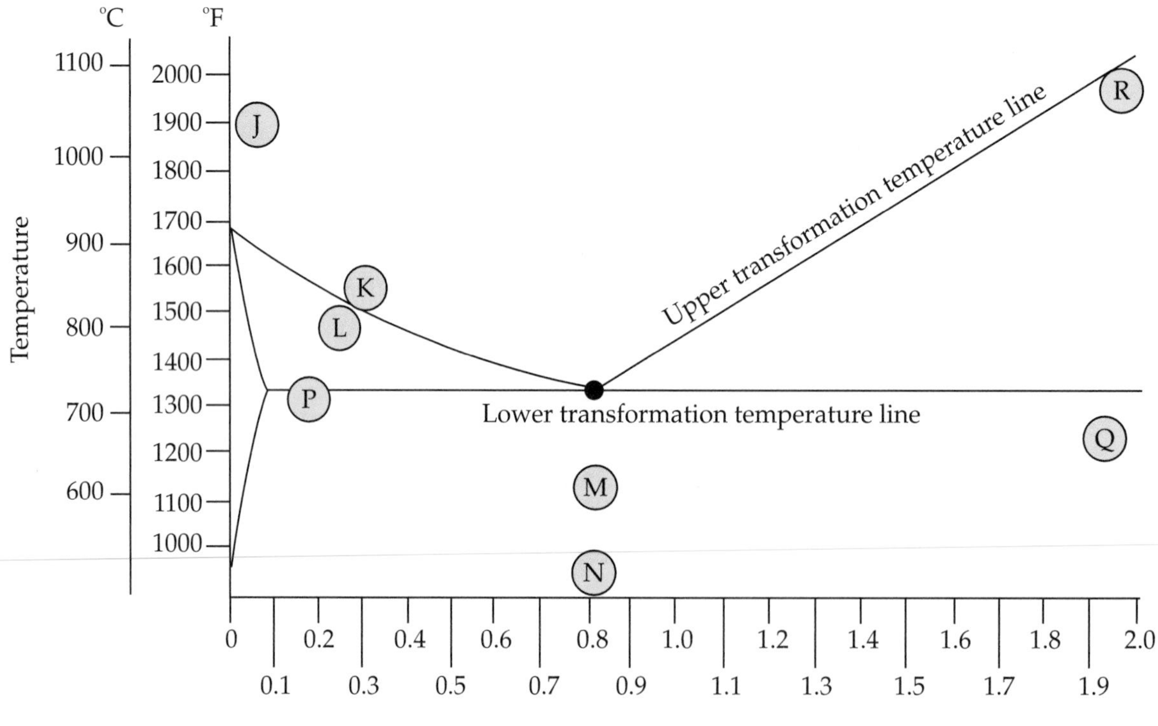

Figure 9-17. *Using these iron-carbon phase diagrams for Question 20, match each of the points to the corresponding structure of steel.*

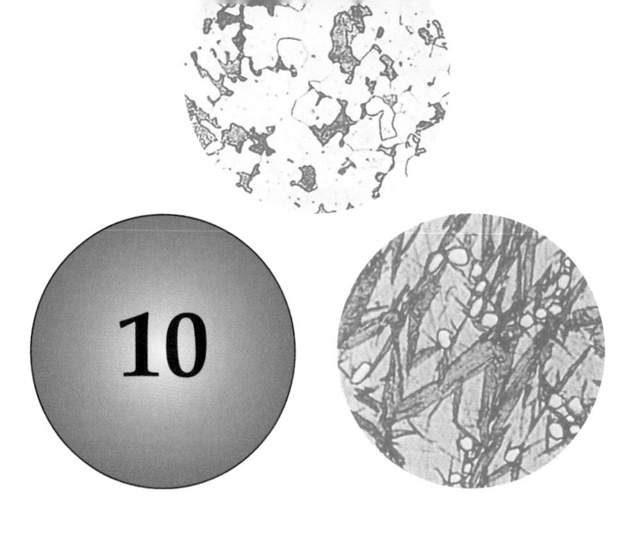

Microstructural Analysis

After studying this chapter, you will be able to:

- Compare different structures of steel under a microscope.
- Describe the microscopic appearance of ferrite, pearlite, cementite, austenite, and martensite.
- Recognize ferrite, pearlite, cementite, austenite, and martensite by looking at their microstructures.
- Prepare a sample of metal for microscopic observation.

When steel is magnified, it has a much different appearance. This can be seen by looking at steel through a microscope, Figure 10-1. *Photomicrography* is the process of creating a photograph from a microscope image. The resulting photograph is called a *photomicrograph.*

Many types of microscopes are equipped with photomicrographic attachments, Figure 10-2. A photomicrograph of steel magnified 500 times (500X) is shown in Figure 10-3.

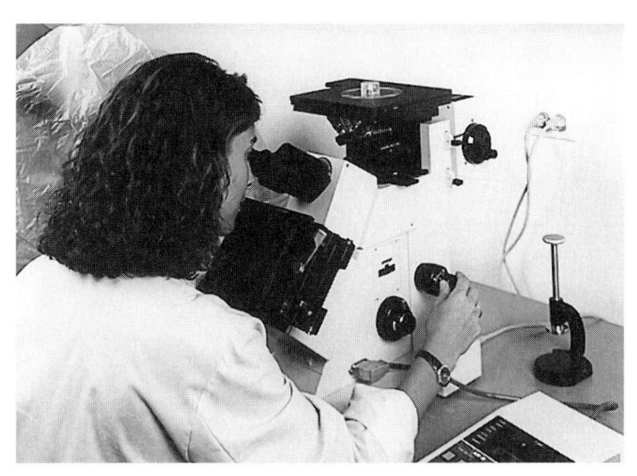

Figure 10-1. *Microscopes are used to examine metal samples. (Buehler, Ltd.)*

Microscopic Appearances

Ferrite, pearlite, cementite, martensite, and austenite look very different from each other under a microscope. A ferrite structure resembles patches. Pearlite looks like a series of ridges. Cementite resembles white country roads. Martensite has the appearance of needles. Austenite looks like broken concrete slabs. The appearance of each type of structure is discussed in the following sections.

Ferrite

Ferrite appears white in color in the microscope, Figure 10-4. The small, dark portions are the pearlite structures containing some carbon. If only ferrite appeared in the structure, the microstructure would be solid white.

Pearlite

Pearlite is made up of ferrite and cementite. This was covered in Chapter 9. At 0.8% carbon, the content of ferrite and cementite is

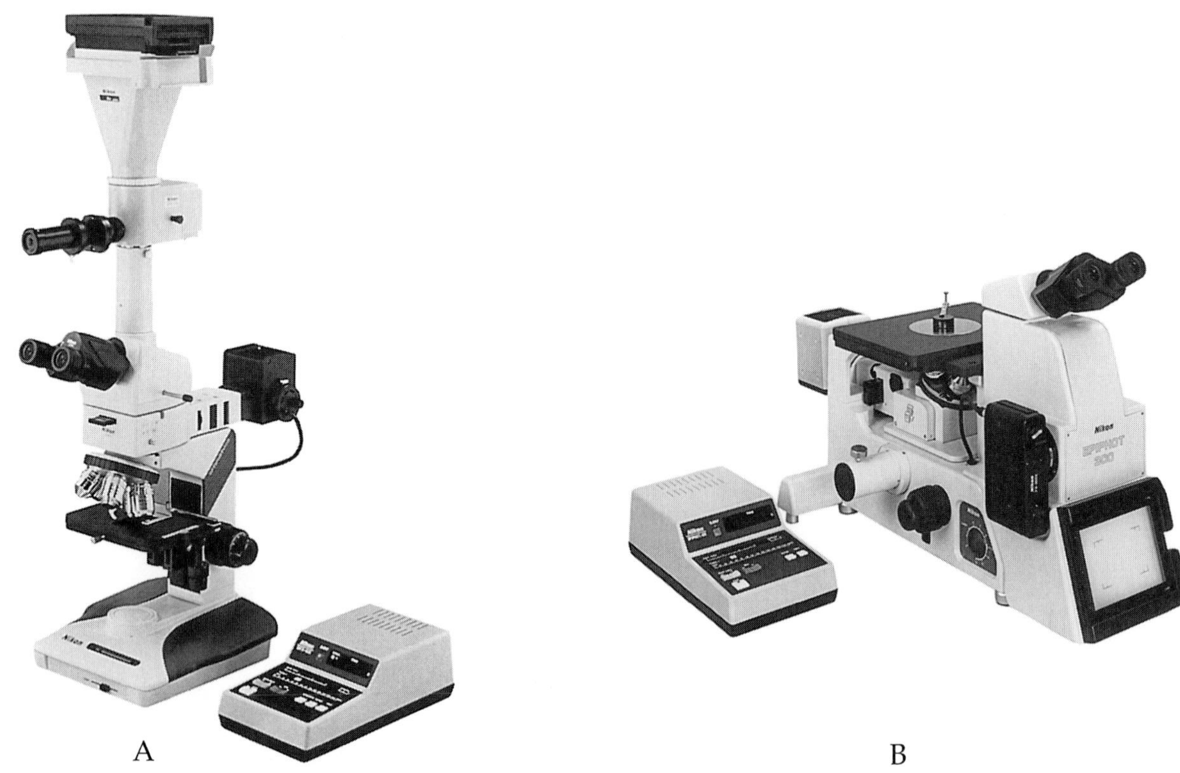

A B

Figure 10-2. *Any of the steel structures shown in this chapter could be seen and photographed by using these metallurgical microscopes. A—An upright microscope with a photographic attachment. B—An inverted microscope. (Nikon, Inc.)*

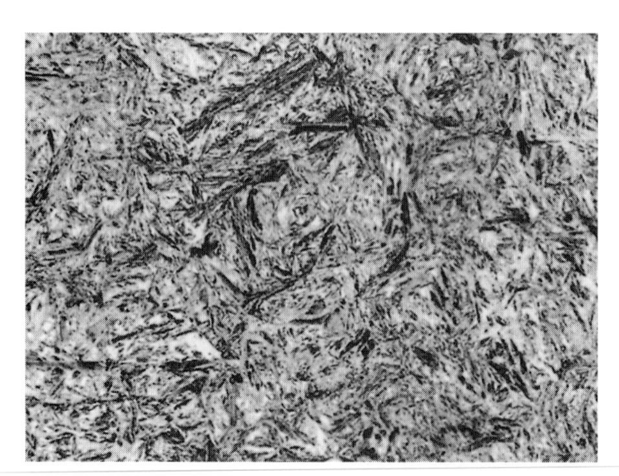

Figure 10-3. *A photomicrograph of a martensite and bainite structure in 1095 steel. Magnification is 500X. (Buehler Ltd.)*

balanced and the entire metallurgical structure appears as pearlitic ridges. At higher percentages of carbon, a combination of pearlite and pure cementite exists. The dark lines are cementite. The light-colored ridges are ferrite. See Figure 10-5.

Ferrite-Pearlite Structure

A composition of ferrite and pearlite in steel produces a very striking image. The ferrite appears white. The pearlite appears dark or laminated. The amount of pearlite present is proportional to the carbon content of the steel. That is, as the carbon content increases, the amount of pearlite in the structure also increases.

It can be very interesting to compare similar structures of different types of steel. In comparing 1018 steel and 1045 steel, 1045 steel has a higher percentage of pearlite because it contains more carbon. There is approximately 0.45% carbon in 1045 steel

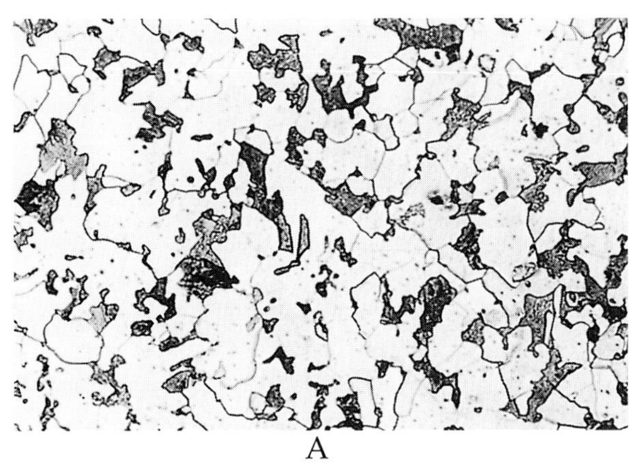

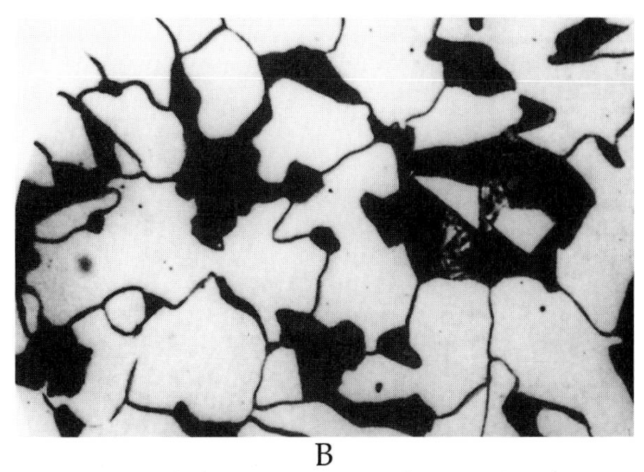

Figure 10-4. *Ferrite is present in steel samples with carbon content below approximately 0.86%. The lower the carbon, the greater the amount of ferrite. Ferrite appears as white patches in photomicrographs. The black areas are pearlite. A—A 1020 steel ferrite structure with some pearlite. Magnification is 100X. (Buehler Ltd.) B—A photomicrograph of steel with 0.15% carbon content viewed at 700X magnification. (Struers, Inc.)*

and approximately 0.18% carbon in 1018 steel. Due to its lower carbon content, 1018 steel has a lighter microstructure. Remember, as the carbon content increases, the amount of pearlite increases. This produces a darker microscopic view.

The photomicrographs in Figure 10-6 illustrate the different appearances of ferrite-pearlite structures. As the carbon content increases, the white ferrite portions diminish and change to darker, ridge-like pearlitic structures.

In Figure 10-6A, at just 0.06% carbon, the steel contains almost all ferrite. As the carbon content increases to 0.20%, as shown in Figure 10-6B, small colonies of pearlite start to appear as ridges even though the ferrite structure is still predominant.

Since 0.36% carbon is almost exactly halfway between pure ferrite (0% carbon) and pure pearlite (0.80% carbon), approximately half of the structure in Figure 10-6C is white ferrite and the other half consists of pearlitic ridges. When the carbon content reaches 0.53%, as shown in Figure 10-6D, the pearlite is predominant and the white ferrite is reduced to thin sections. At 0.86% carbon, the ferrite has disappeared, Figure 10-6E.

Figure 10-5. *A 1095 steel pearlitic structure with some cementite. Magnification is 500X. (Buehler Ltd.)*

Cementite-Pearlite Structure

A cementite-pearlite structure is shown in Figure 10-7. The portion of cementite resembles small, white country roads. The pearlite still appears in its characteristic ridge pattern. As the percentage of carbon increases, the amount of cementite increases.

Martensite

Martensite can have many different microscopic appearances. However, it always has a

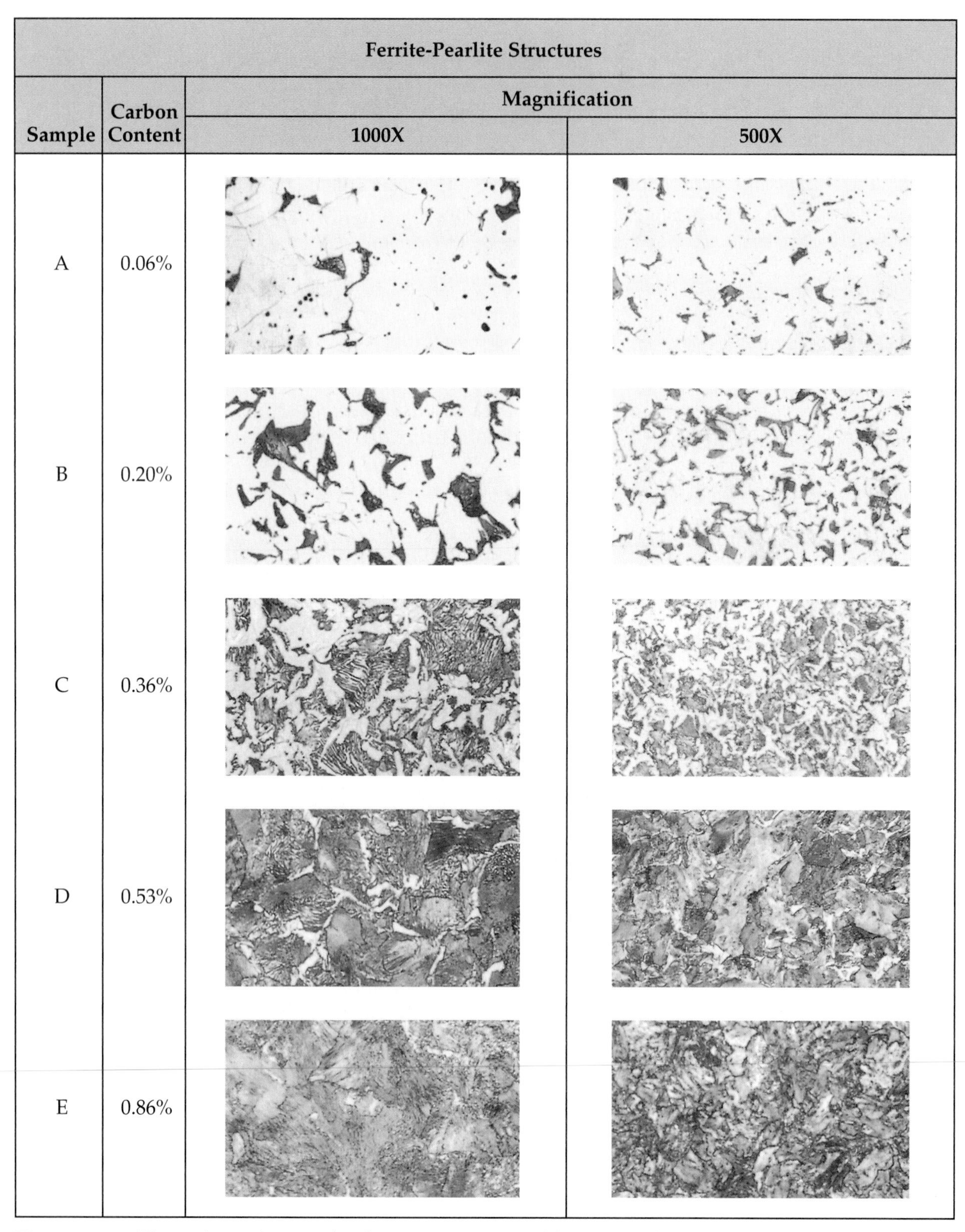

Ferrite-Pearlite Structures			
		Magnification	
Sample	Carbon Content	1000X	500X
A	0.06%		
B	0.20%		
C	0.36%		
D	0.53%		
E	0.86%		

Figure 10-6. *These photomicrographs show various types of ferrite-pearlite structures. Notice that the amount of ferrite (white) decreases as the carbon content increases. (LECO Corporation)*

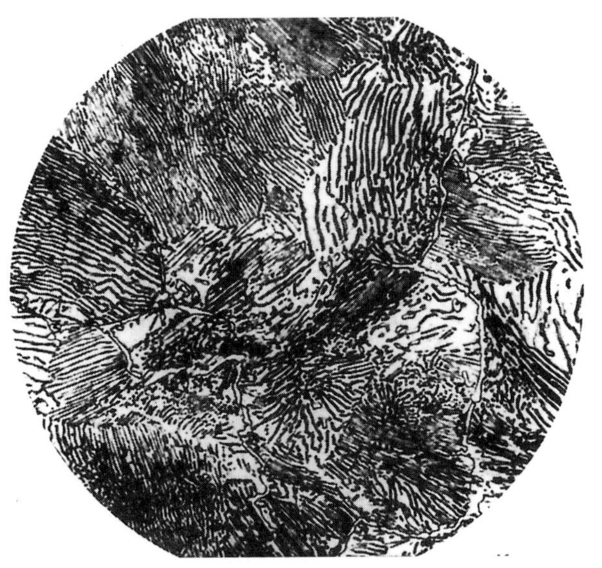

Figure 10-7. *A cementite-pearlite structure. Magnification is 700X. (Struers, Inc.)*

fine, needlelike appearance. The sight of martensite under a microscope gives the impression of pointed lines, Figure 10-8. Those that look like small needles in Figure 10-8A and Figure 10-8B would look similar to the larger needles in Figure 10-8C if the magnification of the microscope's lens were increased.

Austenite

It is often important to observe the grain size and grain boundaries of austenite, even if it is going to be transformed to martensite. You may wonder how austenite can be photographed under a microscope, since it occurs only at elevated temperatures. It does not seem possible to operate a microscope at 1700°F (930°C).

Austenite can be retained at room temperature by adding special alloying elements to the material or by using special etching techniques. A heat treatment technique that uses carbide to "decorate" the austenitic boundaries can also be used. This makes the grain boundary lines visible.

A photomicrograph of austenitic stainless steel is shown in Figure 10-9. The structure resembles the broken slabs of an old concrete

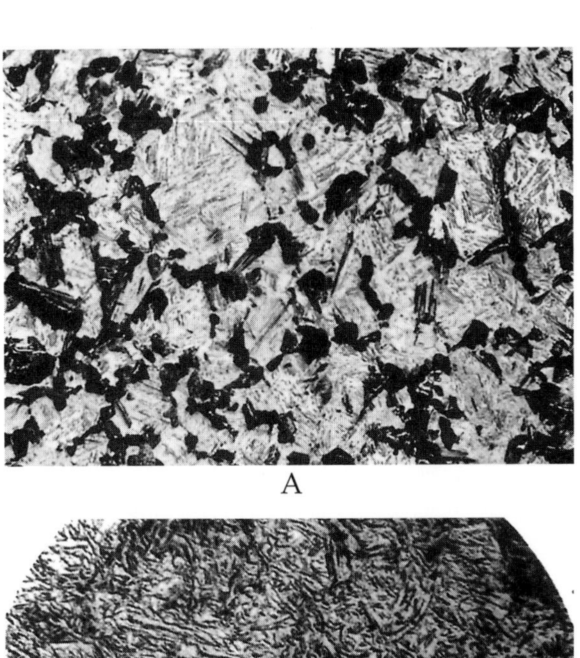

A

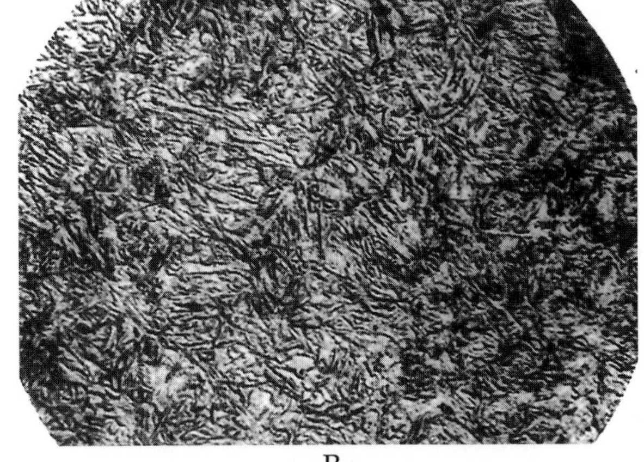

B

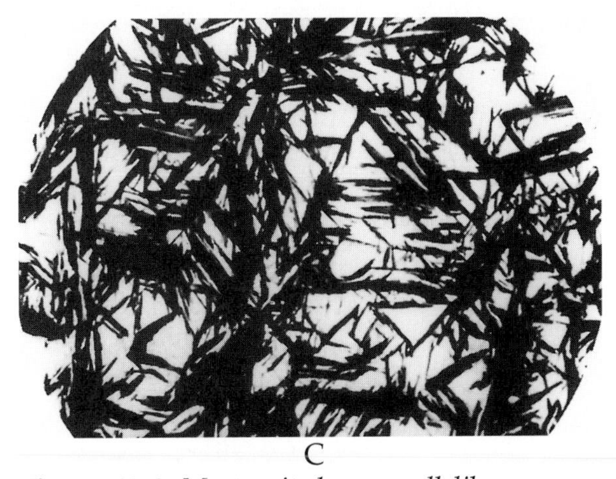

C

Figure 10-8. *Martensite has a needlelike appearance. A—A 1045 steel martensitic structure with some bainite. Magnification is 500X. (Buehler Ltd.) B—A martensitic structure viewed at 700X magnification. C—The martensitic "needles" can be clearly seen at this greater magnification. (Struers, Inc.)*

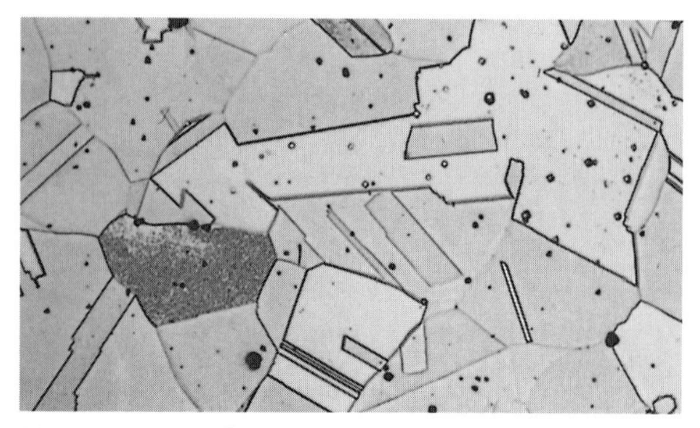

Figure 10-9. *This 303 stainless steel has an austenitic structure. Magnification is 100X. (Buehler Ltd.)*

highway. These "slabs" have a slight resemblance to ferrite. However, a ferrite formation has a rounder and more continuous curvature than austenite. Austenitic lines appear straighter and more abrupt. Although ferrite and austenite look similar when they are viewed through a microscope, the physical behavior of the two structures is still far different.

Structural Combinations

Many times in metallurgy, a heat-treated material may have a combination of structures. An example of this is shown in Figure 10-10. The following observations can be made:

- The white areas are ferrite.
- The lighter gray areas are martensite.
- The laminated portions are pearlite.
- The very darkest sections are *bainite*. This structure will be discussed in Chapter 13.

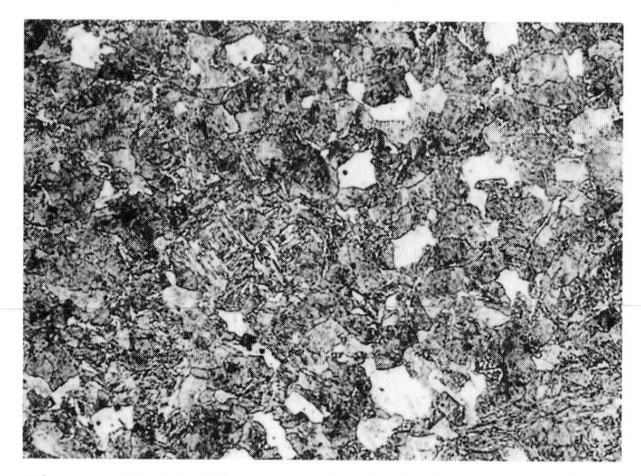

Figure 10-10. *This sample of 1045 steel reveals a martensite-bainite structure with some ferrite and pearlite. Magnification is 500X. (Buehler Ltd.)*

Sample Preparation Procedure

Before a sample of steel can be viewed and photographed, the metal must be carefully prepared. The surface viewed under the microscope must be completely flat and smooth. Any irregularity will appear as a dark surface and confuse attempts to analyze the structure. In order to obtain a smooth, flat surface, several preparatory steps are required. These steps include grinding, molding, polishing, and etching.

Grinding

To prepare a sample for viewing, the surface is ground to remove the rough scale and any gross imperfections from the surface of the material. See Figure 10-11. After rough grinding, fine grinding is performed on

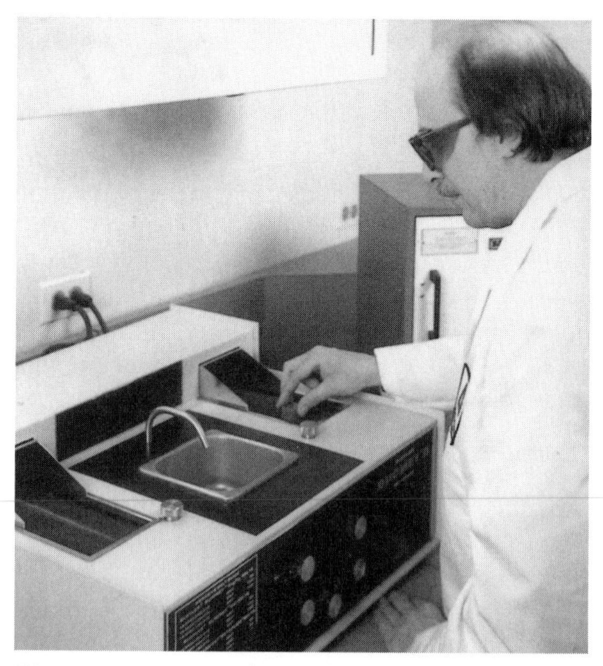

Figure 10-11. *Rough grinding is the first step in preparing a metal sample for microscopic examination. (Buehler Ltd.)*

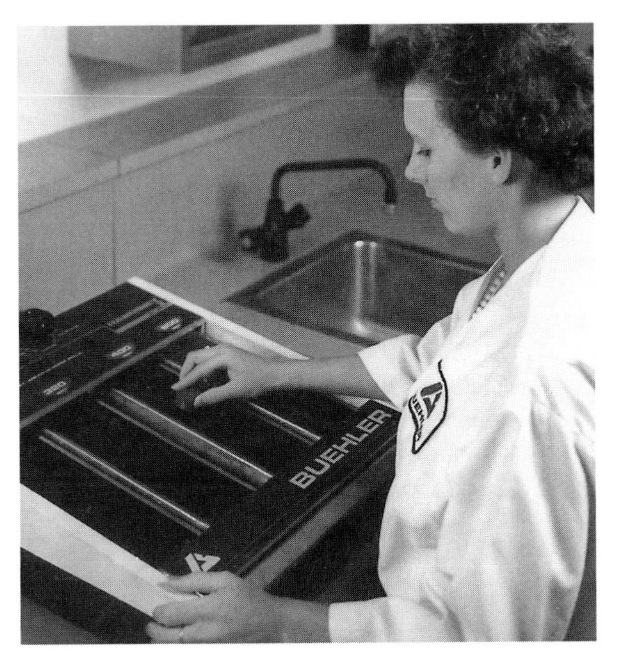

Figure 10-12. *Fine grinding further improves the surface of the sample. (Buehler Ltd.)*

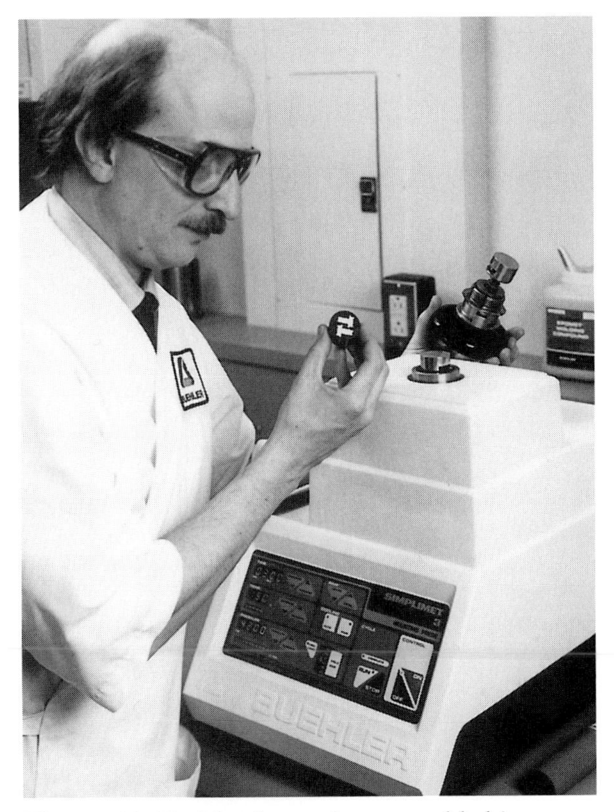

Figure 10-13. *Metal samples are molded in plastic during preparation of the sample using this press. (Buehler Ltd.)*

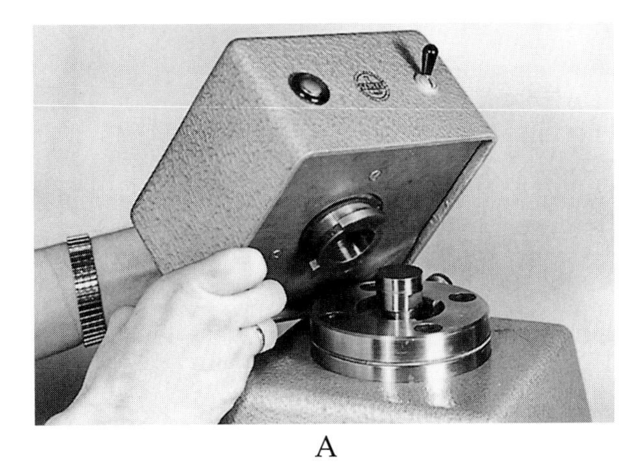

A

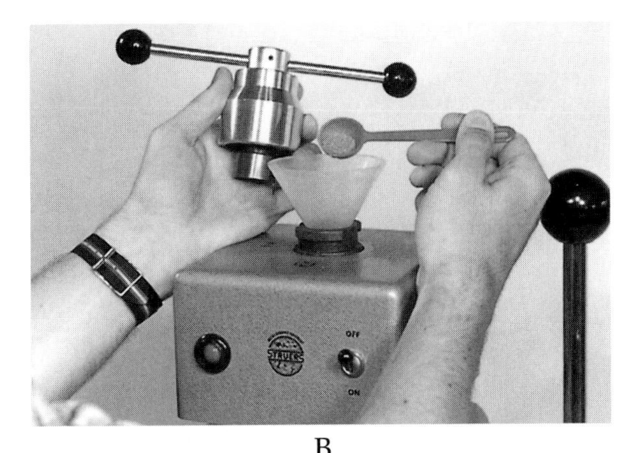

B

Figure 10-14. *This hydraulic molding press is used to mount metallurgical samples. A—Removal of the upper assembly reveals the molding area of the press. B—Adding material is one of the first steps for molding the sample in plastic. (Struers, Inc.)*

the sample. This improves the surface until it begins to shine and slightly reflect light, Figure 10-12.

Molding

The metal specimen is usually molded in plastic after rough grinding. See Figures 10-13 and 10-14. This makes the sample easier to hold throughout the polishing procedure. Generally, it is best not to mold before grinding the sample. The sample can overheat during grinding if your fingers are not touching it.

Polishing

Metallurgical samples are commonly polished in two stages. First, rough polishing removes the imperfections that grinding has left. Then, fine polishing produces a mirror-like finish on the surface of the steel with all scratches removed. See Figure 10-15.

Etching

Etching involves the application of acid to a smooth, polished metal surface to make details of the microstructure more visible. Etching is the final step in preparing samples for microscopic analysis. When acid is applied to a smooth surface, some metallic structures will be eaten away by the acid more rapidly than others. The areas that are dissolved most rapidly by etching appear as dark shadows under a microscope. The surfaces that react slowly to the acid have a lighter appearance.

It takes practice to etch a metallurgical sample properly. The procedure involves the following steps:

1. Clean the surface with alcohol, Figure 10-16. Let the sample air dry. Alcohol evaporates rapidly. Wiping it off may leave smear marks on the surface.
2. Apply acid to the sample. Acid can be applied by dripping acid onto the sample, by submerging the sample into a dish

A

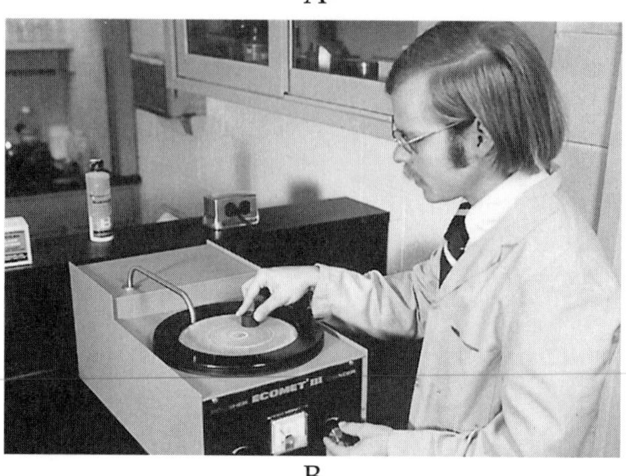

B

Figure 10-15. *Polishing metallurgical samples is a two-step process. A—Rough polishing removes imperfections from grinding. B—Fine polishing removes all scratches. (Buehler Ltd.)*

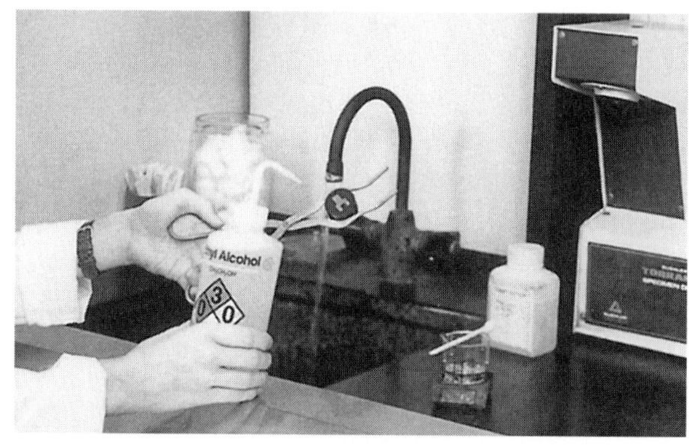

Figure 10-16. *An alcohol rinse prepares the surface for etching. (Buehler Ltd.)*

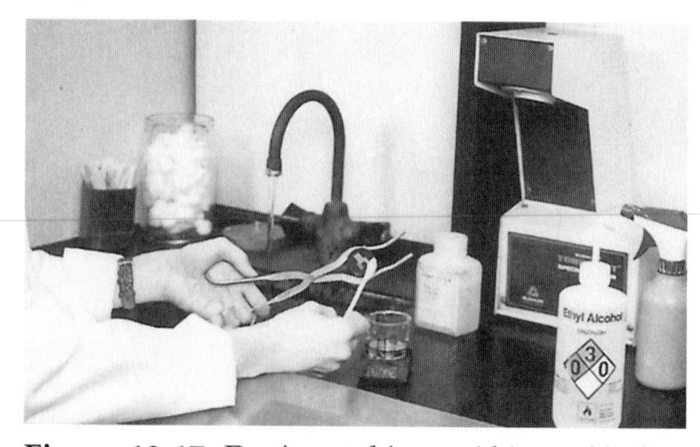

Figure 10-17. *During etching, acid is swabbed onto the sample. (Buehler Ltd.)*

of acid, or by swabbing an acid solution onto the sample. See Figure 10-17. The amount of time the acid should be kept in contact with the sample varies considerably for different materials. The time required may vary from a few seconds to as long as a few minutes.

3. After the acid has had enough time to erode the surface, wash it away with water. It is best to run water over the sample to remove all traces of acid.
4. As soon as the sample is rinsed with water, wash it again with alcohol. This will help prevent water marks. A nitric acid solution in methyl or ethyl alcohol is often used.

After etching is completed, the sample is ready to be viewed under a microscope. A properly etched surface will allow you to identify the microscopic structure of the steel. A comparison of surface finishes through each step of etching is shown in Figure 10-18.

Comparing Light and Dark Structures

There is no absolute rule to use in identifying the microscopic structures that appear light (white) and dark in steel. However, there are a few general rules that can usually be applied in microstructural analysis.

The structures that have the greatest percentage of iron or ferrite will generally appear white. The structures that contain a greater percentage of carbon will generally appear dark. There are exceptions to this rule, however.

Cementite, which has a higher percentage of carbon than pearlite, appears white despite its carbon content. The pearlite in this structure appears darker.

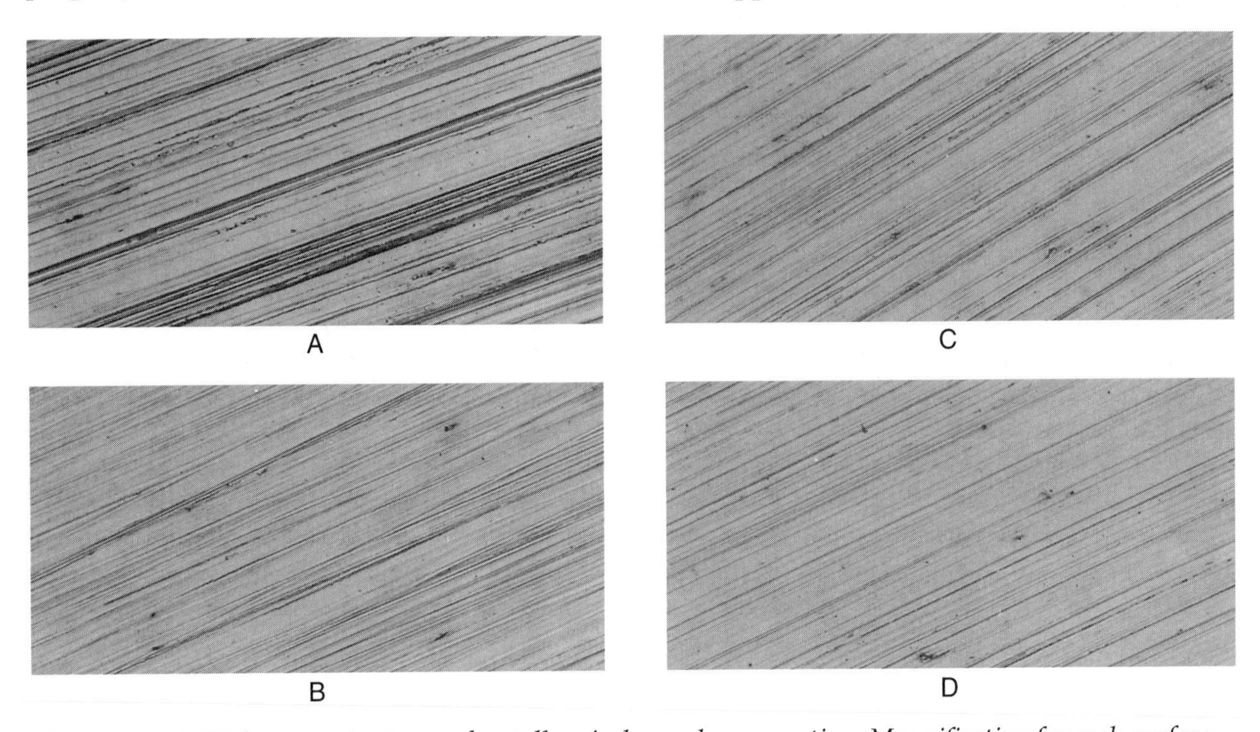

Figure 10-18. *Eight separate stages of metallurgical sample preparation. Magnification for each surface is 250X. A—Surface after grinding with 180 grit paper. B—Surface after grinding with 240 grit paper. C—Surface after grinding with 320 grit paper. D—Surface after grinding with 400 grit paper. E—Surface after grinding with 600 grit paper. F—Rough polishing removes the imperfections produced by grinding. This surface has been polished with a 6 micron diamond abrasive on nylon cloth. G—The surface is ready for etching after fine polishing with a 0.05 micron aluminum oxide on microcloth. H—Etching with acid reveals details in the microstructure of the sample. (Buehler Ltd.)*

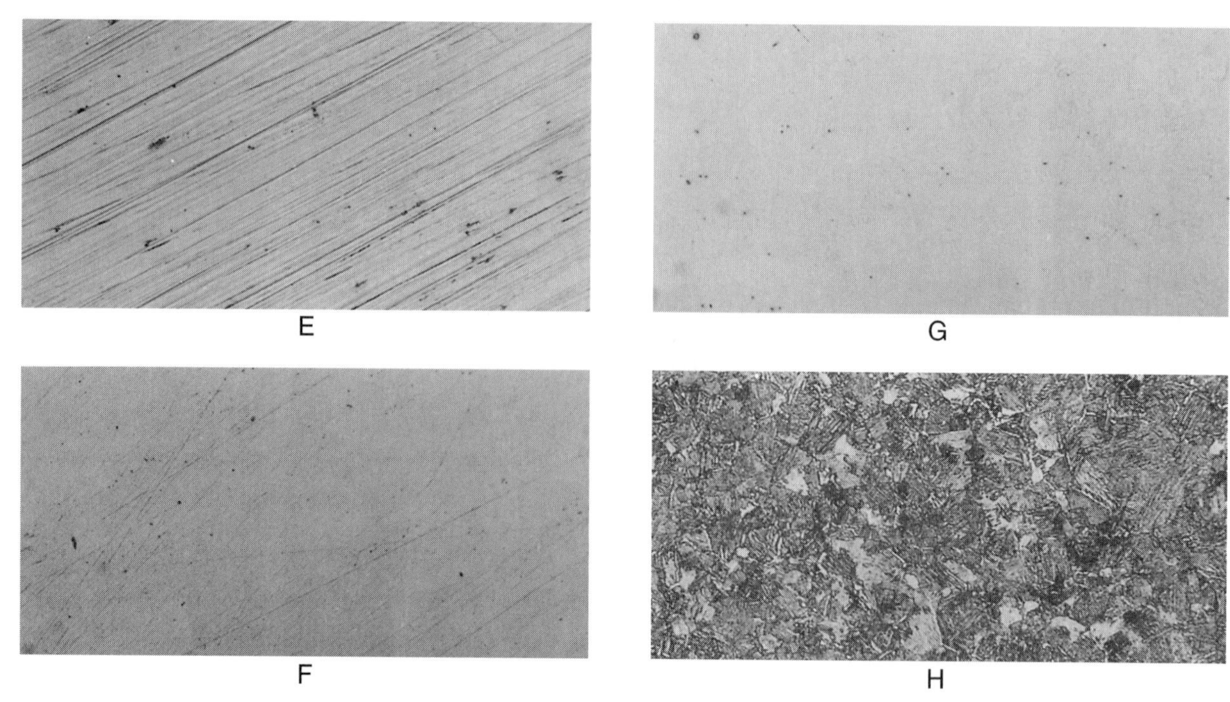

Figure 10-18. *(continued)*

Test Your Knowledge

Write your answers on a separate sheet of paper. Do not write in this book.

1. What does 400X mean?
2. Which would show more detail and a closer view of a piece of metal, 50X or 200X?
3. Define *photomicrography.*
4. Which steel structure appears as patches of white in a photomicrograph?
5. Describe the microscopic appearance of a ferrite-pearlite structure.
6. Which percentage of carbon would generate a lighter photomicrograph, 0.18% carbon or 0.36% carbon?
7. Describe the appearance of a ferrite-pearlite structure that contains 0.4% carbon.
8. Which steel structure appears to be made up of white country roads and ridges?
9. Which steel structure has a needlelike appearance?
10. How can austenite be retained at room temperature for microstructural analysis?
11. When a metal sample is prepared for microstructural analysis, four steps are required. List the steps in correct order.
12. List three methods used to apply acid to a metal sample during etching.
13. Make a freehand sketch of a microscopic structure that would result from slow cooling of 1045 steel. Sketch the structure as it would appear at approximately 500X magnification.
14. Sketch a microscopic structure of 1045 steel that has been slowly cooled after heat treatment. Sketch the structure at approximately 1000X magnification.
15. Sketch a microscopic structure of 1.8% carbon steel that has been slow cooled. Sketch the structure at approximately 500X magnification.
16. Sketch a microscopic structure of 1060 steel that has been quenched in cold water after being heated to 1650°F. Sketch the structure at approximately 500X magnification.

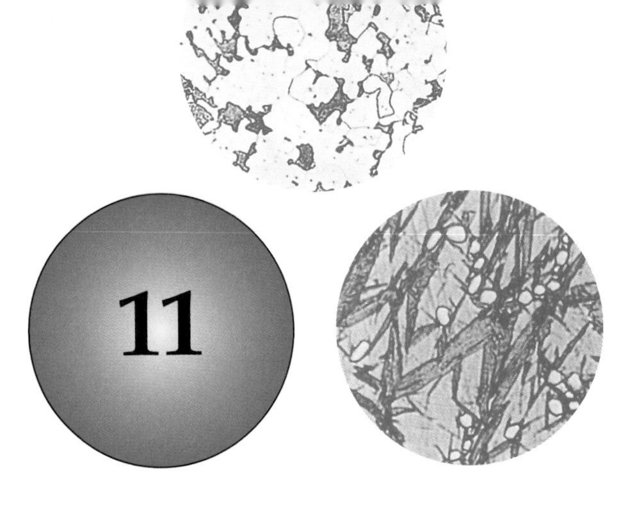

Heat Treating and Quenching

After studying this chapter, you will be able to:

- Explain heat-treating processes.
- Describe the effects and purpose of quenching.
- Discuss the four stages that metal goes through as it is quenched.
- Identify common quenching mediums and techniques and describe how they affect metal.
- Point out the advantages and disadvantages of quenching a material more rapidly.
- Apply some practical quenching techniques.

Heat Treating

Heat treating can be defined as any metallurgical process that involves heating or cooling. It is widely used in industry. Among the many heat-treating processes are such operations as heating, quenching, annealing, normalizing, tempering, and surface hardening.

When a metal goes through several heating and cooling processes, the entire procedure, or "recipe," is often referred to as the *heat treat* for that material.

Heat-treating operations are commonly performed in an oven or furnace. There are many different sizes, styles, and types of heat-treating furnaces. See Figure 11-1 and Figure 11-2.

Quenching

Quenching is a controlled cooling process that causes metals to harden. Before quenching is performed, the material must be heated to a high temperature. Quenching can be done from any elevated temperature. However, if hardness is important, the material should be heated to a temperature above the transfer temperature range. This process was discussed in Chapter 9.

In quenching, parts may be inserted and removed individually from the oven or furnace with tongs, Figure 11-3. Parts to be quenched may also be carried in groups and lowered into a quenching tank. See Figure 11-4.

The parts are plunged quickly into the quenching medium and submerged until they are cool. During the cooling operation, it is advisable to agitate or vibrate the parts very rapidly so cooling takes place as quickly as possible.

There are several different quenching mediums used to cool metals, including water, brine, oil, air, molten salt, and sand. The most common quenching mediums will be discussed later in this chapter.

Four Stages of Quenching

Quenching a metal is a complex process. A typical procedure is shown in Figure 11-5. Metal goes through four separate quenching stages as it cools from an elevated temperature

Figure 11-1. *Furnaces are widely used in the heat treatment of steel. A—This electrical heat-treating furnace is used to heat treat strip steel. B—The interior of this roller hearth heat-treating furnace has cast heating elements on the top, bottom, and side walls. (The Electric Furnace Company)*

Figure 11-2. *This heat-treating furnace is used for annealing coils of tubing, rod, strip, or wire and for annealing and heat-treating motor and transformer laminations. (The Electric Furnace Company)*

to room temperature. Each stage is listed and discussed below:

1. Vapor formation stage.
2. Vapor covering stage.
3. Vapor discharge stage.
4. Slow cooling stage.

In the *vapor formation stage,* the metal starts to cool. As soon as the metal is plunged into the quenching medium, the liquid making contact with the metal boils, forming a vapor film of bubbles that coat the metal. The cooling process slows down as this vapor film continues to form. See Figure 11-5A.

During the *vapor covering stage,* the film of vapor acts like a blanket covering the metal. The bubbles stick to the metal and insulate it from the quenching medium, Figure 11-5B. Unless this vapor film is removed quickly, the metal will not cool rapidly enough.

To reduce the effect of this vapor and speed up the quenching process, the metal should be agitated as much as possible. Agitation causes some of the bubbles to "fall off," allowing the metal to start cooling rapidly again. The longer the vapor is allowed to insulate the metal, the slower the cooling process will be, resulting in less hardness and strength. Unless the film is removed completely and evenly, soft spots, warpage, and cracking can also occur.

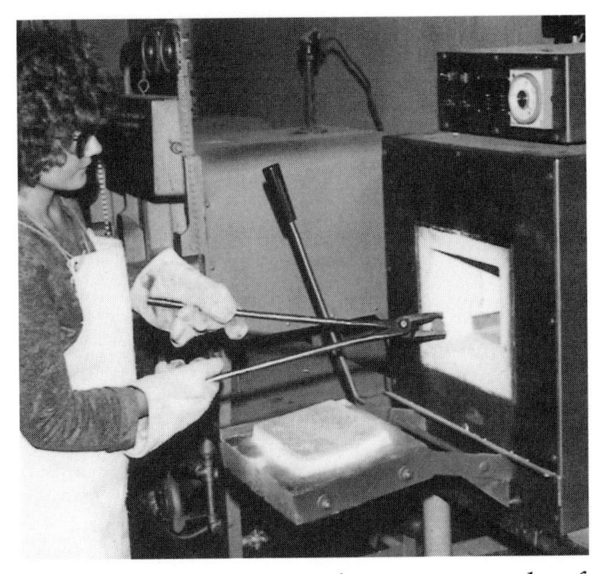

Figure 11-3. *Tongs are used to remove samples of metal from a heat-treating furnace.*

The *vapor discharge stage* occurs next, Figure 11-5C. It is a violent process. At this point, the vapor film begins to collapse and the metal starts to make contact with some of

Figure 11-4. *These cylinders, held by a special fixture, are being lowered as a group into a quenching tank. (J.W. Rex Company)*

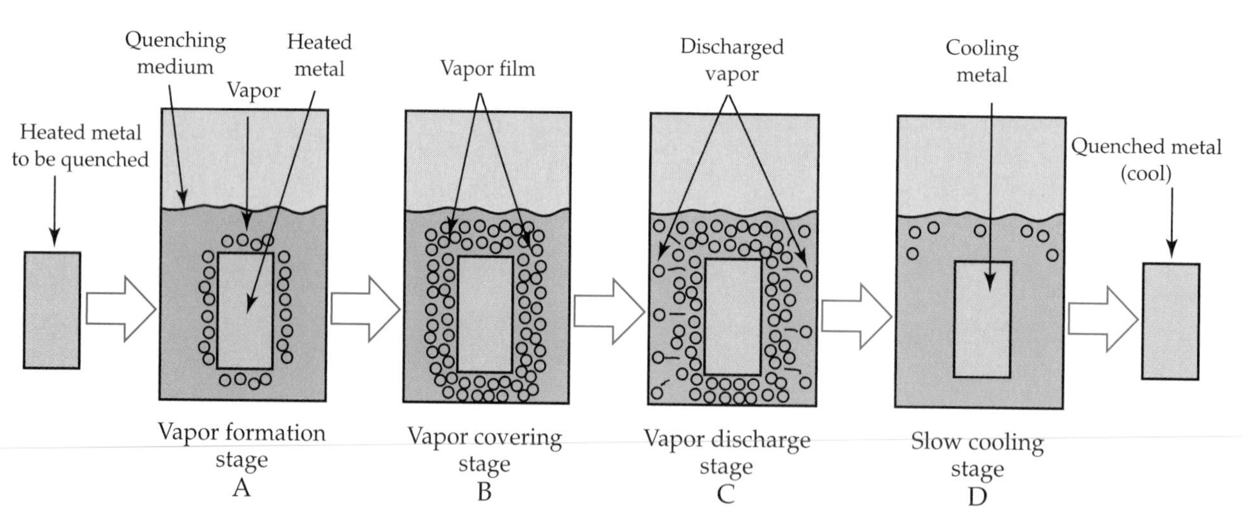

Figure 11-5. *Metal goes through four different stages after its entrance into the quenching medium. A—The metal starts to cool during the vapor formation stage. B—Bubbles of vapor surround the metal in the vapor covering stage. C—The vapor film explodes during the vapor discharge stage. The greatest amount of cooling occurs during this stage. D—The last stage is the slow cooling stage.*

the liquid again. As the film collapses, it tends to "explode" off the surface and more boiling occurs. This explosive action is violent enough to "rip off" the outer scale of the metal.

The vapor discharge stage will generally occur quite quickly, sometimes in less than one second. The "fireworks" in this stage are loud enough to be clearly heard. The discharge of vapor can be recognized as a cracking or sizzling sound. At one time, this sound was referred to as the "water biting the steel." The greatest amount of cooling takes place during this quenching stage.

After the vapor discharge stage ends, the *slow cooling stage* takes place. See Figure 11-5D. During this stage, the metal cools at a slower rate until it reaches room temperature. No further vapor film is formed. The cooling process in this stage is much more gentle than the cooling in the vapor discharge stage.

Figure 11-6 plots the cooling of a typical material as it passes through all four cooling stages. Note from the curve that the most rapid cooling occurs during the third stage (the vapor discharge stage). During this stage, the most drastic drop in temperature occurs.

Quenching Mediums and Techniques

A *quenching medium* is the liquid or material into which metal is plunged during the quenching process. Since not all quenching mediums are liquid, the term "medium" is used rather than "liquid." Some gases, such as air, are used in quenching. There are also some solid materials (such as sand) that are occasionally used in quenching. Some of the

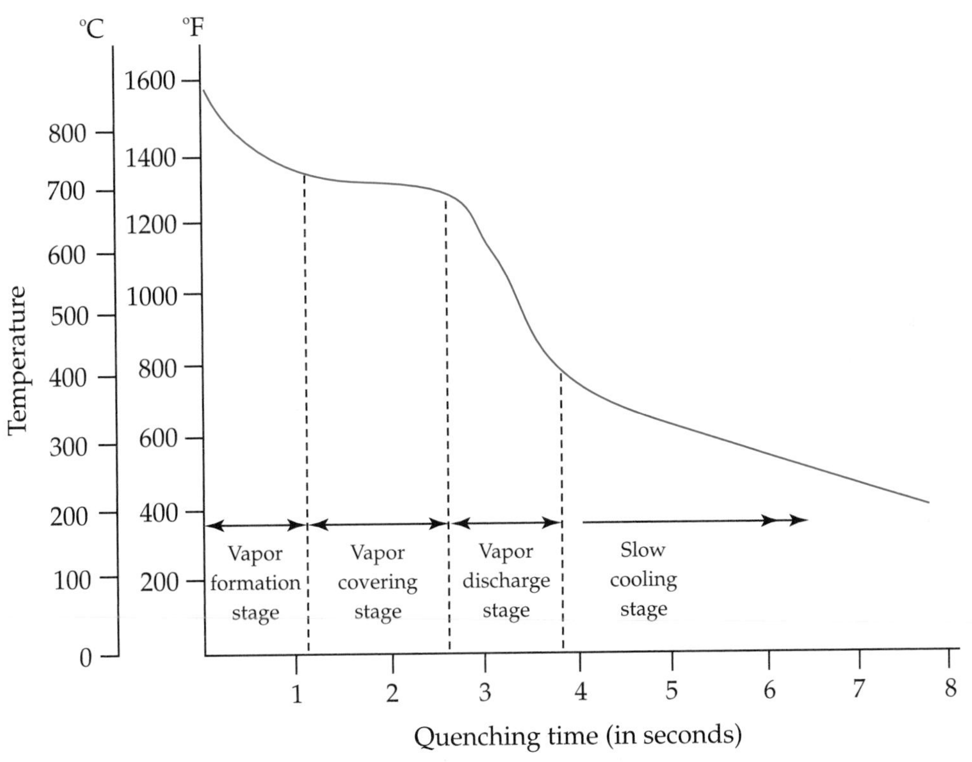

Figure 11-6. *This graph shows metal temperature during quenching. Changes in temperature occur over a period of seconds during the four stages of quenching.*

most common quenching mediums and techniques are discussed next.

Water Quenching

Water is the most commonly used quenching medium. It is inexpensive and convenient to use, and provides very rapid cooling.

Water quenching is especially used for low-carbon steel, which requires a very rapid change in temperature in order to obtain good hardness and strength. Water provides a very sudden, drastic quench. This can cause internal stresses, distortion, or cracking.

Brine Quenching

The effects produced by *brine quenching* are very similar to those that result from water quenching. *Brine* is salt water. It cools material slightly more rapidly than water, so the quenching action is a bit more drastic. The difference in the results from the two processes, however, is very slight.

A typical brine quenching medium contains 5% to 10% salt (sodium chloride) in water. The action of the salt accounts for the slightly increased cooling rate. The salt particles reduce the amount of time spent in the vapor discharge stage. The salt precipitates out and causes the outer surface of the metal to "explode" off the part. These tiny explosions disrupt the film of vapor and remove it rapidly. A brine quench will cause more scale to leave the metal than a water quench.

Quenching a part in a molten salt bath should not be confused with brine quenching. In this method, parts are submerged in a molten salt bath to produce a rapid rate of heat transfer. The quenching furnace shown in Figure 11-7 employs molten salt at temperatures of 350°F to 750°F (180°C to 400°C).

Oil Quenching

Oil is more gentle than water or brine when used as a quenching medium. Therefore, it is used for more critical parts, such as parts that have thin sections or sharp edges. For example, metal used for razor blades, springs, and knife blades is typically quenched in oil.

Since oil quenching is a less violent technique than water quenching, the chance of producing internal stresses, distortion, or cracking is reduced. However, because it is more gentle, oil generally does not produce steel that is as hard or strong as steel quenched by water. Therefore, the metallurgist must decide which is more important: hardness and strength, or eliminating cracking and distortion.

Oil quenching is a more effective technique when the oil is heated slightly above room temperature to 100°F or 150°F (40°C or 65°C). This may seem strange; normally, the colder the quenching medium, the harder and stronger the quench. However, oil is very

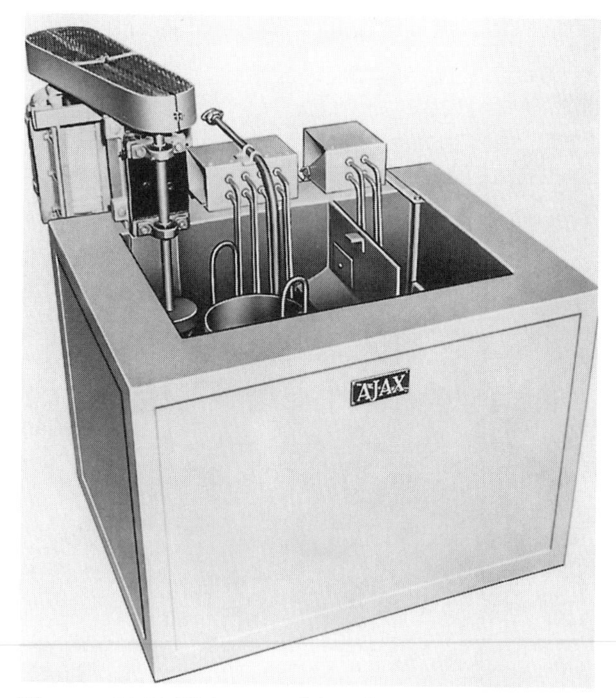

Figure 11-7. *This quenching furnace employs molten salt at temperatures of 350°F to 750°F (180°C to 400°C). It is heated internally by electric resistance heating elements. (Ajax Electric Company)*

thick. Heating it to 100°F reduces its viscosity (makes it thinner), which permits it to circulate more easily as it quenches the part. Hence, the use of warm oil will result in a more rapid quench than the use of cold oil.

Air Quenching

Air quenching results in less drastic quenches than those produced by oil, water, or brine. The physical procedure involved in air quenching is illustrated in Figure 11-8. The heated sample is placed on a screen. Cool air is blown at high speed from below. It passes through the screen and strikes the hot metal parts, Figure 11-9. Air quenching can be conducted in a chamber unit after the metal leaves a furnace. See Figure 11-10.

Air quenching does not cool parts as rapidly as oil or water quenching can. This has advantages and disadvantages. Due to the slower rate of cooling, there is less chance that internal stresses, distortion, or cracking will occur. However, because the cooling rate is slower, the resulting strength and hardness are not as high, unless special alloys are used in the metal. As a result, air quenching is generally used only on steels that have a very high alloy content. Special alloys (such as chromium and molybdenum) are selected because they are known to cause materials to harden even though a slower quenching method is used.

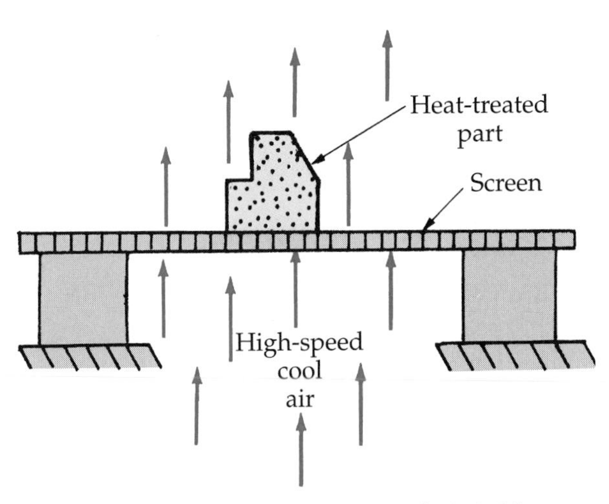

Figure 11-8. *In air quenching, cool air is blown at high speed through a screen and over the hot part.*

Figure 11-9. *These high-temperature alloy cones, just removed from the furnace, are being air quenched. (J.W. Rex Company)*

Summary of Quenching Methods

The effects produced by different quenching techniques can vary widely. Water, brine, oil, and air are used to generate specific results when they are used as quenching media. The next section of this chapter compares the advantages and disadvantages presented by these four quenching media.

The rate of speed at which a quench occurs can have various effects on the cooling process. When comparing the speed of quenching media, the following principles apply:

- Water and brine quenching are fastest.
- Oil quenching is next fastest.
- Air quenching is the slowest.

Quenching is designed to impart hardness and strength to materials. The following compare the hardness and strength produced by quenching media:

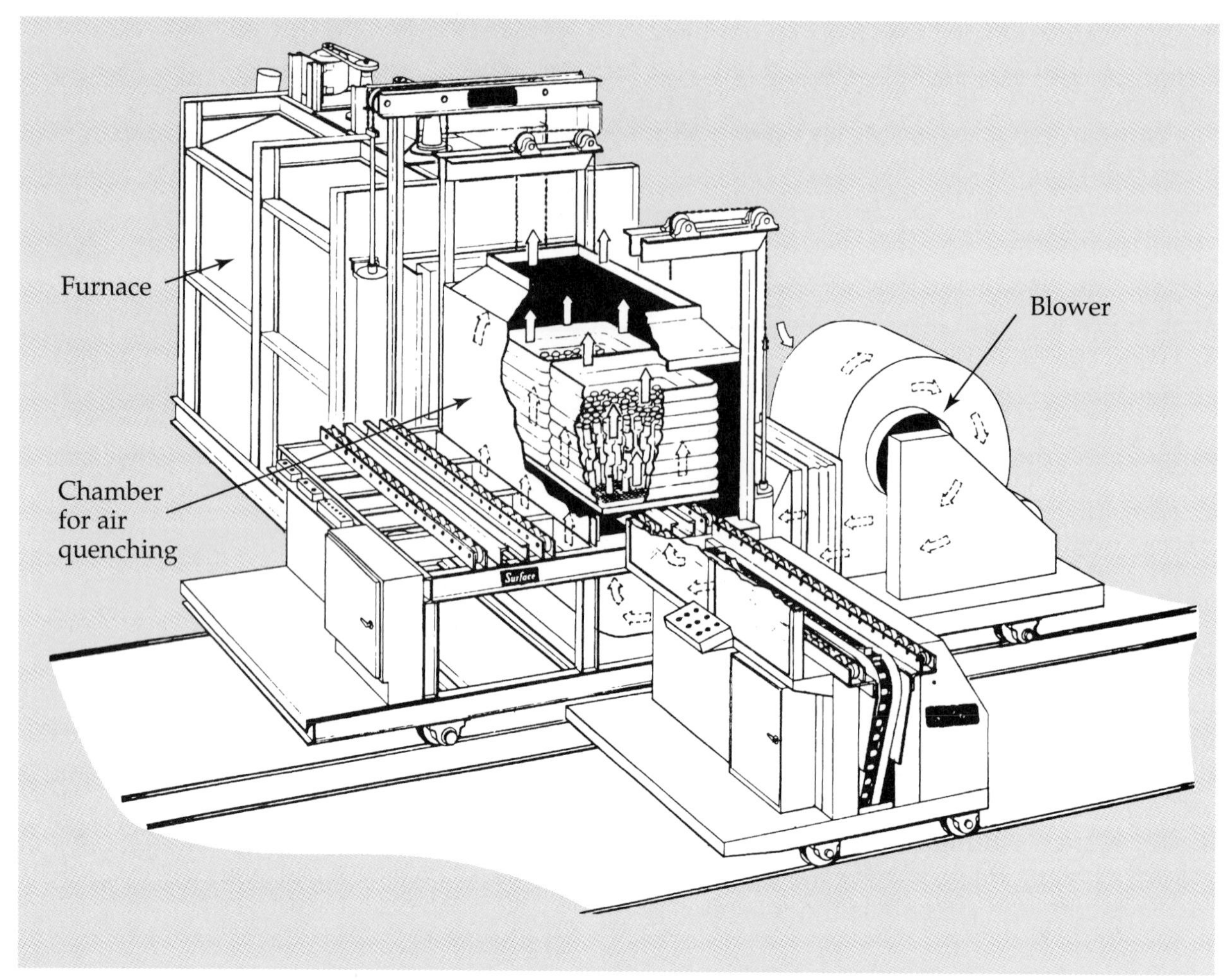

Figure 11-10. *A cutaway view of an air quenching chamber. Loaded boxes or trays are transferred by conveyor into the furnace shown at the left rear of the drawing. After the parts are heated, the trays are moved on a conveyor from the furnace to the air quenching chamber. The parts are then quenched by a forced blast of air from a blower. (Iron Castings Society)*

- Water and brine produce the greatest strength and hardness.
- Oil produces less strength and hardness.
- Air produces the least strength and hardness. However, the addition of expensive alloys to a metal can cause air-quenched steel to attain a hardness that is comparable to that of materials quenched in oil, water, or brine.

Due to the nature of quenching, the danger of producing internal stresses, distortion, and cracking must be considered. Each quenching media is rated as follows:

- Air quenching is best because it is the least drastic.
- Oil quenching is more likely to cause internal stresses, distortion, or cracking.
- Water and brine quenching media are most likely to cause internal stresses, distortion, and cracking.

The criteria outlined thus far can be used to select a quenching medium based on the application of the part. If you need to apply the best quenching medium for a job, your first questions would be:

- Is it important that the part be hard and strong?
- Would a certain amount of distortion or cracking be very detrimental to the part?

A comparison of these qualities and each quenching technique is shown in Figure 11-11. If hardness and strength are the most important qualities, water or brine quenching would be selected. If resistance to distortion and cracking are the most important characteristics, air quenching would be used. Oil quenching can be considered as a compromise because it is a "happy medium" between water quenching and air quenching. If both strength and elimination of distortion are extremely important, it would be best to use a more expensive alloy steel, and then air quench the steel.

Quenching Medium Temperature

The temperature of the quenching medium affects the results of quenching. For example, water can be used at lower temperatures to reduce the time for quenching. As shown in Figure 11-12, water used at 70°F (21°C) can cool a metal part in almost half the time that water used at 120°F (49°C) requires. In many cases, this can make the difference in attaining a metal with satisfactory hardness.

Quenching times for brine, water, oil, and air are illustrated in Figure 11-13. Note that brine and water quenching are the fastest. *Normalizing* (allowing the part to cool in still air) cools more slowly than any quenching method. Oil quenching, as shown, is faster than air quenching but occurs at a slower rate than water or brine quenching.

Practical Techniques Used in Quenching

Many factors must be kept in mind when selecting a quenching method. Metallurgists must be very practical in deciding which media to use and how to apply it. The following are some of the many questions to consider before starting a quenching operation:

- *How long should the material be left in the quenching tank?* The rule of thumb is to leave the sample in the quenching medium until it is cool enough to be touched by hand.
- *How long must the part be left in the oven for heat treatment before removing it for quenching?* The rule of thumb is to leave the part in the oven one hour for every inch of thickness (once the oven reaches the desired heating temperature). Thus, if a metal cube to be heat treated measures 2" × 4" × 10", it would be left in the oven for at least two hours. If a part measures 1/4" × 6" × 10", it would need to be left in the oven for only 15 minutes before quenching. See Figure 11-14 for other examples.

Comparison of Quenching Media		
Water or Brine Quenching	**Oil Quenching**	**Air Quenching**
Most drastic	Less drastic	Least drastic
Most hardness and strength	Less hardness and strength	Least hardness and strength
Least resistance to distortion and cracking	More resistance to distortion and cracking	Most resistance to distortion and cracking

Figure 11-11. *Quenching techniques are selected based on the results associated with each method.*

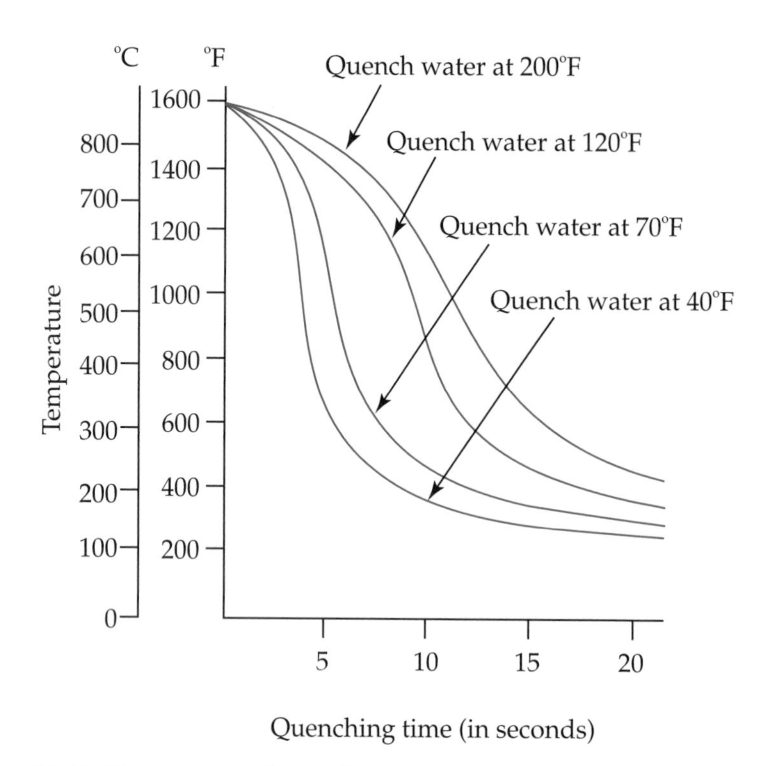

Figure 11-12. *Temperature of quenching medium has an effect on the quench time.*

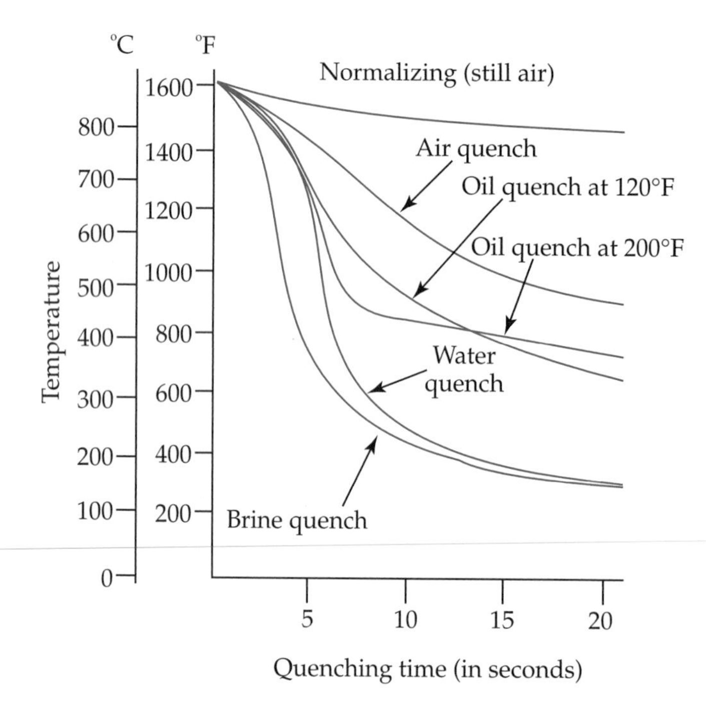

Figure 11-13. *Quenching rates for different quenching mediums.*

Self-Demonstration
Comparing Effects of Quenching Mediums

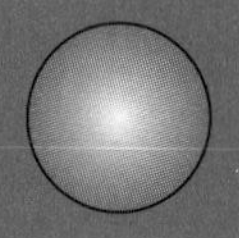

Obtain at least five small samples of 1045 steel. If 1045 is not available, 1095 or 1018 steel may be substituted. If a different steel is used, the temperature values given below will need to be changed. It is likely, however, that 1045 steel will be in inventory in your metallurgical laboratory stock room or machine shop stock room.

First, test the hardness of each of the five samples on the Rockwell C scale and record the hardness values.

Load all five samples into the metallurgical furnace and heat it to 1600°F (if 1045 steel is used). For 1095 steel, the temperature should be 1500°F. If 1018 steel is used, heat the samples to 1700°F. If any other type of steel is used, refer to an iron-carbon diagram and heat the metal to a temperature above its upper transformation temperature.

After holding this temperature for one-half hour, remove the samples one at a time and quench each of them in a different quenching medium. Quench one sample in ice water, one in water at room temperature, one in salt water (brine), and one in oil. Hold the fifth sample in a fast stream of air from a strong fan until it is cooled to room temperature. If you have more than five samples available, you might try different types of oil or several different temperatures of water for quenching.

Special quenching-type oils are available, but in order to conserve costs for your demonstration, you can use any commercial grade of motor oil.

Be sure to immediately plunge the samples into the quenching medium as soon as they are removed from the furnace, so that the samples have no opportunity to precool. If the samples are located near the rear of the furnace, and if the door is opened and closed quickly, there is less chance of precooling the last samples to be quenched.

After each sample has cooled, record the hardness value. Compare the hardness values corresponding to each quenching medium.

Compare the results of each quench to the quenching times in Figures 11-12, 11-13, and 11-15. Why can we expect to see some discrepancies between your results and the graphs?

- *How important is agitation?* Sometimes, agitation is not practical because there is a large quantity of parts or a part is awkward to handle. When practical, however, the effects of agitation on quenching are significant. Heavy or violent agitation reduces cooling time considerably, as shown in Figure 11-15.

  Agitation is not as effective in oil quenching as it is in water quenching. Oil bubbles tend to cling more tenaciously to the metal surface. In a brine or water quench, a second or two of agitation is often sufficient to dislodge the vapor bubbles. Since the vapor film stages last only a second or two, there is little benefit in agitating or shaking the part after it has been in the quenching medium for more than a few seconds.
- *How can distortion be avoided in thin parts?* When a part has a thin section or *web* between heavier sections, Figure 11-16, there is a good chance that distortion will occur during quenching. To reduce the

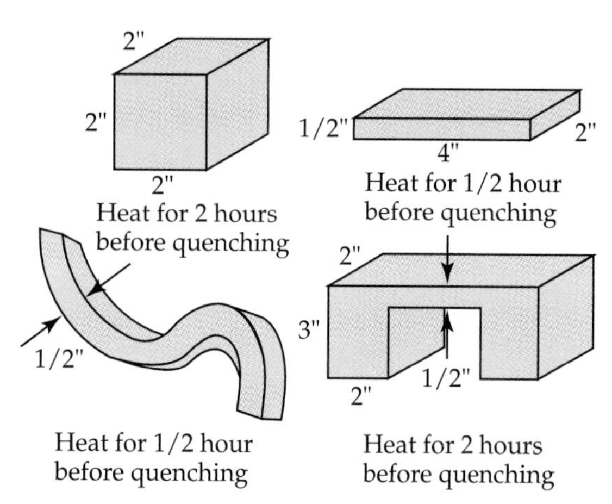

Figure 11-14. *A part should be left in a heat-treating oven for at least one hour for every inch of thickness before quenching.*

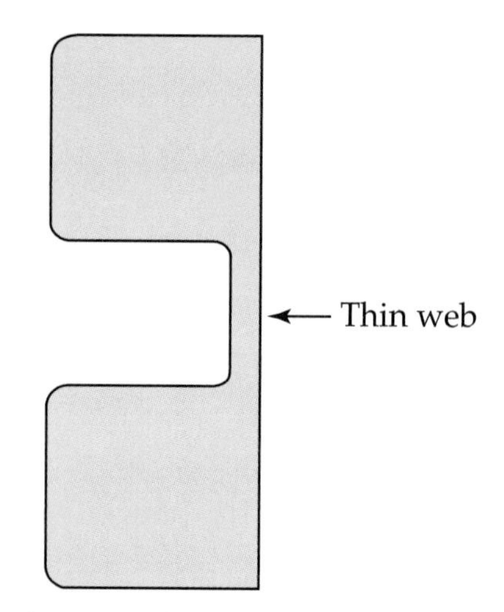

Figure 11-16. *A part with a thin web between heavier sections is susceptible to distortion during the quenching process.*

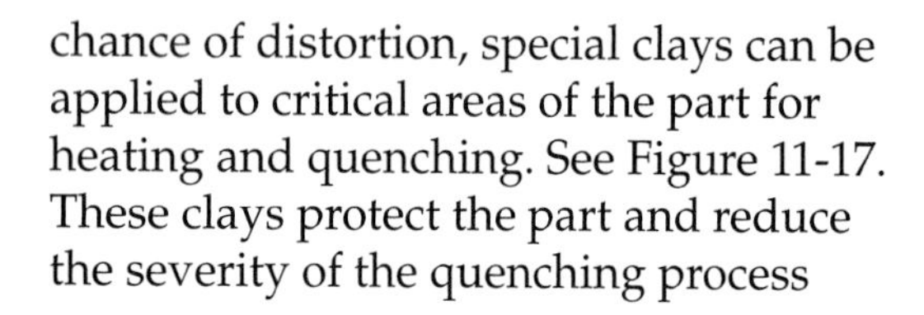
chance of distortion, special clays can be applied to critical areas of the part for heating and quenching. See Figure 11-17. These clays protect the part and reduce the severity of the quenching process when the metal first enters the liquid. Special clays can also be used to reduce the chance of cracking and distortion in parts that have thin projections or sharp corners, Figure 11-18.

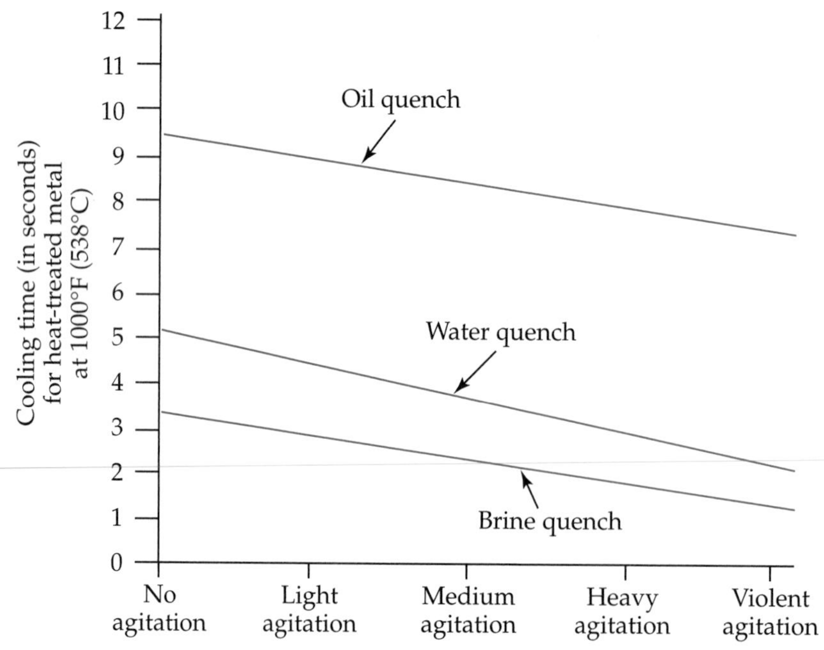

Figure 11-15. *Agitation of part being quenched reduces cooling time.*

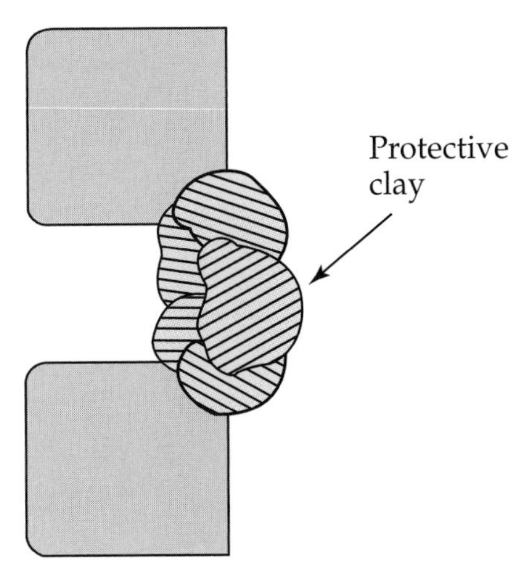

Figure 11-17. *Special clays can be used to protect thin areas of a part during heating and quenching to reduce the chance of distortion.*

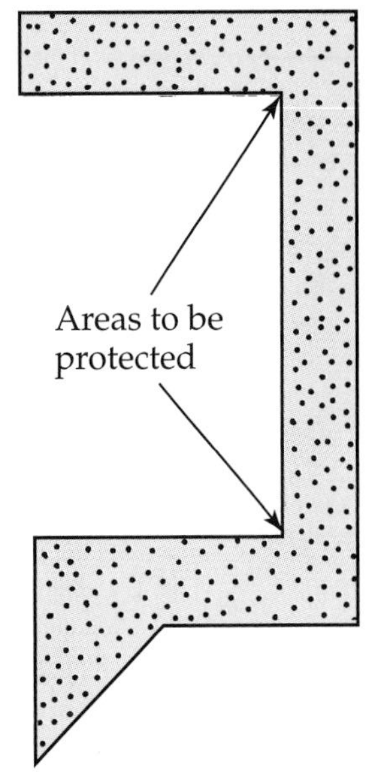

Figure 11-18. *Special clays can be added to sharp corners to reduce the chance of cracking and distortion.*

Test Your Knowledge

Write your answers on a separate sheet of paper. Do not write in this book.

1. What is the purpose of quenching a metal?
2. In addition to water, name four other quenching mediums.
3. Of the four stages of quenching, which one is the most violent?
4. During which of the four stages of quenching do bubbles form a blanket around the sample?
5. What is *brine* and why is it effective in quenching?
6. Name the most commonly used quenching medium.
7. Which type of quenching is the most drastic?
8. Which type of quenching is the most gentle?
9. Why will heated oil result in a more rapid quench than oil at room temperature?
10. Which will provide the faster quench, water at 75°F or water at 120°F?
11. What is the minimum amount of time that a sample should be left in an oven before quenching if the sample measures 3" × 4" × 6"?
12. Sketch a part that would distort considerably as a result of quenching.
13. Sketch a part shape that would not become greatly distorted due to quenching.
14. Sketch a very complex part designed for use in an aircraft engine that could be damaged severely by a harsh quenching operation. How might hardness be attained for this part if quenching is considered to be too severe?

Some processing methods for nonferrous metals are similar to methods used to process steel. Top—A 100′ continuous hot rolling mill for aluminum. Bottom—Technicians monitor the operation of the mill. (Alcoa)

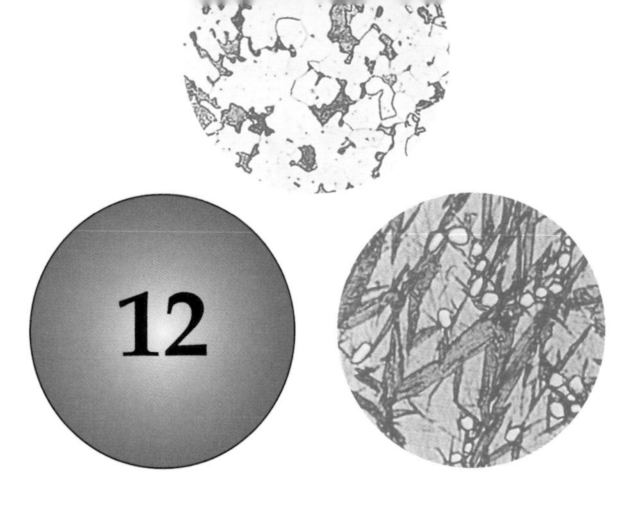

Annealing and Normalizing

After studying this chapter, you will be able to:

- Describe the purpose of annealing and normalizing metal.
- Explain how annealing and normalizing affect the hardness, strength, and brittleness of metal.
- Describe how annealing and normalizing affect the crystal structure of metal.
- Identify three different types of annealing processes.
- List the different ways in which annealing and normalizing affect metal compared to quenching.

Overview of Annealing and Normalizing

Quenching is a form of heat treatment that uses rapid cooling to make metal harder and stronger. This process, introduced in Chapter 11, also makes material more brittle and less ductile. In metallurgy, it is sometimes more important for a material to be ductile rather than hard and strong. Ductility can be improved through annealing and normalizing. These two heat treatment processes are used to make steel softer and more ductile, and they are similar in nature. Both techniques involve *slow cooling.* They produce steel that is easier to machine, easier to bend and form, and less prone to cracking and distortion.

Annealing and *normalizing* can both be defined as heating a metal above a critical temperature and slowly cooling it to room temperature in order to obtain a softer and less distorted material. Just as a fast quench produces a hard part, slow cooling processes produce a softer and more ductile part. The differences between the two methods are discussed below.

Annealing

Annealing is a controlled cooling process that involves heating a metal and then slowly reducing the cooling temperature to attain a high degree of softness in the metal. It is commonly performed in an oven or furnace, Figure 12-1.

In annealing, the metal is first heated above its upper transformation temperature (where it becomes austenite) or some other high temperature. It is left to soak at that temperature for a period of time. Next, it is placed in an annealing oven, where the temperature is slowly lowered to cool the metal. The process is completed when the metal is cooled to room temperature.

The oven may be the same one in which the steel is heated, or the steel may be transferred to a second oven. When the oven temperature has finally been lowered enough for the material to reach room temperature, the part is said to have been *annealed.* Annealing sometimes takes as long as several days in order to make the material as soft as possible.

Figure 12-1. *The metal tubes shown on the conveyor are being simultaneously annealed in this roller hearth furnace. (The Electric Furnace Company)*

Normalizing

Normalizing involves a more rapid cooling process than annealing. In this technique, the material is heated above its upper transformation temperature and is then cooled slowly at room temperature. In other words, when the material is removed from the furnace or oven, it is set out to cool slowly on a bench or a floor. Large parts may even be left to cool slowly outside.

Annealed steel cools at a slower rate than normalized steel. Normalized steel merely cools to room temperature with no external influence. In normalizing, after the material has had enough time to cool, the entire part will be at room temperature and the material is said to be *normalized.*

Effects of Annealing and Normalizing

Annealing and normalizing generally produce a material that is less hard and less strong. They make a material less brittle and more ductile.

Annealing and normalizing also reduce the amount of internal stresses in a material. This, in turn, reduces the tendency of the material to distort and crack.

A comparison of the purposes of annealing and normalizing and quenching is shown in Figure 12-2.

Purposes for Annealing and Normalizing

In the study of metallurgy, there is so much emphasis on attaining hardness and strength that one starts to consider hardness and strength as the "good guys" and softness and ductility as the "bad guys." This is not always true. There are many purposes in making a material softer and less strong:

- To make the material easier to machine.
- To produce material that is easier to form.
- To relieve internal stresses.
- To refine the crystal structure of a part.

Comparison of Cooling Methods	
Annealing and Normalizing	**Quenching**
• Slow cooling process	• Rapid cooling process
• Softens and weakens metal	• Hardens and strengthens metal
• Produces ductility	• Produces brittleness
• Reduces internal stresses	• Causes internal stresses
• Helps prevent cracking and distortion	• Increases chances of cracking and distortion

Figure 12-2. *A comparison of the different purposes for annealing, normalizing, and quenching.*

Machinability

A softer and more ductile material is easier to machine in the machine shop. Often, a material in an annealed or normalized condition will receive as much machining as possible, even for a part that requires strength. Then, after most of the machining is completed, the part is hardened by heating and quenching. Only the last finishing cut is taken with the material in a hardened condition. This saves considerable machining time and costs and reduces tool wear.

Forming

A part that has been annealed or normalized will respond better to forming operations. Processes such as spinning, rolling, bending, and deep drawing require considerably less machine time and energy if the part has been softened. A softer, more ductile part will also be much less likely to crack or distort if it has been annealed or normalized before forming.

Internal Stress Relief

Internal stresses can build up in metal as a result of processing. Stresses may be caused by previous processing operations such as welding, cold working, casting, forging, punching, drawing, extruding, or machining. If internal stresses are allowed to remain in a metal, the part may eventually distort or crack.

Annealing and normalizing help relieve internal stresses and reduce the chances for distortion and cracking. Just as a human being who feels tension can take a hot shower to relieve stress and relax nerves, metal can be relaxed by being annealed or normalized.

Refinement of Crystal Structures

Another reason for annealing or normalizing is to change the crystal shape. After some types of metalworking (particularly cold working), the crystal structures are elongated. Annealing or normalizing can change the shape of the crystals back to the desired form.

First, the material is heated to a temperature above the upper transformation temperature (where it becomes austenite). Then, upon cooling, the crystal grains change and a more desirable shaped crystal is redeveloped.

Types of Annealing

There are three basic types of annealing processes used in the heat treatment of metal:

- Full annealing.
- Process annealing.
- Spheroidizing.

Each type of annealing is based on a specific heating and cooling process. These processes are discussed next.

Full Annealing

Full annealing is the most basic of the annealing processes and is often simply referred to as "annealing." In this technique, the steel is first heated to a high temperature, normally 50°F–100°F (28°C–55°C) above the upper transformation temperature. See Figure 12-3. Upon reaching this temperature, the steel changes to austenite and its crystal structure becomes face-centered cubic. The steel is held at this temperature for a long time. A rule of thumb is to soak the material for one hour at the annealing temperature for every inch of thickness.

The material is then slowly cooled in the oven. A cooling rate of 100°F per hour is typical for full annealing. A much slower cooling rate is used when time is not important, and when the elimination of internal stresses is especially critical.

It is best not to heat the material much above the upper transformation temperature in order to prevent the grain size from becoming excessively large. This will weaken the metal. The higher the steel is heated above the upper transformation temperature, the larger the grain size. A temperature from 50°F–100°F above the upper transformation temperature is normally sufficient to produce austenite without generating a large grain size.

Process Annealing

One major problem with full annealing is the time it takes. In order to attain completely stress-free material, the part may need to be cooled slowly for an entire day or, occasionally, several days. Not only does this delay the availability of the parts, but it also ties up ovens for long periods of time.

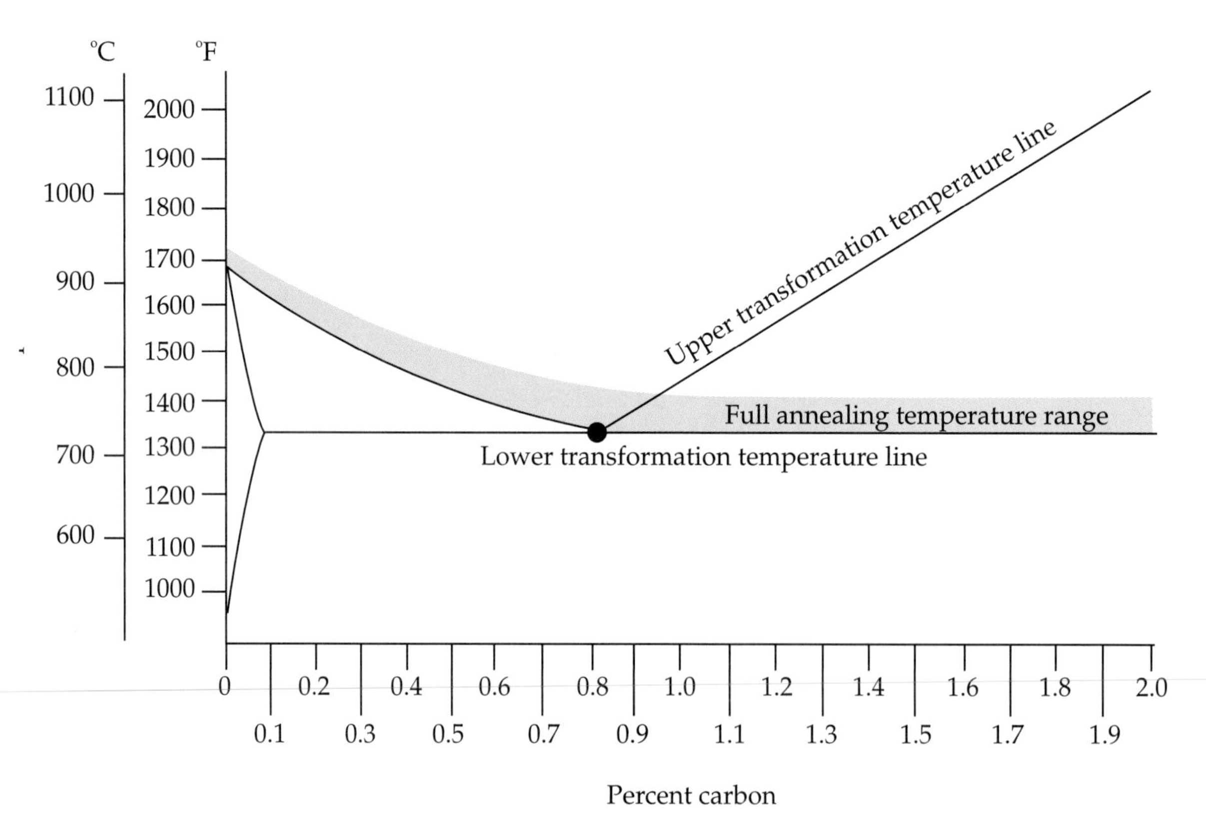

Figure 12-3. *In full annealing, steel is heated to a temperature above the upper transformation temperature before it is very slowly cooled.*

In *process annealing,* parts are not as completely softened as they are in full annealing, but the time required is considerably lessened. Thus, process annealing is a compromise between full annealing and no annealing at all. Process annealing is used when time is an important factor.

During manufacture, internal stresses build up in parts. This is a result of some of the manufacturing discussed earlier (such as welding or cold working). It is not always necessary to remove all of these stresses if the time required to do so is excessive. If most of the stresses can be eliminated in a short period of time, process annealing is a practical and economic solution.

In process annealing, the part is heated to only 1050°F–1300°F, Figure 12-4. The part is then cooled. This will *stress relieve* the part. Most of the internal stress will be removed, but not all of it.

Process annealing is frequently used as an intermediate heat-treating step during the manufacture of a part. A part that is stretched considerably during manufacture may be sent to the annealing oven three or four times before all of the stretching is completed. Thus, a part may be stretched a little, and then process annealed to eliminate some internal stresses. It may then be stretched a little more and process annealed again.

After a rough machining cut is made, a part may be process annealed to remove most of the stresses caused by the cutting action. After stress relieving, a thin, final cut may be taken, producing little additional stress. Thus, the complex part will end up in a nearly unstressed condition. See Figure 12-5.

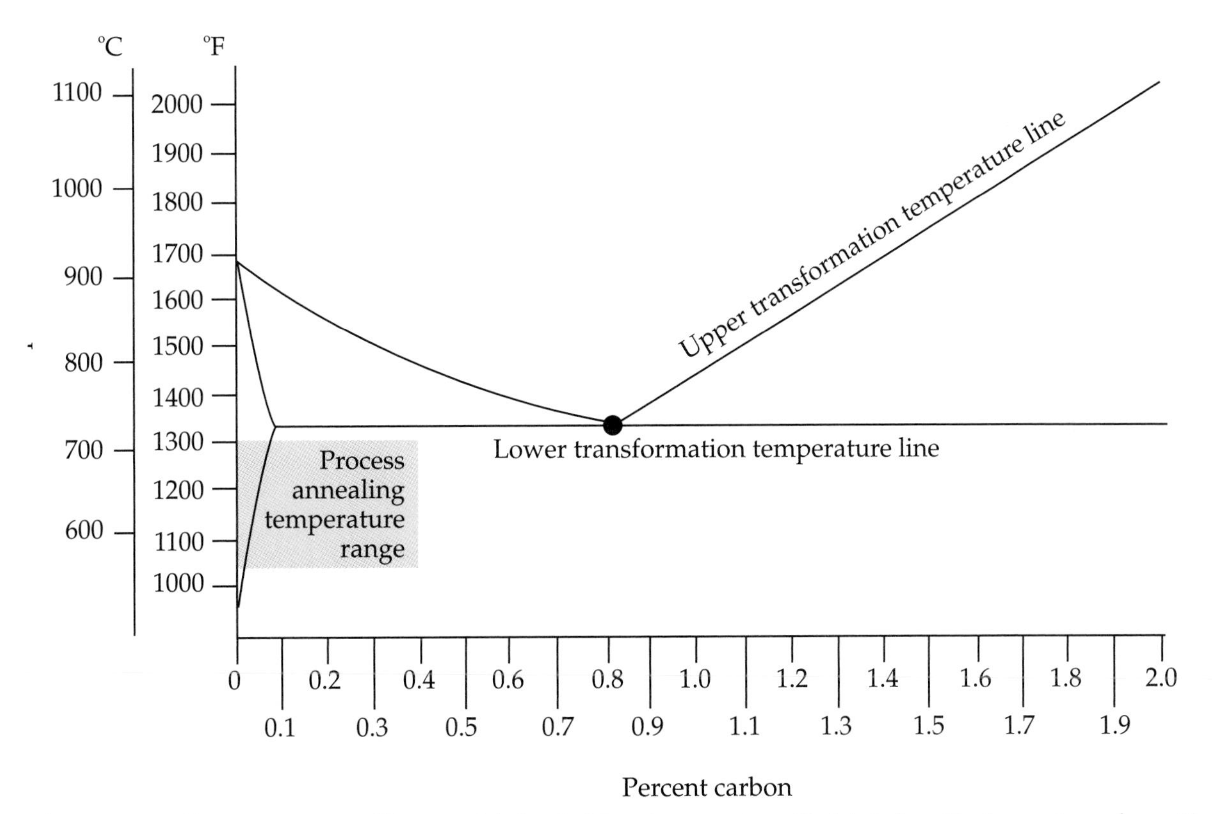

Figure 12-4. *In process annealing, steel is heated to a temperature that is below the lower transformation temperature. This process relieves some internal stresses without taking a great deal of time.*

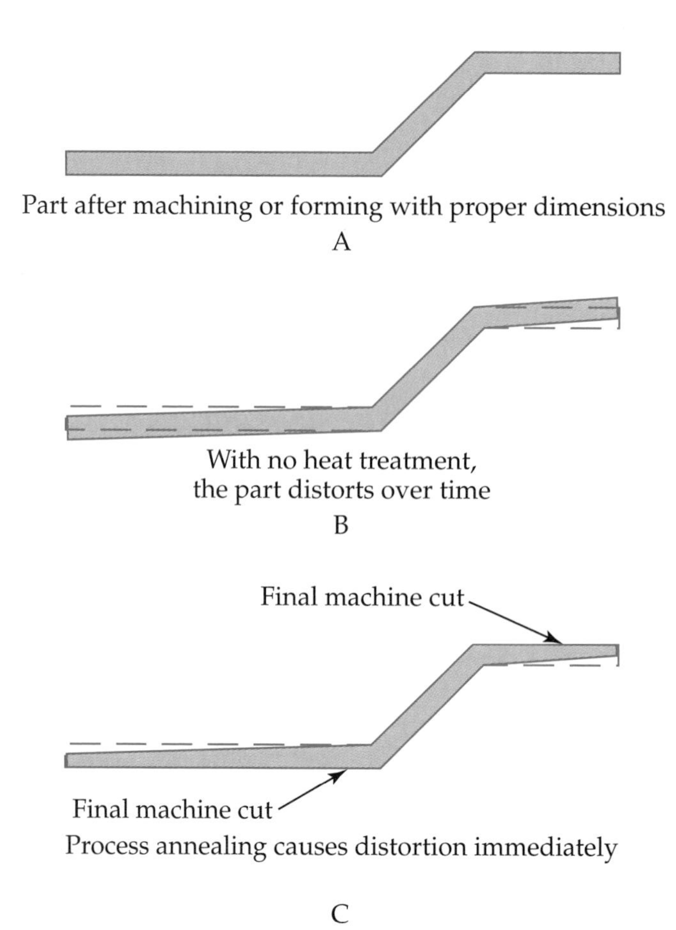

Figure 12-5. *Process annealing can eliminate the hazard of distortion taking place after a long period of time. A—After machining or forming, the part appears to be holding at its proper dimensions. B—After a long period of time (perhaps weeks), internal stresses may cause the part to distort. C—Process annealing the part after machining or forming will relieve internal stresses. Some distortion will be eliminated and the rest of the distortion will take place immediately. A thin, final machine cut will remove the effects of this final distortion.*

Spheroidizing

Spheroidizing is a rapid annealing method that is almost identical to process annealing. Parts are heated to a temperature near the lower transformation temperature and are then slowly cooled, Figure 12-6.

The term "spheroidizing" comes from the microscopic appearance of the steel after it has been annealed. Many tiny spherical forms occur throughout the microscopic structure of cementite when it is process annealed. See Figure 12-7.

Spheroidizing is used for high-carbon steel. When cementite (high-carbon steel) is process annealed and the globular forms appear in the microscopic pattern, the annealing process is referred to as "spheroidizing." When ferrite (low-carbon steel) is annealed and globular forms do not appear, the annealing process is referred to as "process annealing."

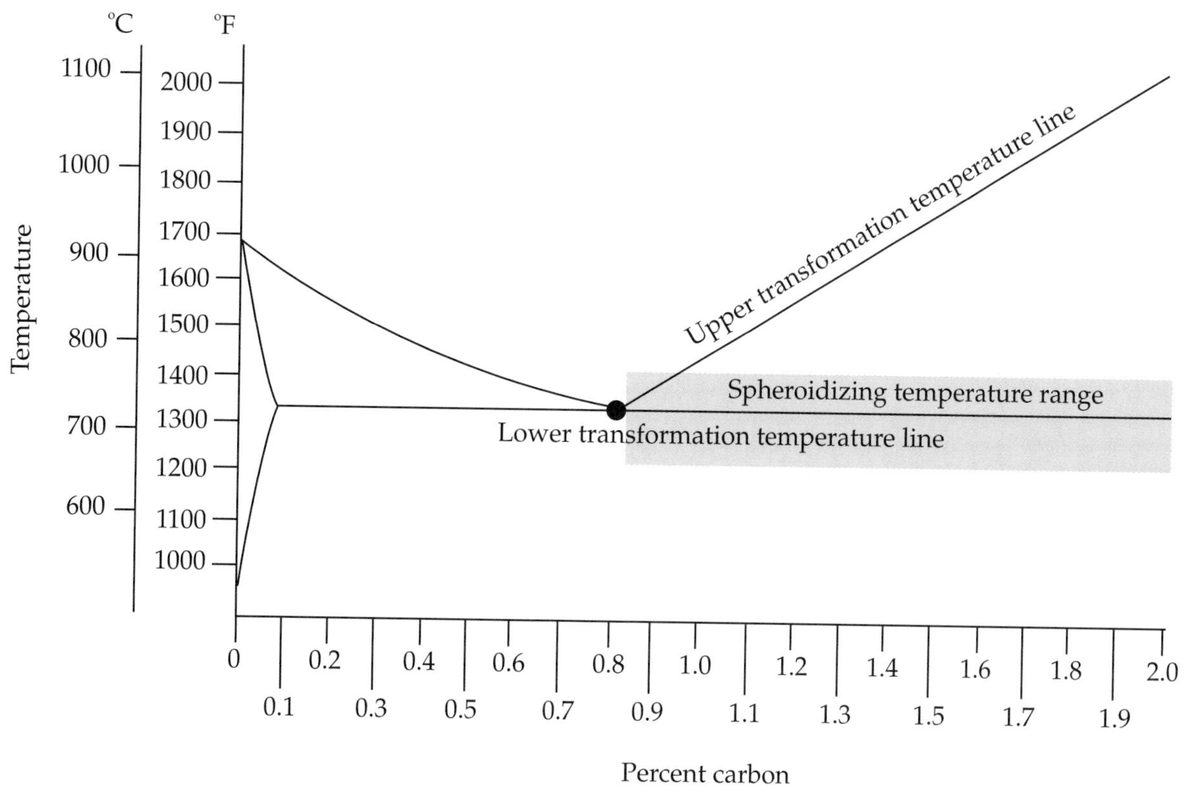

Figure 12-6. *In spheroidizing, parts are heated to a temperature near the lower transformation temperature before very slow cooling.*

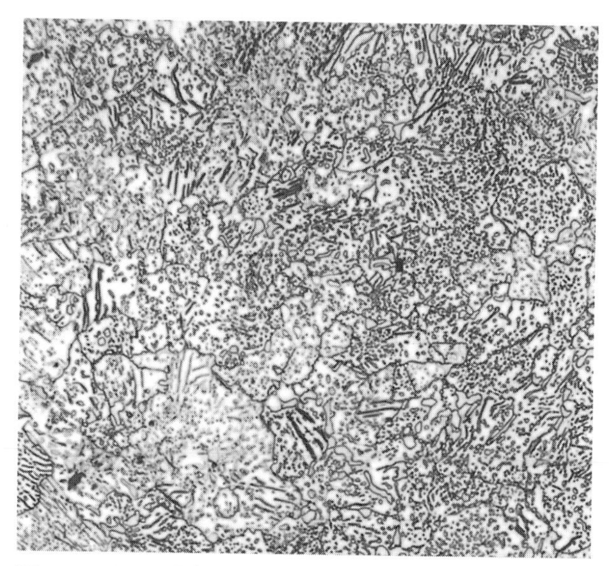

Figure 12-7. *The spherical structures in 1095 steel are a result of the spheroidizing process. (Buehler Ltd.)*

Normalizing

Normalizing is a heat treatment method that involves slow cooling. The name "normalizing" comes from the original intended purpose of the process, which was to return steel to the "normal" condition it was in before it was altered by cold working or other processing.

The heating process in normalizing is similar to that of full annealing. In normalizing, the steel is heated above the upper transformation temperature (where it changes to austenite). See Figure 12-8. It is then cooled in still air at room temperature. See Figure 12-9.

Normalizing does not soften the material as much as full annealing does. The cooling process does not leave the material as ductile

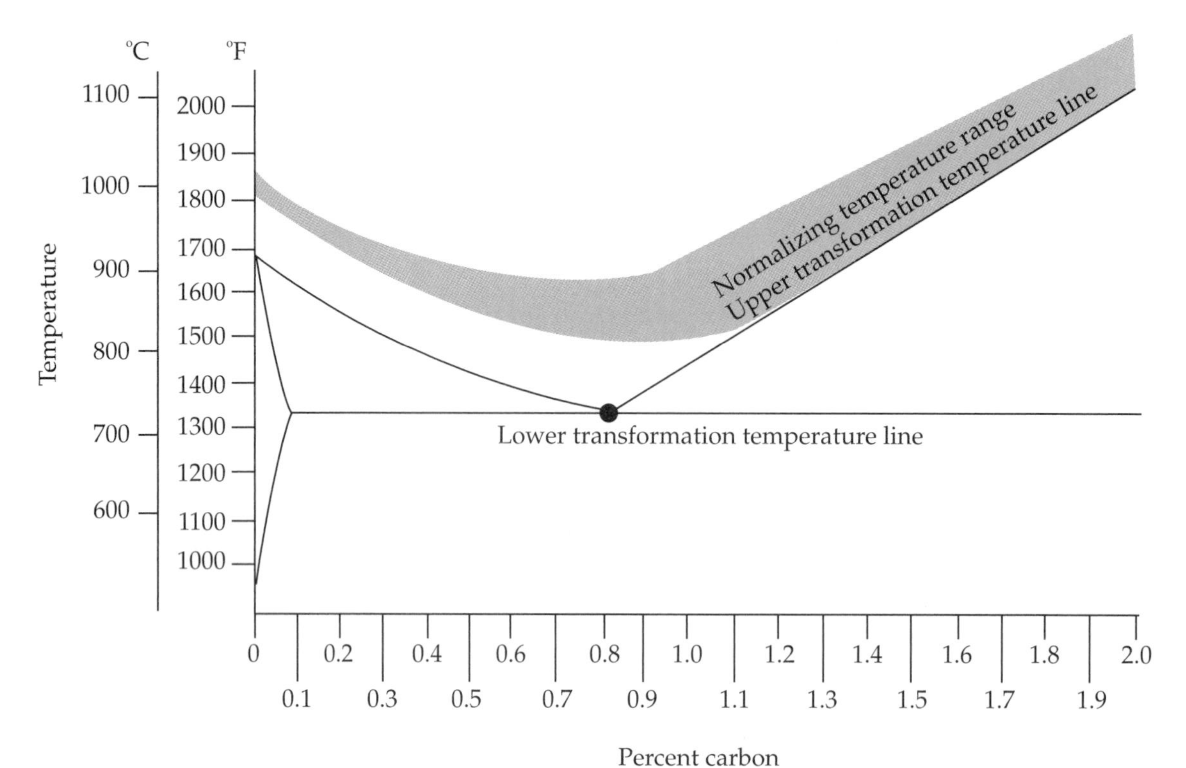

Figure 12-8. *Normalized steel is first heated to a temperature above the upper transformation temperature, where it changes to austenite. It is then cooled slowly at room temperature.*

Figure 12-9. *These drill rods are being transferred to an air-cooling station after being removed from this 22′ high bottom-opening gantry furnace. (J.W. Rex Company)*

or as internally stress-free. A normalized part will usually be a little stronger, harder, and more brittle than a full-annealed part. A normalized part will have a slightly finer, smaller grain structure.

Comparing Annealing, Normalizing, and Quenching

The table in Figure 12-10 compares annealing, normalizing, and quenching. At the right of the table is the hard and strong region. The left portion of the chart is the soft and ductile region.

Note that full annealing will allow a piece of steel to attain as much softness as possible. A brine or water quench will transform the material into the hardest possible condition.

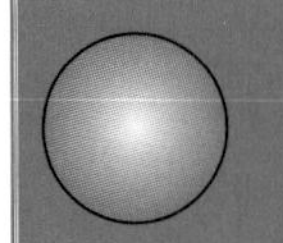

Self-Demonstration
Effects of Heat Treatment Processes

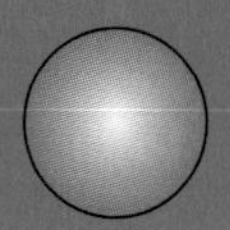

Obtain at least four small pieces of several different types of steel, including 1018, 1045, 1095, 4140, and 52100. If other types of steel are substituted, the temperature value given will need to be adjusted accordingly. Try to obtain an assortment of steel types, including low-carbon, medium-carbon, and high-carbon steels.

First, test the hardness of each of the samples on the Rockwell C scale and record the hardness values.

Load all of the samples into the metallurgical furnace and heat it to 1700°F. If you are using other types of steel, refer to an iron-carbon diagram to determine the upper transformation temperature for each sample. Heat the furnace to a temperature above the highest upper transformation temperature required for the materials.

After holding this temperature for one-half hour, remove the samples one at a time.

Quench one sample of each type of steel in water.

Place one sample of each type of steel on a table and let the samples slowly cool to room temperature. These will be your normalized samples.

Quench one sample of each type of steel in another quenching medium of your choice, such as brine, ice water, or oil.

Leave the fourth samples of each type of steel in the oven for controlled cooling. Slowly lower the temperature of the furnace until the samples are cooled to room temperature. If this is too time-consuming for your laboratory, simply shut the furnace off and allow the samples to cool at their own rate. These will be your full-annealed samples.

After the entire cooling process is completed, test the hardness of all the samples and tabulate the results. Are there any surprises?

Compare your results to the common heat-treating effects listed in Figure 12-10. Why can you expect to see some discrepancies between your results and Figure 12-10?

Effects of Annealing, Normalizing, and Quenching on Metal					
Full Annealing	Normalizing	Air Quenching	Oil Quenching	Water Quenching	Brine Quenching
← Softer, less strong		Harder and stronger →			
← More ductile		More brittle →			
← Less internal stress		More internal stress →			
← Less distortion, cracking		More distortion, cracking →			

Figure 12-10. *A comparison of full annealing, normalizing, and quenching. Full annealing and normalizing produce softer, more ductile, and more stress-free parts. Quenching produces parts that are relatively hard, strong, and brittle.*

Process annealing and spheroidizing cannot be compared to the other processes since these two methods are dependent on the condition of the metal before it is process annealed. Full annealing or normalizing can reduce the hardness of any steel to a very soft condition, even if it is a water-quenched part that measures harder than 60 R_C (on the Rockwell C scale).

Test Your Knowledge

Write your answers on a separate sheet of paper. Do not write in this book.

1. How is material cooled when normalizing?
2. Explain the difference between *annealing* and *normalizing*.
3. What are some of the results of annealing and normalizing?
4. List four purposes for annealing and normalizing.
5. What effect does cold working have on the crystal shapes in steel?
6. How long should a sample be held at the annealing temperature before allowing it to cool?
7. What is the advantage of process annealing over full annealing?
8. To what temperature range is steel heated for process annealing?
9. Which takes longer to cool, a part that has been normalized, full annealed, or spheroidized?
10. In which process is the material heated the hottest?
    a. Full annealing.
    b. Process annealing.
    c. Spheroidizing.
11. After a part has been heated to a high temperature, which of the following three cooling processes would cause the part to become the softest?
    a. Full annealing.
    b. Normalizing.
    c. Water quenching.
    d. Air quenching.
12. After a part has been heated to a high temperature, which of the following three cooling processes would cause the part to become the strongest?
    a. Full annealing.
    b. Normalizing.
    c. Water quenching.
    d. Air quenching.
13. Sketch an irregular-shaped part that would be affected greatly by heat treatment. Select a part shape that would show considerable difference in the results from the different heat treating methods. Make the following sketches of the part:
    a. The original manufactured part before heat treatment.
    b. The part after full annealing.
    c. The part after normalizing.
    d. The part after water quenching.

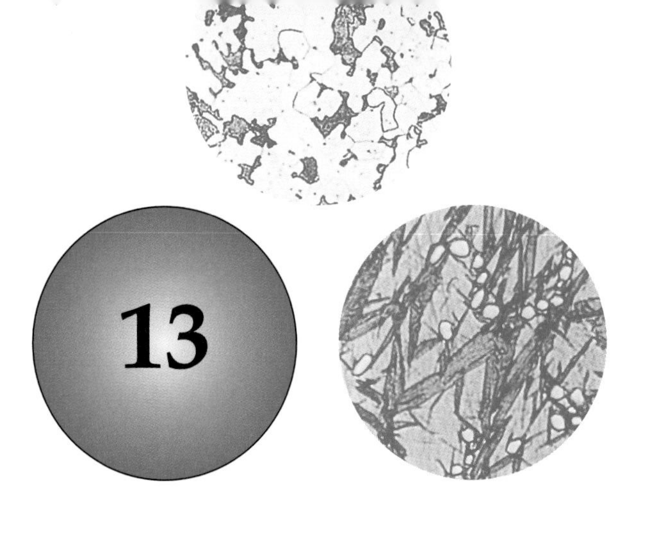

Isothermal Transformation Diagrams

After studying this chapter, you will be able to:

- Explain the purpose of an isothermal transformation (I-T) diagram and describe how it is used.
- Recognize the difference in usage between an I-T diagram and an iron-carbon phase diagram.
- Determine whether a material will become stronger during the heat-treat process by evaluating its I-T diagram.
- Compare different industrial I-T diagrams.
- Plot a temperature-time line on an I-T diagram.

Introduction to Isothermal Transformation Diagrams

The iron-carbon phase diagram is a very useful tool in metallurgy. When the percent carbon and the temperature are known, the structure of a particular steel can be established. When the percentage of carbon and the transformation temperatures for a particular type of steel are known, the basic structure of the material can be established.

One important element is missing from the iron-carbon phase diagram, however. It does not indicate the amount of *time* involved in the cooling process. If a certain type of steel is heated above its upper transformation temperature until it becomes austenite and is then cooled *slowly*, it becomes ferrite, pearlite, or cementite. If the steel is heated above its upper transformation temperature and is quenched *rapidly*, it becomes martensite.

The iron-carbon phase diagram does not differentiate between slow and rapid cooling rates in terms of time, so there is no way to know the cooling process or the *type* of steel structure that is produced. If steel is cooled from a high temperature to room temperature in six seconds, is it considered *rapid* cooling or *slow* cooling?

If a material is not cooled *very* rapidly, but is not cooled slowly, what percentage of martensite occurs and what percentage of ferrite, pearlite, or cementite occurs? The iron-carbon phase diagram cannot be used to determine the resulting steel products.

The isothermal transformation (I-T) diagram provides this information. An *I-T diagram* identifies the different steel structures that are transformed from austenite over a period of cooling time. This diagram is a very useful tool and is the focus of this chapter.

Basic Elements of the I-T Diagram

An I-T diagram shows the rate of cooling for a type of steel when the steel is heated to its transformation temperature and cooled at isothermal (constant temperature) conditions. The diagram consists of a graph of temperature versus time when steel is cooled and transformed from austenite. From this diagram, the final structure of the steel can be predicted and identified.

I-T diagrams are commonly called *time-temperature-transformation (T-T-T) diagrams.* They are also called *C curves* or *S curves* because of the shapes of the curves.

The basic structure of an I-T diagram is shown in Figure 13-1. Note that temperature is plotted against time. Temperature is measured in degrees Celsius (°C) and degrees Fahrenheit (°F) and are graphed along the y-axis. The two horizontal lines near the top of the diagram represent the *upper transformation temperature* and the *lower transformation temperature.* To be transformed completely to austenite, steel must be heated above both of these temperature lines.

The cooling time is measured in seconds and is graphed along the x-axis. Cooling time is measured in *logarithmic* values. This means that time is measured in increasingly larger (or exponential) intervals along the x-axis. The time interval that is represented by the space between 1 and 10 seconds, for example, designates much less time than the space between 10 and 100 seconds. This reflects the disproportionate amount of transformation that occurs in steel during the earliest stages of quenching.

In quenching, most of the transformation from austenite takes place in the first ten seconds. Therefore, the amount of time represented by the first two intervals on the graph is much smaller than the incremental times farther to the right on the x-axis. After the first few seconds of quenching, less transformation will take place.

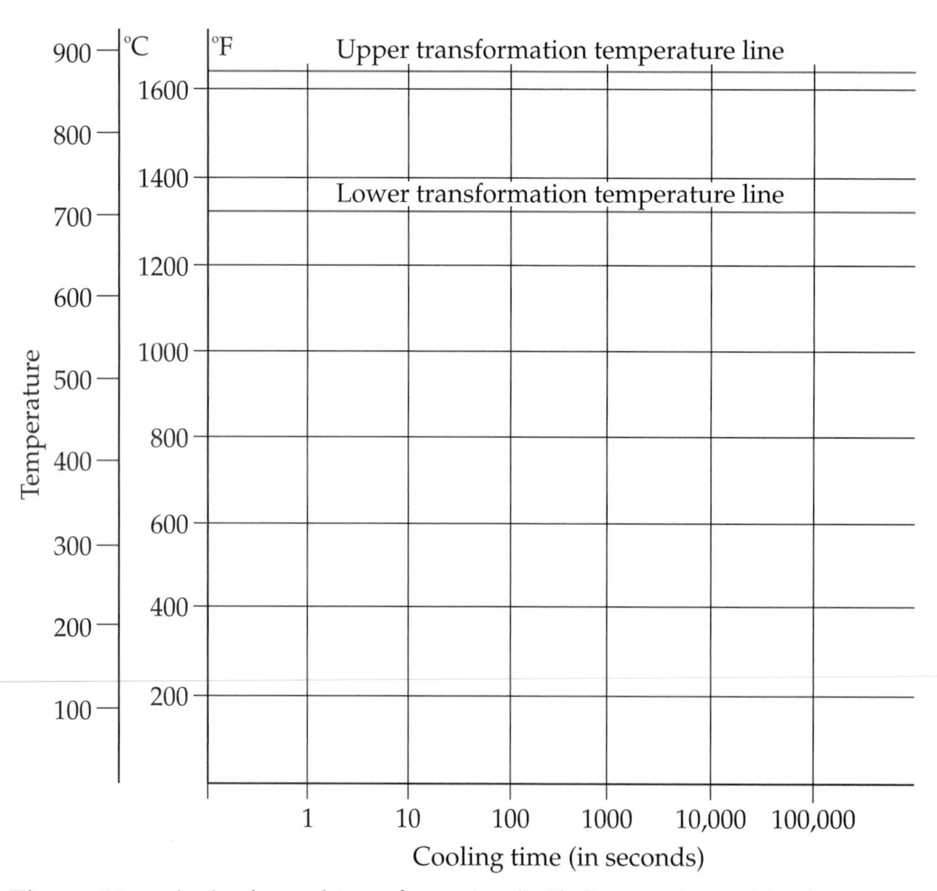

Figure 13-1. *An isothermal transformation (I-T) diagram is used to plot temperature versus time.*

Temperature-Time Line

A *temperature-time line* (also called a *timeline*), shown on the I-T diagram in Figure 13-2, follows the path of the temperature of the steel after quenching begins. This line shows the progress of each quenching stage. First, the steel is heated above the upper transformation temperature to 1700°F (927°C) at point A. It is then quenched. Quenching from point A to 700°F (370°C) at point B takes 10 seconds. When the steel reaches point B, it is held at that temperature.

The steel is held at 700°F for 90 seconds, until it reaches point C. Finally, it is rapidly cooled to room temperature at point D.

A comparison of rapid quenching and slow cooling is shown in Figure 13-3. Timeline A shows steel that has been quenched rapidly. It reaches room temperature in two seconds. Timeline B shows steel that has been cooled very slowly. It takes 10,000 seconds (almost three hours) before it reaches room temperature.

By comparing the two quenching techniques, the following conclusions can be made. Steel represented by line A would be transformed into a hard, strong steel (probably martensite). Steel cooled at a much slower rate, such as the rate represented by line B, would be transformed into ferrite, pearlite, or cementite.

Limitations of the I-T Diagram

The main advantage of the I-T diagram over the iron-carbon phase diagram is the ability to show the transformation of steel over

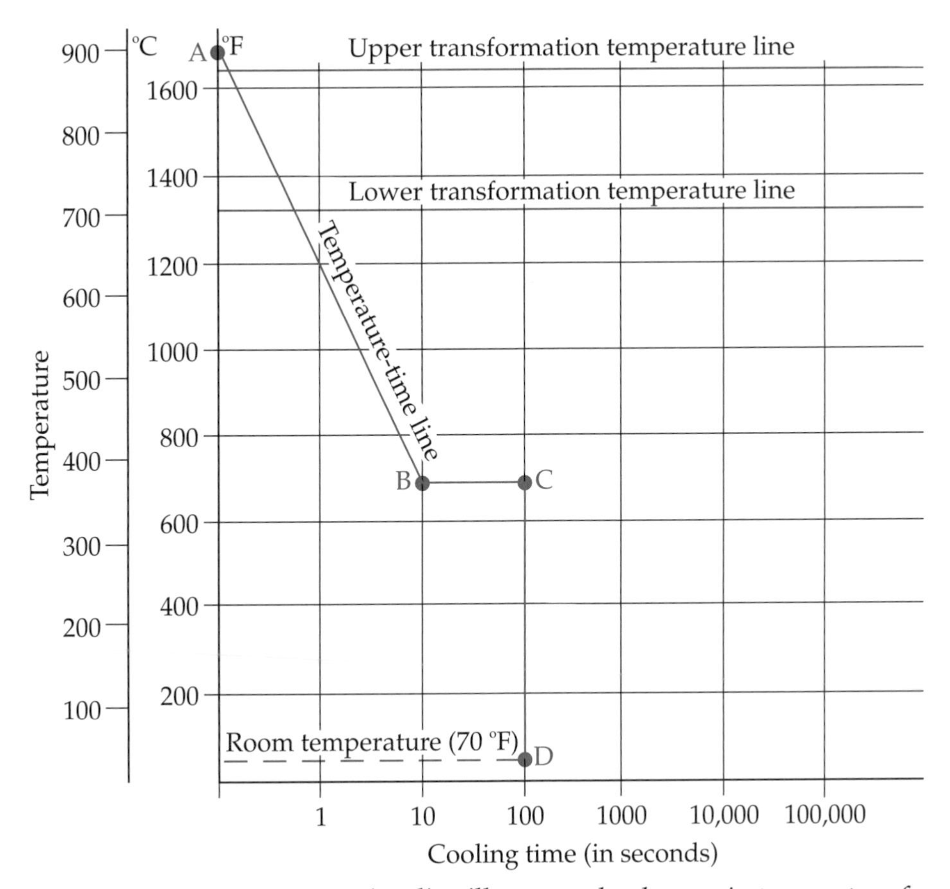

Figure 13-2. *A temperature-time line illustrates the changes in temperature for steel as it is cooled from 1700°F (927°C) to room temperature.*

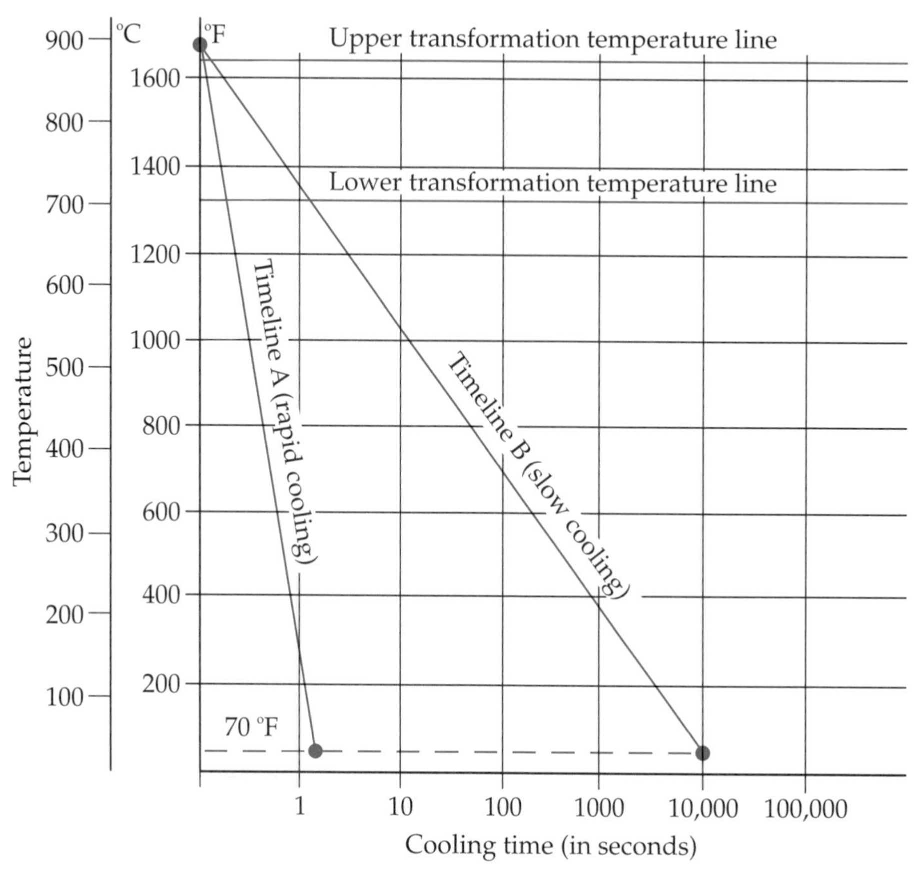

Figure 13-3. *This I-T diagram illustrates temperature-time lines for rapid quenching and slow cooling.*

time. However, there is also a limitation to its use. The I-T diagram does not plot the percentage of carbon present in the steel. For this reason, a new I-T diagram must be used every time the percentage of carbon changes. In fact, a different I-T diagram must be plotted for every type of steel.

Using I-T Diagrams

I-T diagrams are used to determine how the steel structure is effected by different cooling procedures. A simple I-T diagram is shown in Figure 13-4.

The left C curve indicates the beginning of the structural transformation from austenite. Therefore, when the temperature-time line is to the left of both C curves, the steel is 100% austenite and has not begun its transformation. When the steel cools and its temperature-time line crosses over to the right of both C curves, it has been completely transformed; it no longer contains any austenite.

The region between the left C curve and the right C curve is the *region of transformation.* This is where structural changes take place and where austenite is converted to some other structure, such as pearlite or martensite.

The dotted line between the C curves is the *50% transformation line.* When a temperature-time line reaches this line, 50% of the austenite has been transformed to another structure and only 50% austenite remains. In Figure 13-5, for example, steel at point A contains 100% austenite; steel at point B contains 50% austenite and 50% pearlite. At point C, the steel is

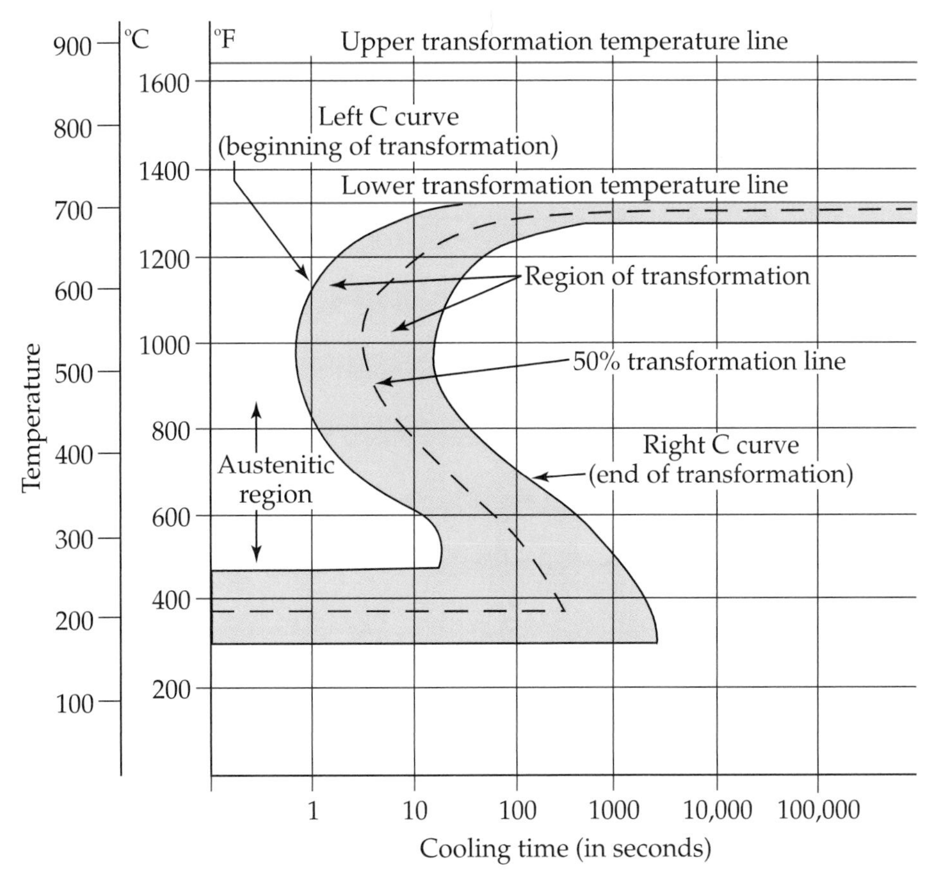

Figure 13-4. *A simplified I-T diagram. Steel passes through several transformation phases as it is cooled from its upper transformation temperature.*

completely transformed from austenite. It consists of 100% pearlite.

Regions of Isothermal Transformation

There are several regions within the I-T diagram. The austenitic region (A) is the area to the left of the region of transformation. Steel occurring in the austenitic region is 100% austenite.

There are four regions of isothermal transformation within the region of transformation. These regions identify different structures of steel as it is transformed from austenite. See Figure 13-6. The four regions are the coarse pearlite region, the fine pearlite region, the bainite region, and the martensitic region. Each is discussed below.

The *coarse pearlite region* (CP) represents coarse pearlite, or large grain pearlite. The abbreviated term "coarse pearlite" is used to describe coarse pearlite, coarse ferrite, coarse cementite, or any combination of these structures. Coarse pearlite is produced when steel is cooled extremely slowly.

The *fine pearlite region* (FP) represents steel that is transformed into fine pearlite. Other steel products that occur in this region are fine ferrite, fine cementite, or any combination of these types of steel and fine pearlite.

The *martensitic region* (M) is the lowest region of transformation on the diagram. If the temperature-time line crosses into the martensitic region, martensite will be formed.

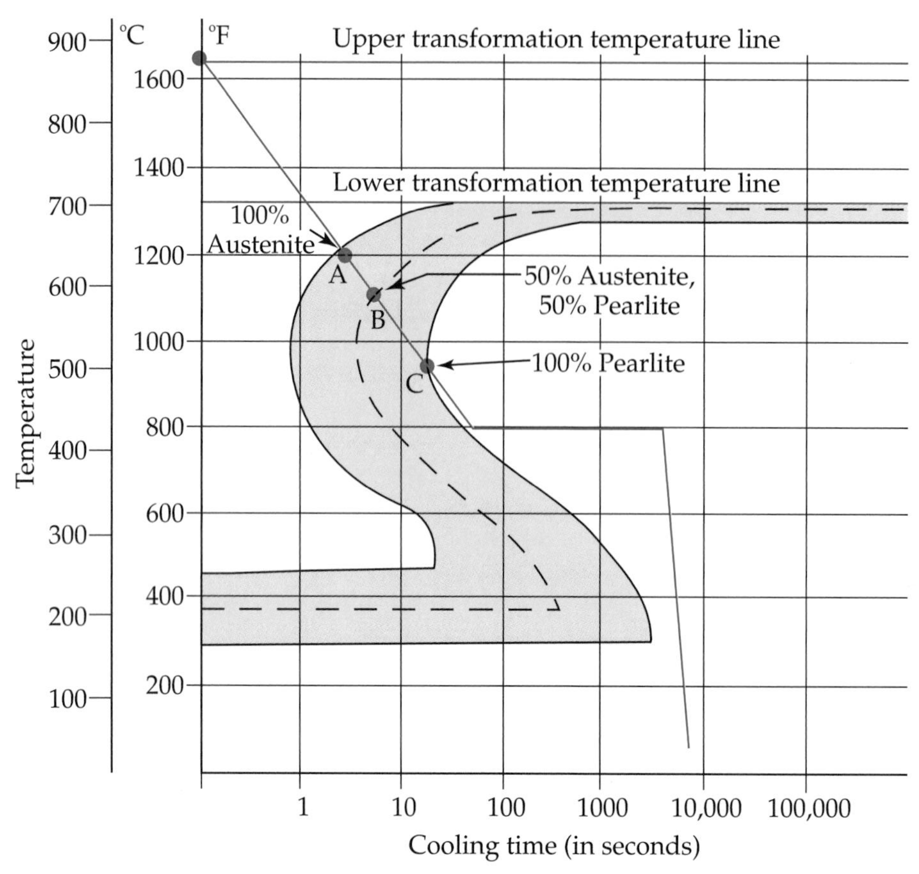

Figure 13-5. *The 50% transformation line is represented by the dotted line shown. This line indicates that 50% of the austenite has been transformed into another structure.*

This type of steel is produced when the material is quenched very rapidly.

The *bainite region* (B) is between the martensitic region and the fine pearlite region. Bainite is named after E.C. Bain, a metallurgist who conducted research on the transformation of austenite.

Bainite is superior to martensite in ductility and toughness, but it is not as hard or strong. It has less ductility than fine pearlite, but it is stronger and harder. Thus, bainite is a "happy medium" between martensite and fine pearlite.

Bainite typically reaches hardnesses of 50 R_C–55 R_C on the Rockwell C scale. It presents an advantage over other transformation products because it combines fairly high strength with fairly good ductility. The structural appearance of bainite when viewed under a microscope is shown in Figure 13-7.

Using I-T Diagrams to Identify Steel

Different quenching methods have a wide range of effects on the isothermal transformation process. The following examples illustrate how to read an I-T diagram and identify different structures of steel that are produced when quenching techniques vary.

Pearlite, bainite, and martensitic products are formed by the quenching processes that correspond to the temperature-time lines in Figure 13-8. Timeline A represents the cooling rate for a rapidly quenched steel. This steel is

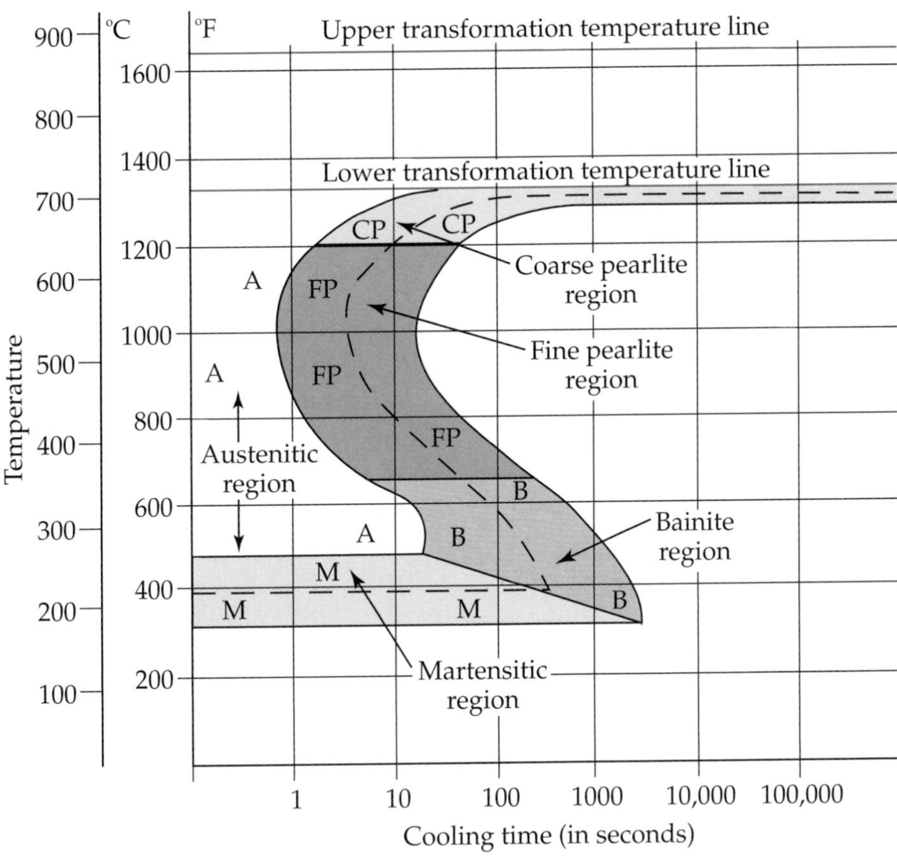

Figure 13-6. *The four regions of isothermal transformation on an I-T diagram. Steel that occurs in the austenitic region is 100% austenite.*

quenched so rapidly that the line never touches the nose of the left C curve. All transformation takes place in the martensitic region. Therefore, this steel structure is martensite.

Timeline B represents steel that is cooled very slowly. The line crosses the region of transformation between points B_1 and B_2 in the coarse pearlite region. Therefore, this material is transformed from austenite into 100% coarse pearlite. Remember, the term "coarse pearlite" can be used to describe any combination of coarse pearlite, coarse ferrite, and coarse cementite. It does not literally refer to 100% pearlite.

Timeline C crosses the region of transformation in the fine pearlite region. Therefore, the steel sample is transformed into fine pearlite.

The steel represented by Line D is transformed into bainite because the line crosses the region of transformation in the bainite region.

A transformation to fine pearlite and bainite occurs in the I-T diagram shown in Figure 13-9. Timeline E crosses two different regions of transformation as the steel is transformed from austenite; it crosses both the fine pearlite and bainite regions. At point E_2, 50% of the austenite has been transformed into fine pearlite.

As the cooling process continues, the line moves farther away from the 50% transformation line. Once the line reaches point E_3, it starts to move closer to the 50% transformation line again. The remaining austenite transforms into bainite between points E_4 and E_5. Thus,

Figure 13-7. *In this photomicrograph of 1045 steel, the bainite areas appear darker than the martensitic areas.*

after it is cooled, this steel will consist of 50% fine pearlite and 50% bainite.

A more complex cooling process is shown in Figure 13-10. When Line F reaches point F_1, the steel is still 100% austenite. When the timeline reaches point F_2, 25% of the material has transformed into fine pearlite. The line then moves away from the 50% transformation line toward the left C curve, so no further transformation takes place between points F_2 and F_3.

When the timeline reaches point F_4 at 25% transformation again, another transformation takes place. Now, since the line has crossed into the bainite region, another 25% of the austenite is transformed into bainite between points F_4 and F_5. At this point, the steel is cooled rapidly. No further transformation takes place until the line again reaches the 50% transformation line at point F_6. Finally,

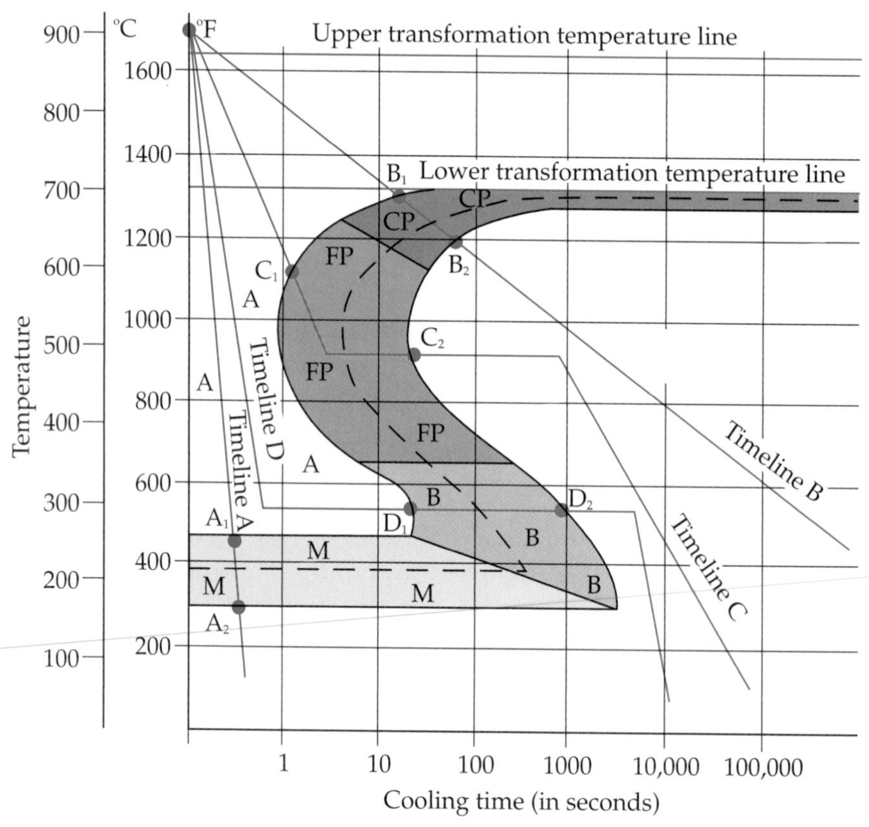

Figure 13-8. *These temperature-time lines show the transformations that occur for various cooling processes.*

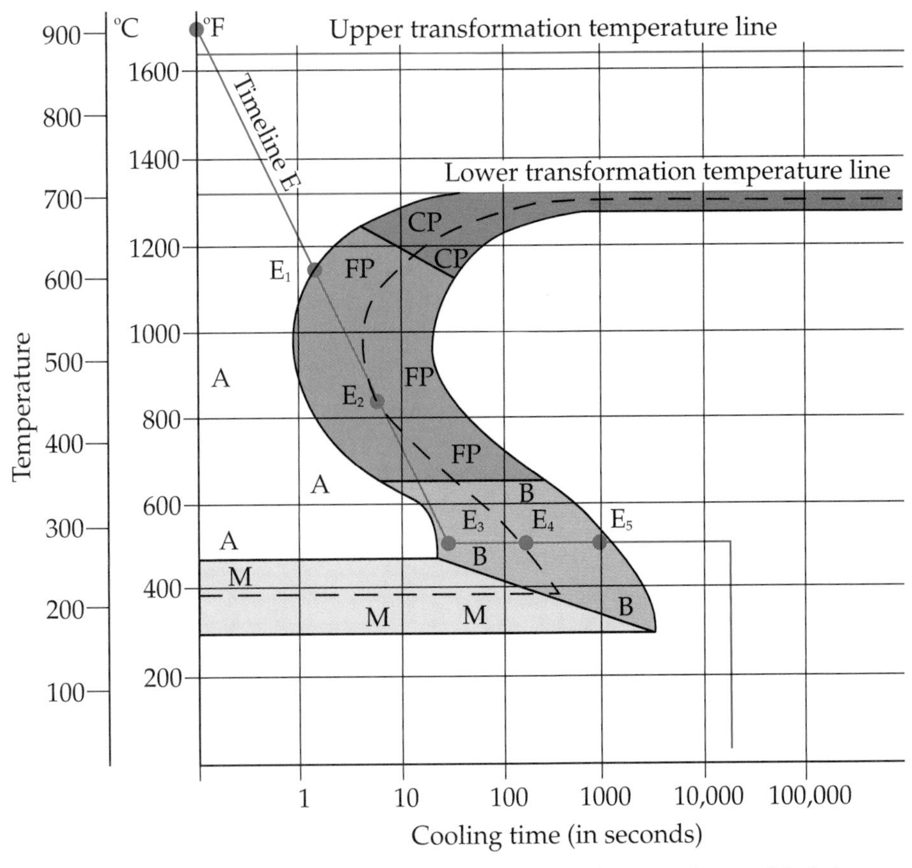

Figure 13-9. *This I-T diagram indicates a transformation to fine pearlite and bainite.*

the last 50% of austenite is transformed into martensite as the line crosses into the martensitic region.

Thus, this steel consists of 25% fine pearlite, 25% bainite, and 50% martensite.

Industrial I-T Diagrams

Industrial I-T diagrams are similar to the basic I-T diagrams previously discussed in this chapter, but they provide more detailed information about steel. See Figure 13-11. The same regions of transformation are used to identify the transformation from austenite into different structures.

A scale of hardness values for different steel products is included on the far right side of industrial I-T diagrams. Hardness values for transformed steel will be discussed later in this chapter. Some industrial I-T diagrams also have a third curve that is used to distinguish between coarse pearlite and coarse ferrite. See Figure 13-12.

Comparing Industrial I-T Diagrams

As is the case with other I-T diagrams, different industrial I-T diagrams are used for different types of steel. Referring to Figures 13-11 and 13-12, the diagram for 1095 steel shows that the nose of the left C curve runs farther into the austenite region and comes much closer to the y-axis than the nose in the diagram for 4140 steel. This means that 1095 steel must be quenched much more rapidly in order to obtain martensite. The 4140 steel, which has more alloys, does not have to be quenched as rapidly to remain outside the upper regions of transformation and become martensite.

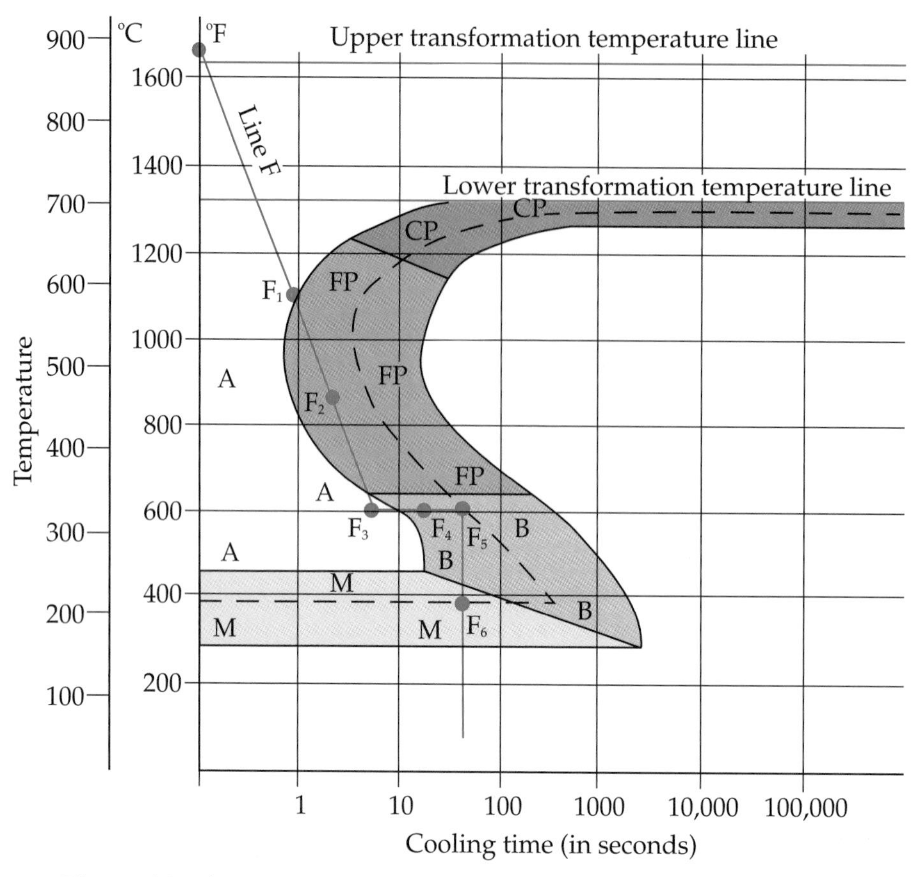

Figure 13-10. *The steel in this I-T diagram is transformed into fine pearlite, bainite, and martensite.*

When 1060 steel is quenched, Figure 13-13, there is no way to obtain 100% martensite. No space exists between the nose of the curve and the y-axis of the diagram. Thus, no matter how fast the quenching speed is, 100% martensite cannot be obtained with 1060 steel.

The shape of the C curve for 9261 steel, shown in Figure 13-14, varies considerably from the curves for the other types of steel. The region of transformation covers a much wider area on the diagram. This illustrates how the use of alloys has different effects on the transformation of steel.

Industrial I-T diagrams can be used to help in the selection of steels for applications. If it is important to obtain martensite, for example, it would be best to use a type of steel whose I-T diagram has a left C curve located far to the right of the y-axis. Referring to Figures 13-11 and 13-12, it is easier to harden 4140 steel than 1095 steel because 4140 steel does not have to be quenched as rapidly. The trick then, in hardening is either to quench the steel very rapidly to miss the nose of the curve, or else use a more expensive, higher alloy steel whose nose has plenty of "breathing room" between it and the left border.

Determining Hardness Values

Hardness values for different structures of steel can be determined by using the scale provided on the right side of an industrial I-T diagram. Hardness is commonly measured

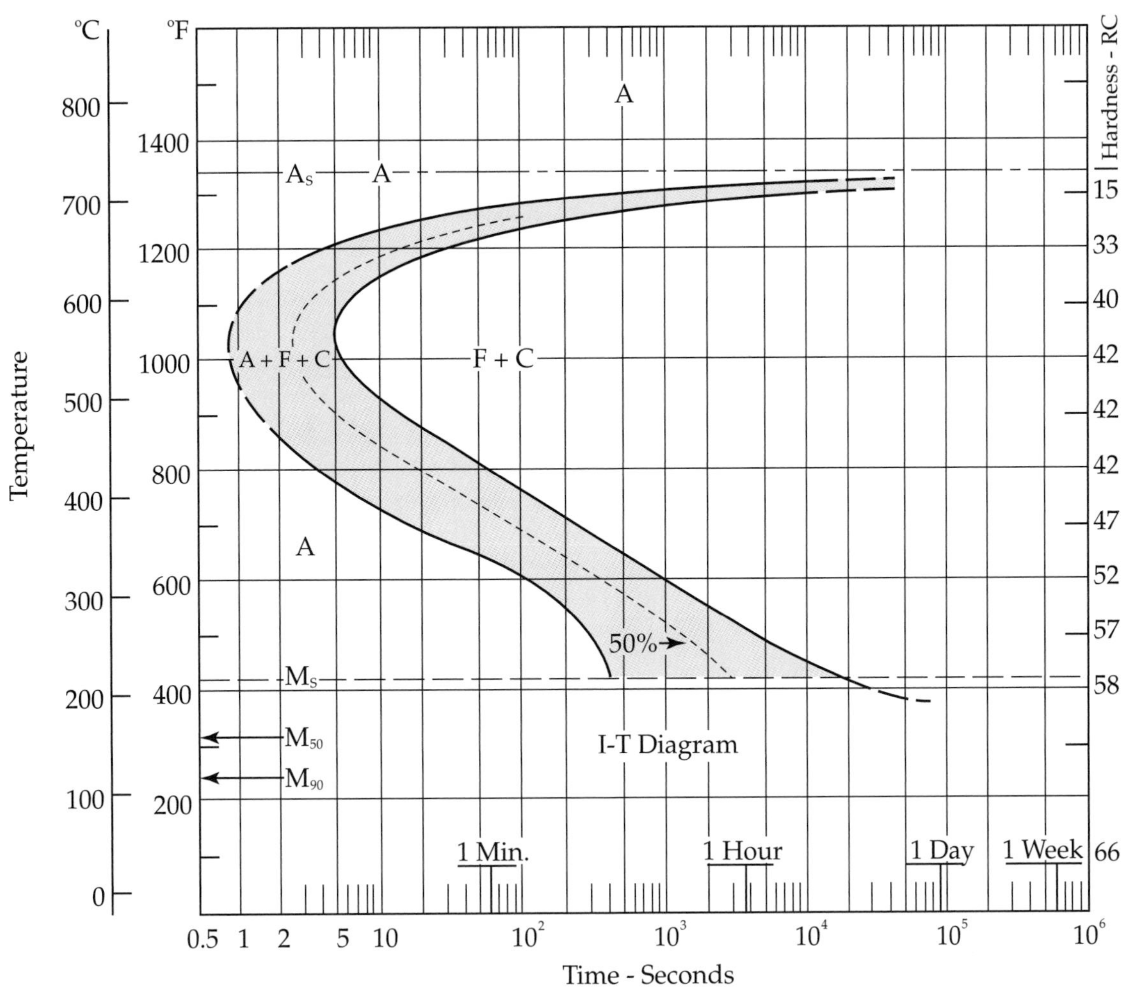

Figure 13-11. *An industrial I-T diagram for 1095 steel.*

with the Rockwell C scale, Figure 13-15. Values are read at horizontal points on the diagram where the temperature-time line crosses the transformation curves as it is quenched.

An exact hardness value generally cannot be determined, but a range can be established by identifying the minimum and maximum hardness values. The minimum possible hardness of a structure corresponds to the hardness at the point where the temperature-time line first crosses the *left* C curve on the diagram. The maximum possible hardness corresponds to the hardness at the point where the timeline crosses the *right* C curve.

For example, referring to Figure 13-15, the hardness of sample A is between 18 R_C and 28 R_C. The hardness of sample B is between 28 R_C and 53 R_C. Sample C has a hardness value between 32 R_C and 66 R_C. The hardness value of sample D is between 58 R_C and 66 R_C.

Plotting I-T Diagrams

Metallurgical and steel corporations that produce I-T diagrams obtain their data by conducting many series of quenching tests on steel. In order to plot an I-T diagram, many

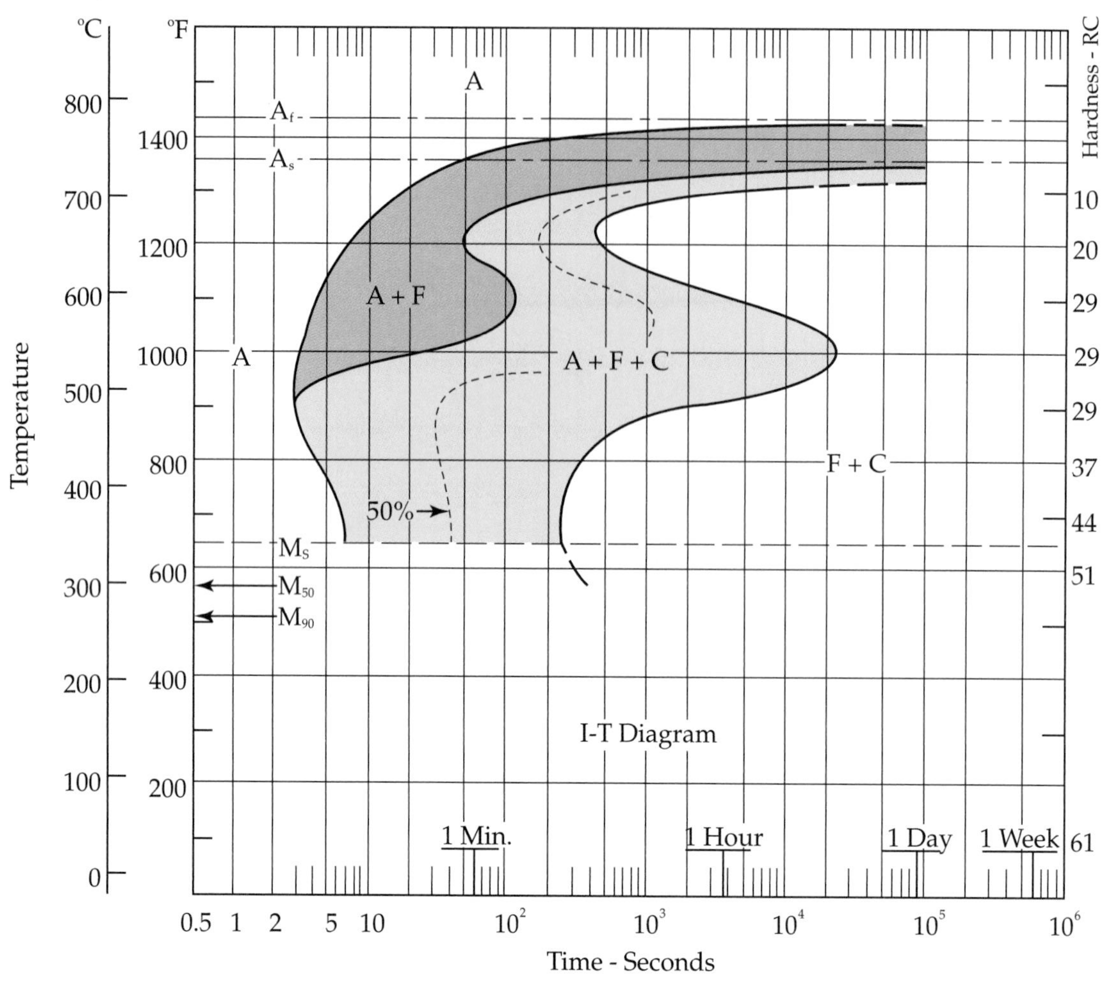

Figure 13-12. *This industrial I-T diagram for 4140 steel uses a third curve to distinguish coarse pearlite products from coarse ferrite.*

small, identical samples (perhaps 100 or more) are heated and quenched. A typical sample is about 1" in diameter and 1/16" thick. Each sample is heated above the upper transformation temperature and quenched using a specific technique. The cooling methods differ slightly for each sample.

In order to establish isothermal conditions, samples may be held at a given temperature during quenching. The samples are plunged into hot molten lead that is maintained at a constant temperature, such as 500°F (260°C), 800°F (427°C), or 1000°F (538°C).

One by one, the samples are heated and removed from the furnace, quenched, and then tested for hardness. The data these samples generate is used to plot the actual shapes of the C curves. The data is also used to determine the hardness values for the hardness scale on the diagram.

Comparing Isothermal Transformations

The results of isothermal transformations vary widely, depending on the type of steel used and the rates at which heating and cooling are performed. When transformation

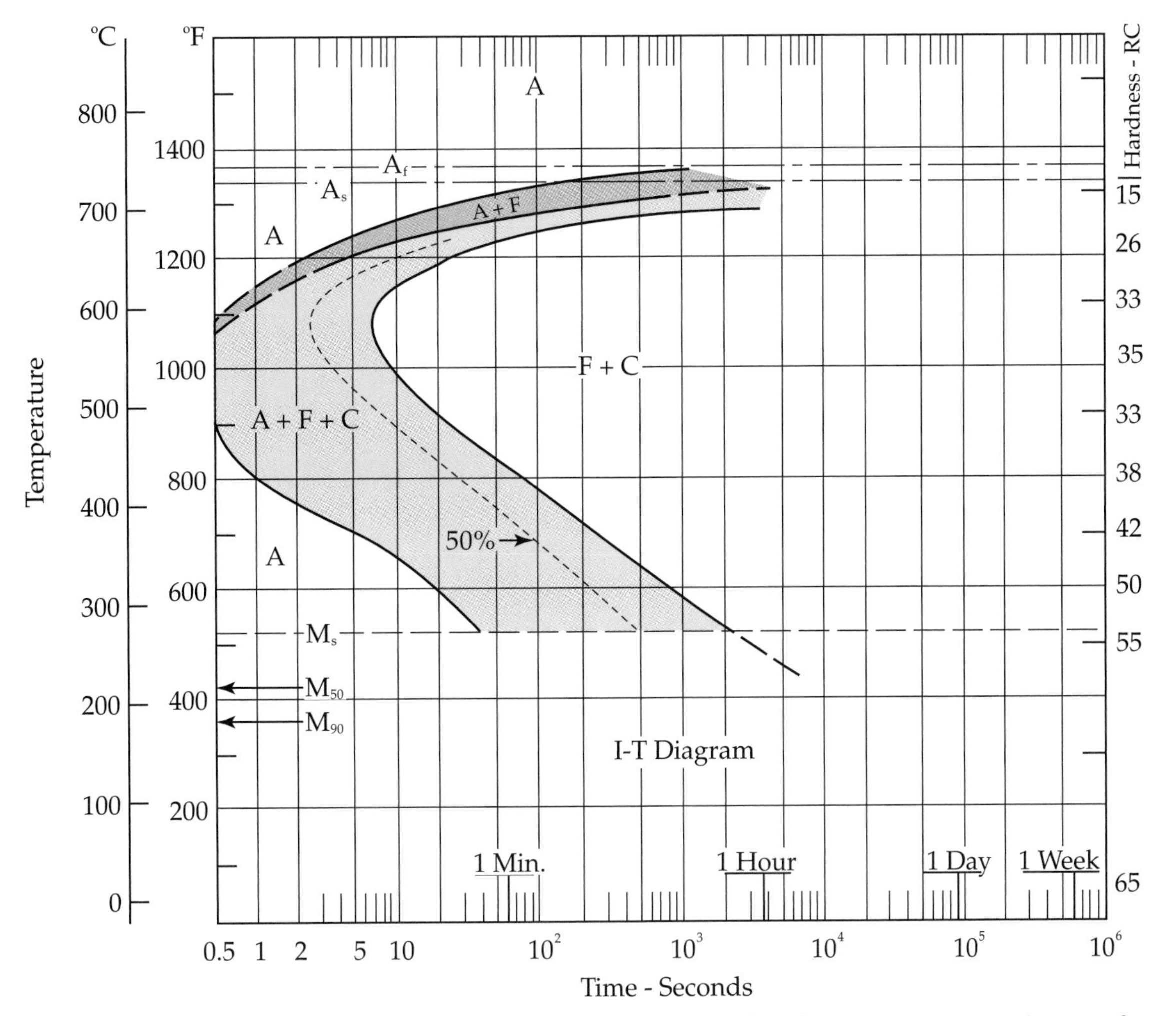

Figure 13-13. *This industrial I-T diagram for 1060 steel shows that the structure cannot be transformed into 100% martensite.*

curves are plotted on I-T diagrams, the structure of the steel can be predicted at various points along the timeline. The following examples can be used to compare the transformations that occur when different types of steel are quenched and when time, temperature, and alloy content are varied.

1340 Steel

Assume a sample of 1340 steel is heated to 1600°F (870°C). It is then gradually cooled to 1200°F (650°C) over a 10-minute period. Finally, it is quenched to room temperature in five seconds.

Based on these conditions, the isothermal transformation for 1340 steel is plotted as shown in Figure 13-16. All transformation takes place in the coarse pearlite region. Therefore, the steel is transformed into 100% coarse pearlite. The hardness value is below 15 R_C. Note that quenching occurs immediately after the steel leaves the region of transformation.

Now assume that a second sample of 1340 steel is heated to 1600°F and cooled to 1000°F in four hours. It is then quenched rapidly to room temperature in five seconds.

A timeline plotted for these conditions is shown in Figure 13-17. Since all transformation

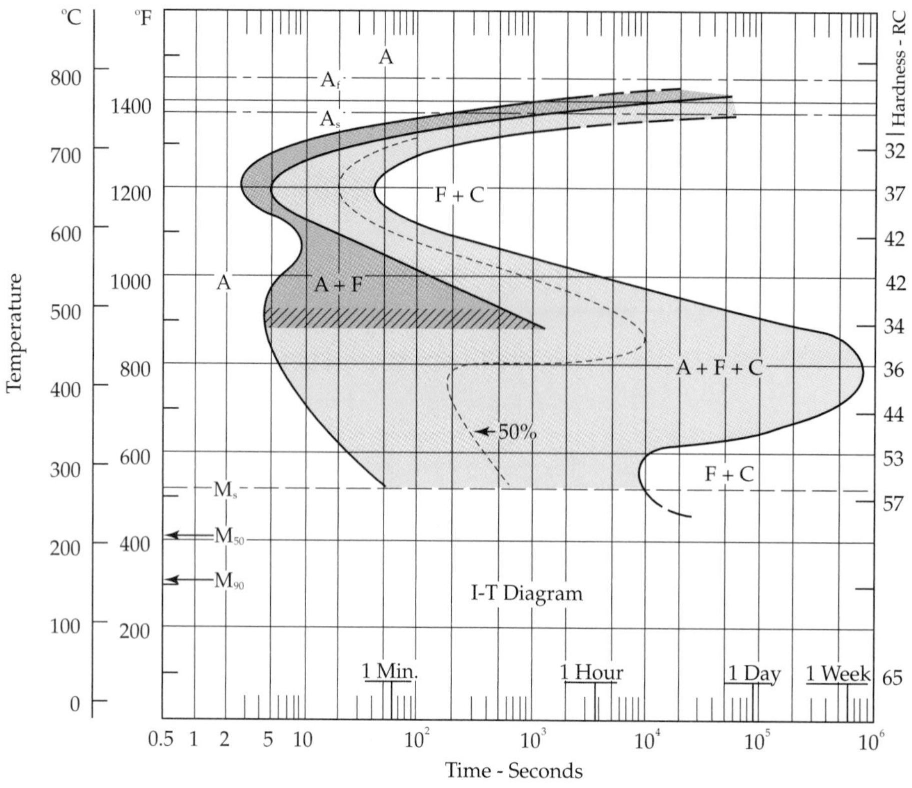

Figure 13-14. *An industrial I-T diagram for silicon 9261 steel.*

takes place in the coarse pearlite region, this structure becomes 100% coarse pearlite. The hardness value is below 15 R_C.

The examples illustrated in Figure 13-16 and Figure 13-17 use the same type of steel. The difference is the last part of the cooling cycle.

On the two diagrams, the timelines cross the region of transformation at almost the same points. Referring to Figure 13-17, once the timeline has reached point B_2, theoretically, all transformation is over. Therefore, the remaining portion of the cooling process (indicated by the two lines between points B_2 and B_3) has nothing to do with determining the final transformation. Both samples attain the same structure, even though the first sample is quenched *immediately* after transformation is completed.

9261 Steel

The following example is based on the use of silicon 9261 steel. Assume one sample of this steel is heated to 1600°F and then quenched rapidly in water in one second. The timeline is shown in Figure 13-18.

All transformation takes place in the martensitic region. Therefore, the new structure is 100% martensite. The hardness value is between 57 R_C and 65 R_C.

A second sample of silicon 9261 steel is then heated to 1600°F and cooled to 600°F (315°C) in one second. Then, it is held at 600°F for 20 seconds in a molten lead bath. Finally, it is quenched rapidly in water from 600°F to room temperature in one second.

The timeline for the second sample is plotted in Figure 13-19. All transformation

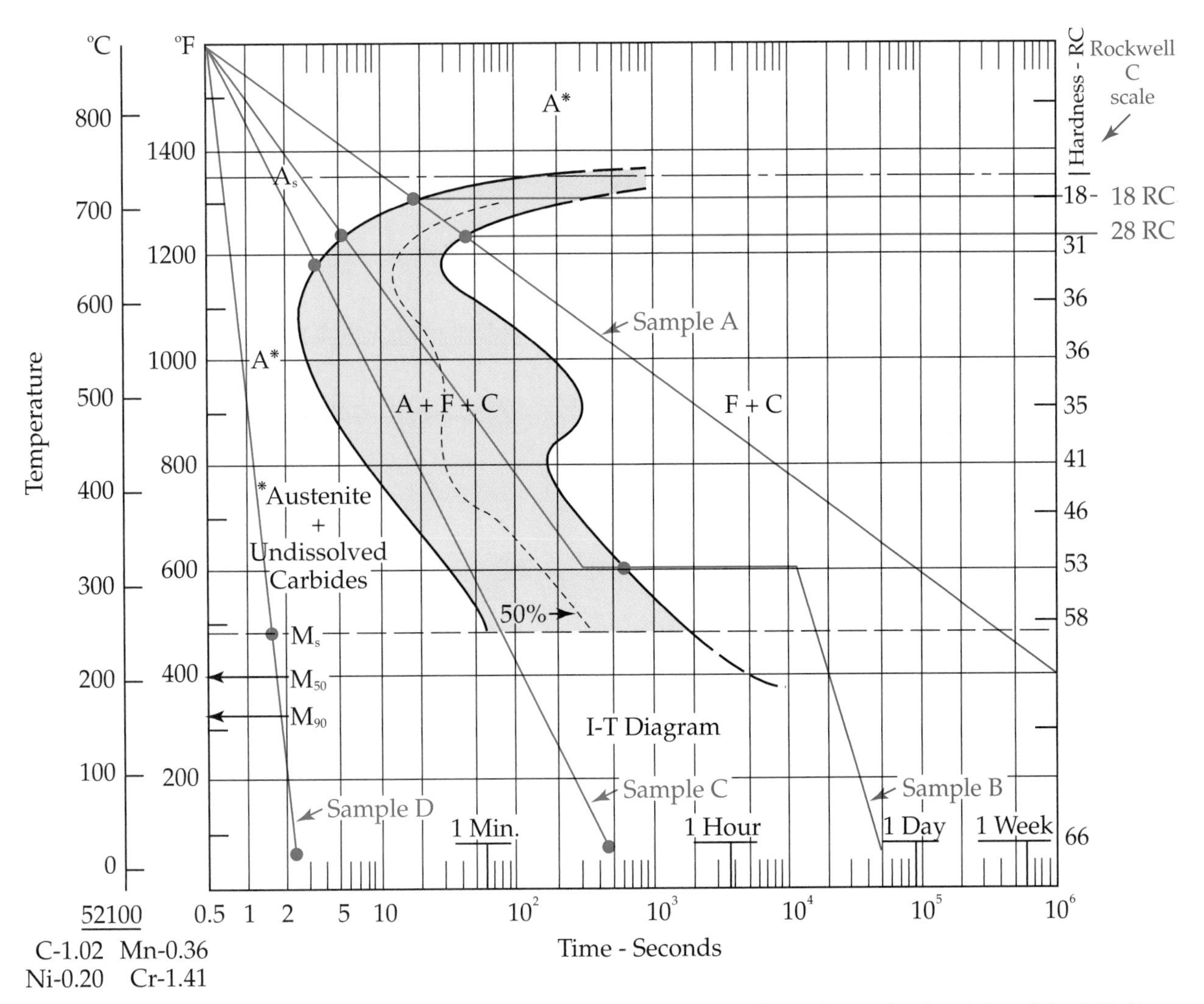

Figure 13-15. *Hardness values for steel can be measured by using the scale on the far right of the I-T diagram. This diagram is for 52100 steel.*

takes place in the martensitic region. Therefore, the second sample is also transformed into 100% martensite. The hardness value of this sample is between 57 R_C and 65 R_C.

Note that both samples are 100% martensite and both have the same hardness value, even though the second sample was given more time to cool. This slightly different cooling technique is known as *martempering*. This process is designed to relieve stress and reduce the chance of cracking and distortion in steel. Martempering is discussed in Chapter 14.

1095 Steel

Assume a sample of 1095 steel is heated to 1600°F. It is then cooled to 600°F in two seconds. Then, it is held at 600°F for one hour. Finally, it is quenched rapidly in water in one second until it reaches room temperature.

The I-T diagram for this example is shown in Figure 13-20. In this transformation, the steel slightly crosses the nose of the left C curve. As a result, approximately

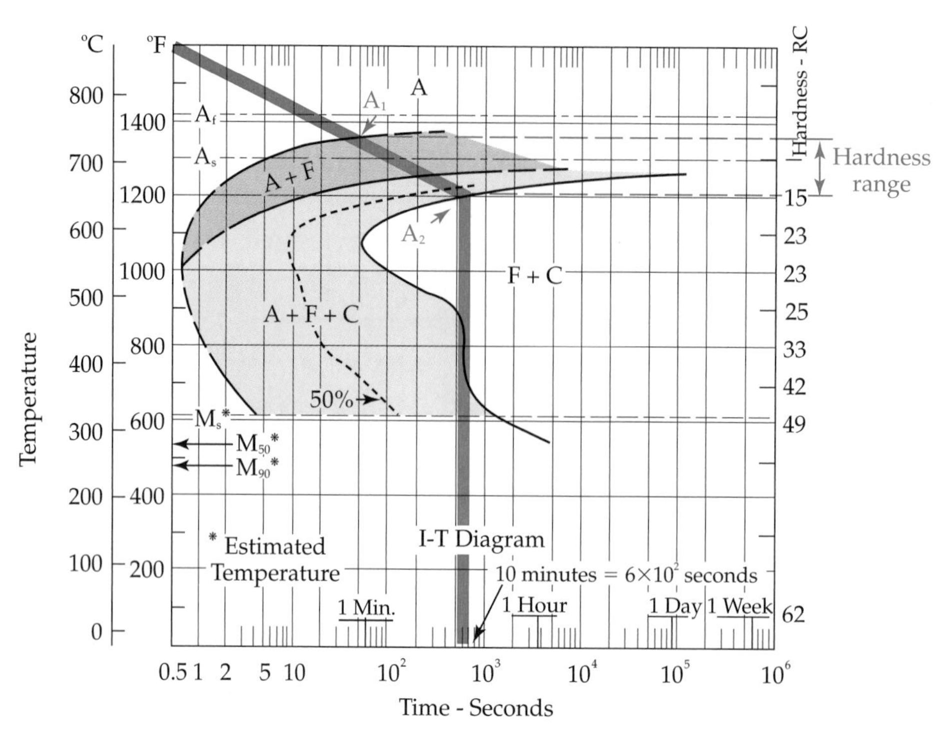

Figure 13-16. *An I-T diagram for 1340 steel. The steel is transformed into coarse pearlite during the quenching process.*

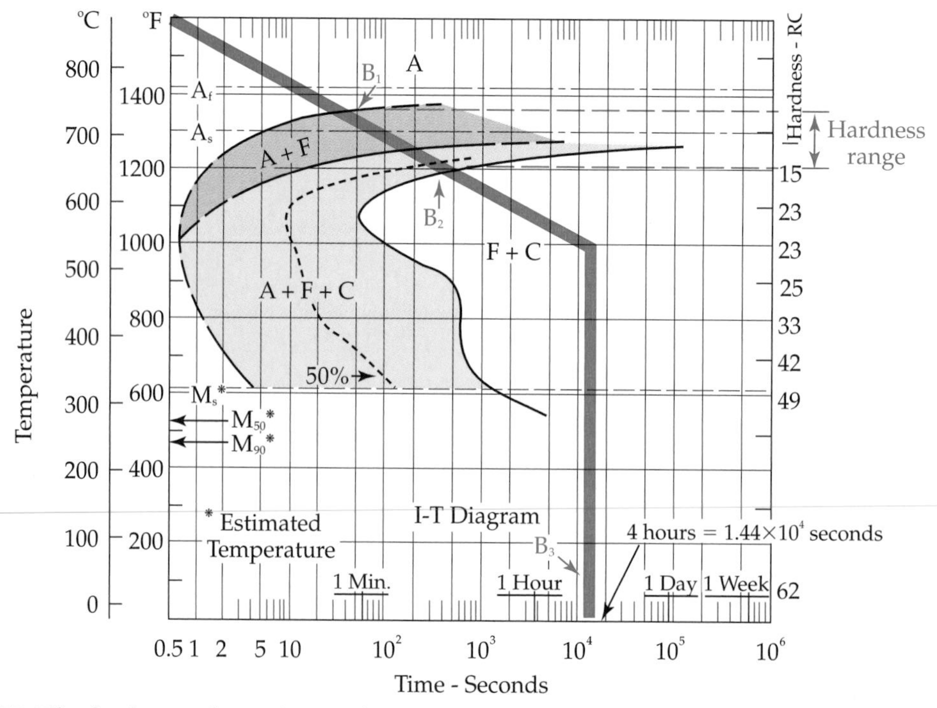

Figure 13-17. *The final transformation in this I-T diagram is 100% coarse pearlite. The structure of the steel has been completely transformed once it reaches point B_2.*

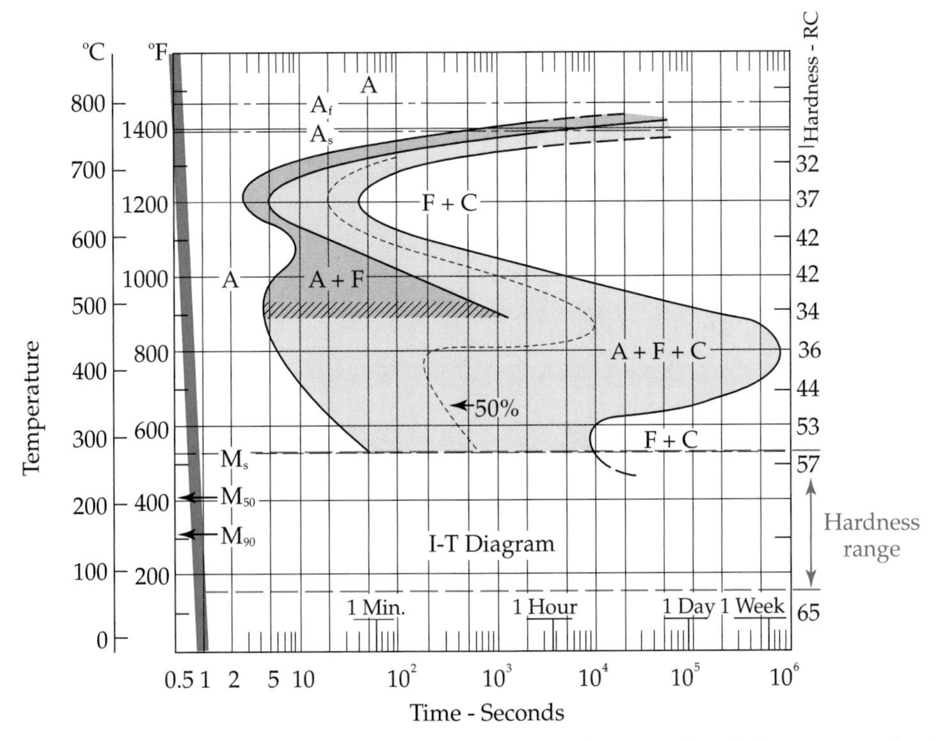

Figure 13-18. *The very rapid quenching shown in this 9261 silicon steel I-T diagram results in 100% martensite.*

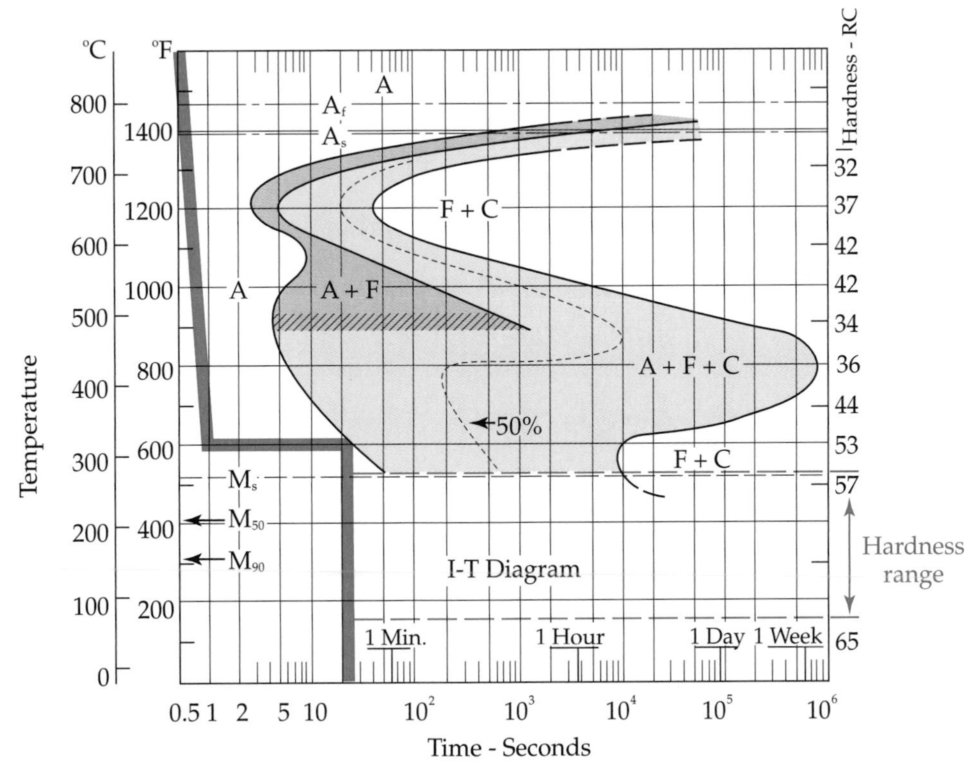

Figure 13-19. *Despite a longer cooling process for 9261 silicon steel, all transformation takes place in the martensitic region and the final product is 100% martensite.*

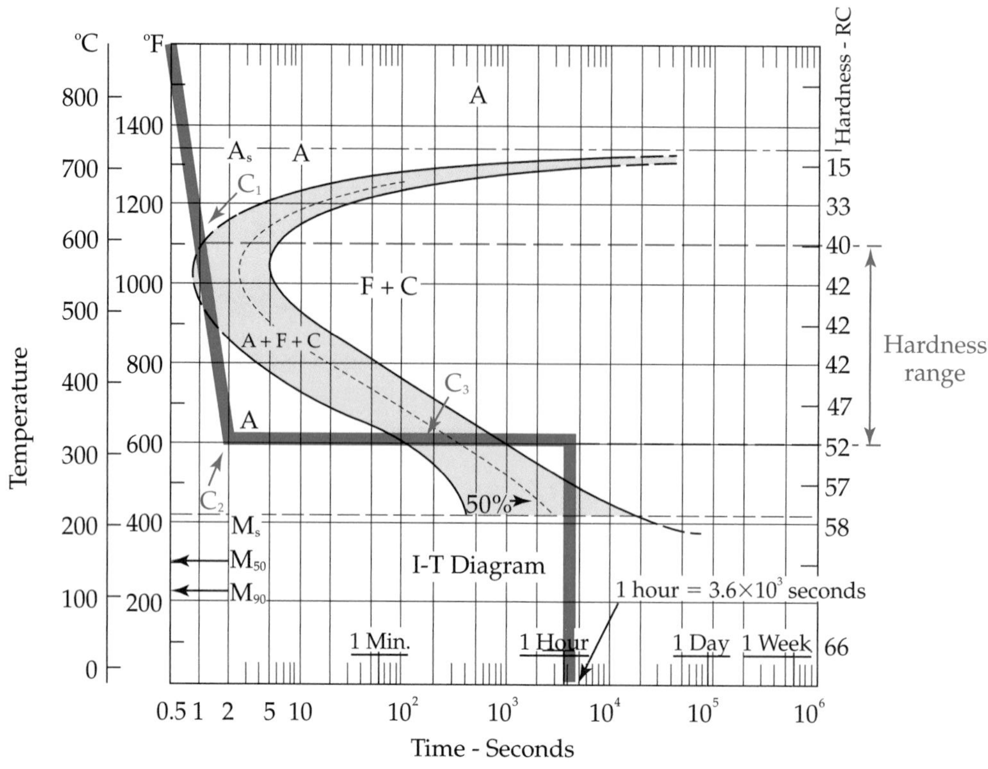

Figure 13-20. *An I-T diagram for a sample of 1095 steel. The timeline crosses into two different regions of transformation as it is quenched, resulting in a product consisting of fine pearlite and bainite.*

10% fine pearlite is formed. Between points C_1 and C_2, the timeline leaves the region of transformation and reenters the austenitic region. Despite this occurrence, the fine pearlite will not transform back to austenite. Further transformation does not occur until the steel reaches point C_3 (at 10% transformation). Then, the remaining austenite transforms to bainite.

Thus, this sample is transformed into 10% fine pearlite and 90% bainite. The hardness value of this sample is between 40 R_C and 52 R_C.

Assume a second sample of 1095 steel is heated to 1600°F and quenched rapidly to 1000°F in one second. It is held at 1000°F for two more seconds and then quenched to room temperature in one additional second.

The timeline for this sample is shown in Figure 13-21. Transformation takes place in both the fine pearlite region and the martensitic region. In the fine pearlite region, 50% of the austenite is transformed into fine pearlite as the steel reaches point D_2. As the timeline moves from point D_2 to point D_3, no additional transformation takes place because the timeline is essentially moving back toward the left C curve and the austenitic region.

Further transformation occurs at point D_3, when the timeline again reaches 50% transformation. Thus, the final structure is 50% fine pearlite and 50% martensite. The hardness value is between 41 R_C and 66 R_C.

1566 Steel

This example is based on the transformation of 1566 steel. Assume a sample of this steel is heated to 1600°F and cooled to 950°F

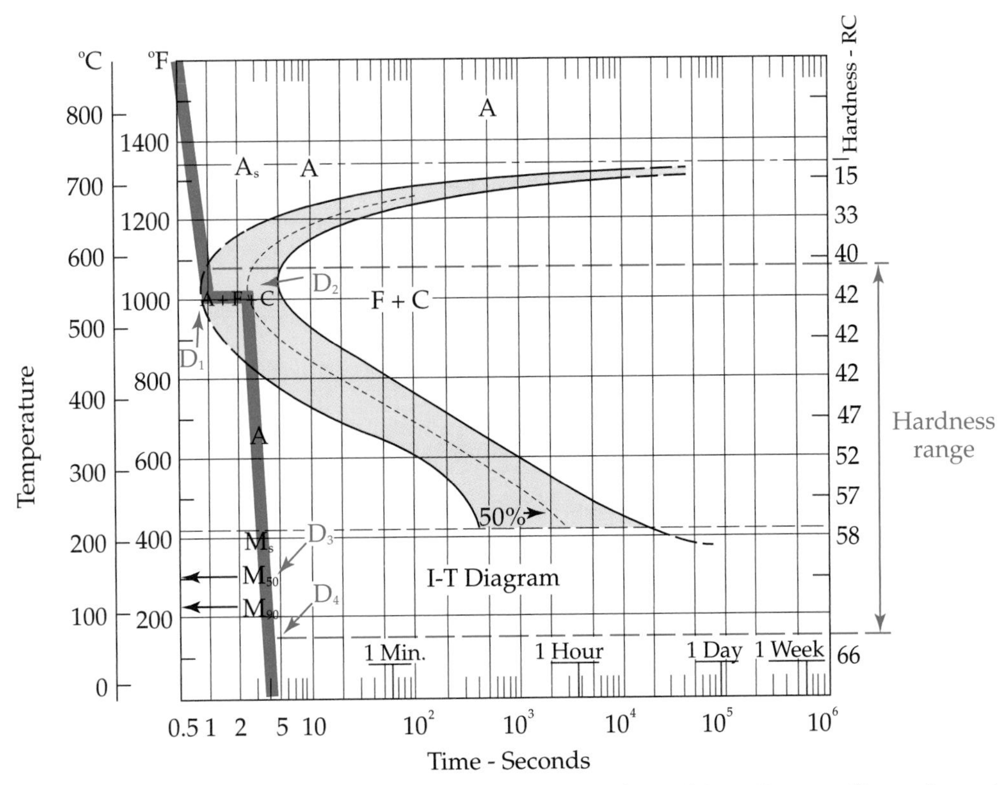

Figure 13-21. *In this I-T diagram, a sample of 1095 steel is transformed into fine pearlite and martensite.*

(510°C) in eight seconds. It is then cooled to 600°F over the next 12 seconds (for a total of 20 seconds). It is held at 600°F for seven minutes and then quenched rapidly in water to room temperature.

The timeline for this sample is shown in Figure 13-22. Transformation will take place between points E_1 and E_2 in the fine pearlite region, between points E_3 and E_4 in the bainite region, and finally between points E_5 and E_6 in the martensitic region. The final structure is 50% fine pearlite and about 20% bainite and 30% martensite.

The 50% fine pearlite is not transformed back to austenite as the timeline moves toward the austenitic region after it reaches point E_2. Note that the martensitic transformation does not begin until the timeline reaches point E_5, since 70% of the material has already been transformed.

Test Your Knowledge

Write your answers on a separate sheet of paper. Do not write in this book.

1. What is the purpose of an isothermal transformation diagram?
2. What are two other names for an I-T diagram?
3. What valuable piece of information does the iron-carbon phase diagram lack?
4. In addition to coarse pearlite, what other types of structures may form in the coarse pearlite region?
5. On any I-T diagram, what steel structure is present before its temperature-time line crosses either C curve?
6. What are the four regions in the region of transformation?
7. Describe what occurs when a sample of steel is quenched and it reaches the 50% transformation line on an I-T diagram.

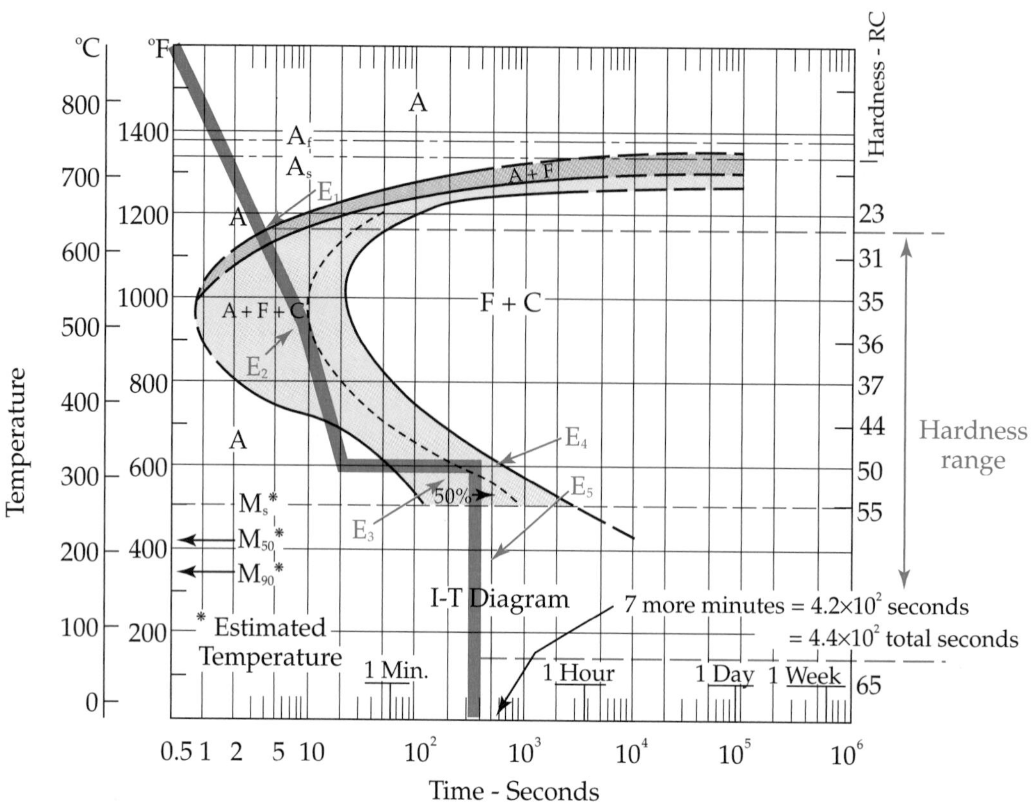

Figure 13-22. *An I-T diagram for a sample of 1566 steel. This sample is transformed into a final product of fine pearlite, bainite, and martensite.*

8. Which transformation product has more ductility, bainite or martensite? Which has greater hardness?

For the following questions, refer to the appropriate industrial I-T diagram provided in the text. Determine the structure and hardness values that would result from each quenching cycle.

9. A sample of 4140 steel is heated to 1600°F (refer to Figure 13-12). It is cooled to 1000°F in two seconds. Then, it is held at 1000°F for 20 minutes. Finally, it is cooled to room temperature in two seconds. What is the resulting structure of steel?
   a. 100% martensite.
   b. 100% bainite.
   c. 100% fine pearlite.
   d. 100% coarse pearlite.
10. What is the range of hardness values for the sample of 4140 steel in the previous question?
11. A sample of 1095 steel is heated to 1500°F (refer to Figure 13-11). It is cooled to 500°F in one second. Then, it is held at 500°F for 10 hours. Finally, it is cooled to room temperature in one hour. What is the final structure of steel?
   a. 100% martensite.
   b. 100% bainite.
   c. 100% fine pearlite.
   d. 100% coarse pearlite.
12. What is the range of hardness values for the sample of 1095 steel in the previous question?

13. A sample of 52100 steel is heated to 1500°F (refer to Figure 13-15). It is cooled to 500°F in one second. It is then held at 500°F for 20 seconds. Finally, it is cooled to room temperature in two seconds. What is the final structure of steel?
    a. 100% martensite.
    b. 100% bainite.
    c. 100% fine pearlite.
    d. 100% coarse pearlite.
14. What is the range of hardness values for the sample of 52100 steel in the previous question?
15. A sample of 52100 steel is heated to 1500°F (refer to Figure 13-15). The sample is cooled to 1300°F in two seconds. Then, it is held at 1300°F for 60 seconds. Finally, it is cooled to room temperature in two seconds. What is the final structure of steel?
    a. 100% bainite.
    b. 100% fine pearlite.
    c. 100% coarse pearlite.
    d. 100% martensite.
16. What is the range of hardness values for the sample of 52100 steel in the previous question?
17. A sample of 9261 steel is heated to 1500°F (refer to Figure 13-14). The sample is cooled to 700°F in one second. It is then held at 700°F for 4 minutes. Next, it is cooled to room temperature in two seconds. What is the final structure of steel?
    a. 50% fine pearlite and 50% coarse pearlite.
    b. 50% fine pearlite and 50% bainite.
    c. 50% bainite and 50% martensite.
    d. 100% martensite.
18. What is the range of hardness values for the sample of 9261 steel in the previous question?
19. A sample of 1095 steel is heated to 1500°F (refer to Figure 13-11). The sample is cooled to 1000°F in two seconds. It is then held at 1000°F for one second. Finally, it is cooled to 200°F in seven seconds. What is the final structure of steel?
    a. 50% coarse pearlite and 50% fine pearlite.
    b. 50% fine pearlite and 50% martensite.
    c. 50% fine pearlite and 50% bainite.
    d. 50% bainite and 50% martensite.
20. What is the range of hardness values for the sample of 1095 steel in the previous question?
21. A sample of 1060 steel is heated to 1450°F (refer to Figure 13-13). The sample is cooled to 450°F in five seconds. It is held at 450°F for 10 seconds, and is then cooled to room temperature in three hours. What is the final structure of the steel?
    a. 50% fine pearlite and 50% martensite.
    b. 30% fine pearlite and 70% martensite.
    c. 100% fine pearlite.
    d. 100% martensite.
22. What is the range of hardness values for the sample of 1060 steel in the previous question?

Using the information given below, plot I-T diagrams that show the different transformation products for each type of steel.

23. Sketch an I-T diagram for 1566 steel. Plot temperature-time lines for four different samples that result in a martensitic transformation. Compare the hardness values for each final product.
24. Sketch an I-T diagram for 52100 steel. Plot temperature-time lines for four different samples that result in a transformation to bainite. Compare the hardness values for each sample.
25. Sketch an I-T diagram and use it to compare the processing costs and time required for producing martensite, bainite and ferrite. Indicate where the most costly transformation products would occur.

Metallurgy plays an important role in the design and construction of airplanes. Relatively light metals, such as aluminum, are used for this type of application. Nonferrous metals are discussed later in the text. (The Boeing Company)

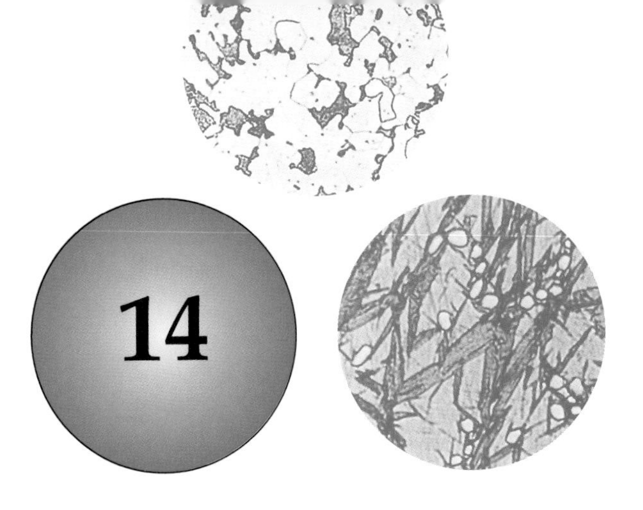

Tempering

After studying this chapter, you will be able to:

- Explain the purpose of tempering.
- Identify characteristics of tempered steel.
- Discuss some practical aspects of tempering.
- Describe common tempering methods used in industry.
- Explain how tempering affects distortion and hardness.

Introduction to Tempering

Tempering can be defined as reheating a steel to a temperature below its lower transformation temperature and cooling it at any rate to increase the ductility and toughness of the steel. After steel has been quenched, it may be hard and brittle. Also, there are internal stresses in the steel. If the steel is immediately reheated after it is quenched, stresses will be relaxed and the material will become less hard. The reheating temperature does not need to be extremely high. Quenched steel is normally given a quick "stress relieving" at a fairly low temperature, such as 800°F (430°C).

Tempering takes place only after quenching and hardening have been completed. The process is sometimes called *drawing*. This term refers to the tendency of tempering to "draw back" the hardness in steel.

Purpose of Tempering

Many people think that tempering *hardens* steel. This is not true. Tempering *softens* steel.

The purpose of tempering is to increase ductility and reduce internal stress. Tempering reduces brittleness and the likelihood of cracking and distortion. Unfortunately, tempering also reduces the hardness and strength of steel.

The amount of hardness and strength lost is not excessive. The benefits of stress relief and the elimination of brittleness generally offset problems caused by the slight loss of hardness and strength. Internal stress relief gives steel several desirable qualities. The material becomes more ductile, it gains more toughness and greater impact resistance, and it is more easily machined and cold worked.

Effects of Tempering

The effects of tempering can be summarized by comparing the process to annealing and normalizing. In tempering, the effects on the properties of the metal are less severe than the effects produced by annealing and normalizing. Tempering also results in less loss of strength. It is somewhat similar to process annealing, but takes less time.

Tempering is a quick heat-treating process that does not remove all internal stress and brittleness, but it removes enough of these to justify its use. A list of the effects of tempering is shown in Figure 14-1.

Practical Tempering

Tempering is designed to impart certain characteristics to steel that has already been heat treated or quenched. To produce the desired results, tempering should be conducted under the proper conditions, and only when the process is suitable for the steel. The following are some questions to consider before tempering steel:

- *At what temperature should tempering be done?* In tempering, metal is generally heated between 300°F and 1200°F (150°C and 650°C) in a furnace, Figure 14-2. If metal is heated to a temperature less than 300°F, essentially no change occurs in the material. By contrast, when steel is heated and it reaches a temperature above 1200°F, structural changes are ready to start taking place. This is not the purpose of tempering. Heating the metal to a high tempering temperature also takes more time than necessary.

Effects of Tempering	
Property of Steel	**Result**
Hardness	Reduced
Strength	Reduced
Toughness	Increased
Brittleness	Reduced
Ductility	Increased
Internal stress	Reduced
Distortion	Reduced
Cracking	Reduced
Machinability	Improved
Formability	Improved

Figure 14-1. *The effects of tempering on material properties of steel.*

Figure 14-2. *This tempering furnace uses molten salt and can be externally heated by gas or oil-fired burners. It uses a brick-insulated steel pot. (Ajax Electric Company)*

Materials that are heated to a temperature near 1200°F will undergo more property changes than those that are heated to temperatures in the 300°F–500°F (150°C–260°C) range. The effects of different tempering temperatures on hardness are shown in Figure 14-3. Note that the change in hardness is considerably greater for metal that is tempered at 1000°F (540°C), as compared to metal that is tempered at 400°F (205°C).

- *How soon should tempering be done after quenching?* Tempering should be done immediately after quenching. If too much time elapses between the two processes, internal stresses have time to work on the metal and cracking or distortion can occur.

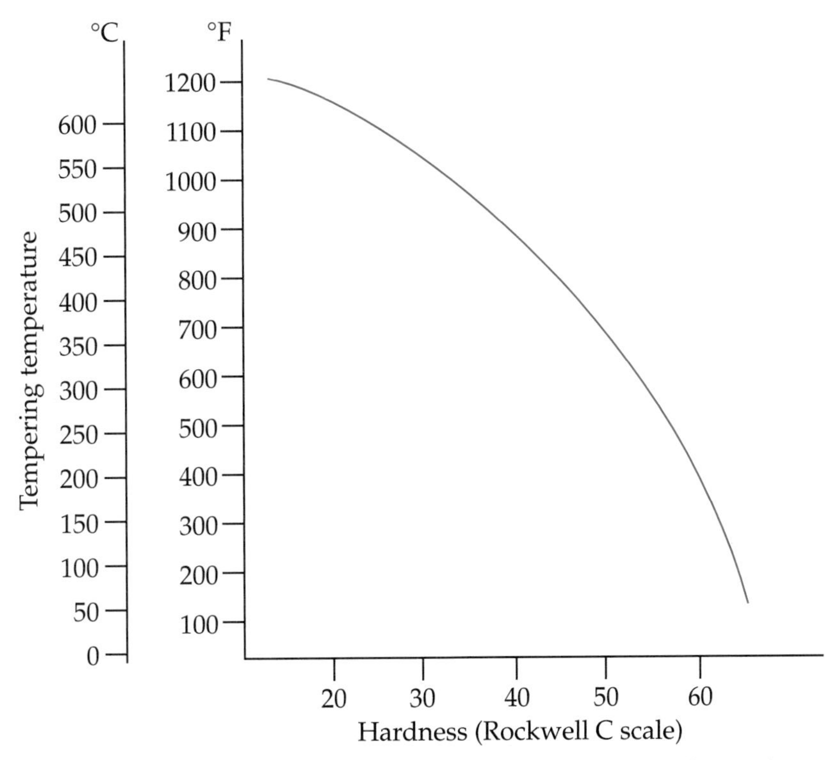

Figure 14-3. *Graph shows effects of different tempering temperatures on hardness of a metal.*

- *How long should the metal be kept at the elevated temperature during tempering?* The effects of soaking the metal at the tempering temperature for a long period of time are apparent, Figure 14-4. Note that when metal is heated to 1000°F, the greatest amount of softening takes place during the first 30 minutes of soaking. Over a longer period of time, such as five hours, tempering has less of an effect on the hardness of metal. Only when the process continues for as long as a day does tempering produce much more of an effect. Steel that is left to soak in an oven for an entire day, for example, softens by about five units on the Rockwell C scale.
- *Is tempering always desirable?* Tempering is not desirable for all applications. When hardness is important, tempering may become a problem. Tool bits, for example, can become very hot during cutting and the metal may reach a tempering temperature. If this occurs, the bit can soften slightly and lose hardness.

Undesired tempering can also occur in welding and other applications. In many types of machines, friction can produce a tempering temperature when two surfaces rub together and build up heat. In such cases, tempering is done unintentionally and can cause the metal to lose hardness and strength.

Special Types of Tempering

Different tempering methods are used in metallurgy to obtain various structures of steel. These methods reduce the chances of cracking and distortion and produce greater ductility and toughness.

There are three special tempering methods that are used to obtain special structures.

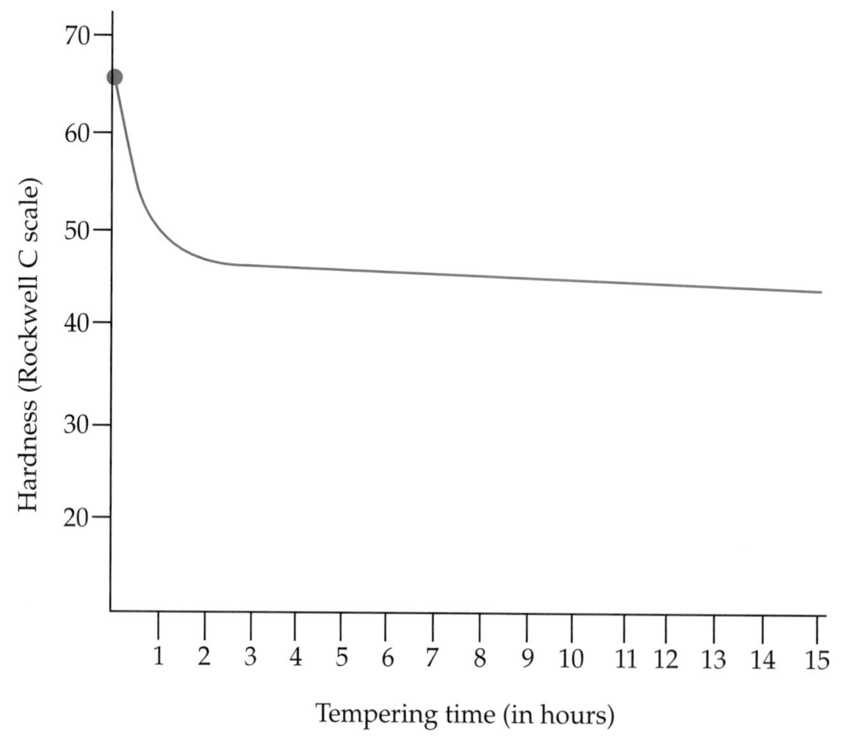

Figure 14-4. *Graph shows effects of soaking time at tempering temperature on hardness of a metal.*

These three methods are designed for applications where the chances for cracking and distortion must be minimized:

- Martempering.
- Austempering.
- Isothermal quenching and tempering.

These methods are covered in the next section of this chapter. They are related to general quenching and tempering, which is discussed next.

General Quenching and Tempering

An isothermal transformation (I-T) diagram for general quenching and tempering is shown in Figure 14-5. This type of diagram was discussed in Chapter 13. It illustrates the transformation of quenched steel into different structures.

Note that the temperature-time line representing the steel does not cross the nose of the C curve as the steel is quenched. The line crosses the martensitic transformation region and the steel becomes 100% martensite.

At point B on the diagram, quenching is completed and tempering begins. The material is quickly reheated to a tempering temperature between 500°F and 1000°F (260°C and 540°C). At point C, the steel reaches the tempering temperature. The steel is soaked at the tempering temperature for several hours and is then quenched back to room temperature.

The resulting structure is known as *tempered martensite.* It is very similar to martensite, but its hardness is slightly "drawn back" by the tempering process. Tempered martensite is slightly less hard, strong, and brittle than martensite. However, it is slightly more ductile, tough, stress-free, distortion-free, and crack-free.

Note that the C curve for the region of transformation in Figure 14-5 pertains only to

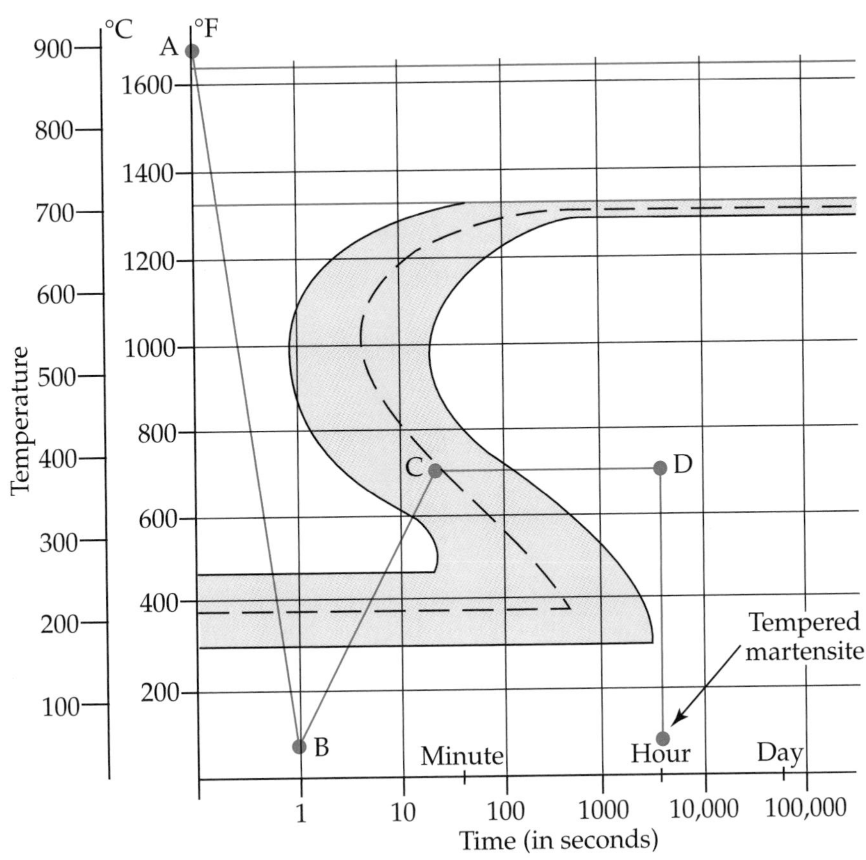

Figure 14-5. *An I-T diagram for general quenching and tempering. This method produces tempered martensite.*

the quenching process. After the material is quenched, it is not transformed into other products by tempering.

Martempering

Martempering is very similar to general tempering. The quenching process is a little more gentle, Figure 14-6. When the steel is quenched, it is cooled rapidly at first; it does not cross the nose of the C curve. However, it is held at about 500°F to 600°F (260°C to 315°C) for a short period of time. As a result, the effects of the quench are less drastic when the steel finally crosses the martensitic transformation region.

After quenching, the material is 100% martensite, but it is slightly more stress-free than the martensite produced by general quenching and tempering. Then, the quenched steel is immediately tempered. It is reheated to an intermediate temperature, held there for a period of time, and then quenched back to room temperature.

Martempering produces tempered martensite, the same type of steel that is attained by general tempering. The final product obtained by martempering, however, is more resistant to cracking and distortion. The disadvantage of martempering versus general tempering is that leveling out at 500°F is time consuming and inconvenient. Therefore, it is

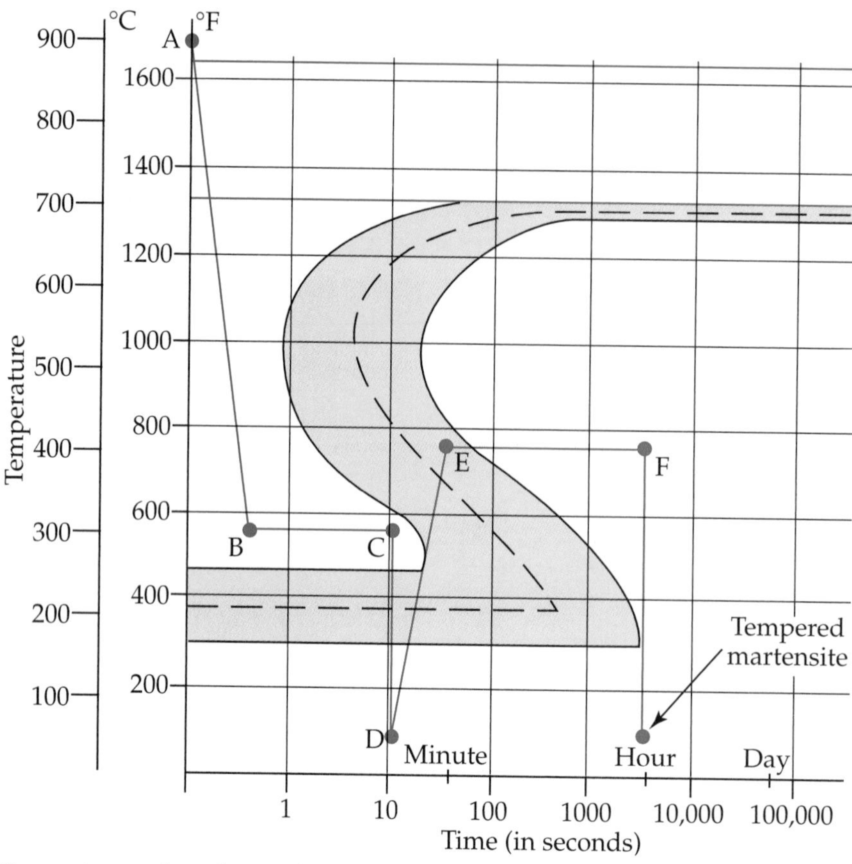

Figure 14-6. *In martempering, the steel is kept at a specific quenching temperature for a short time before it reaches the martensitic region. The resulting structure is tempered martensite.*

done only when the tendency to distort or crack is very critical.

Austempering

In *austempering,* the steel is quenched rapidly until it reaches a temperature between 500°F and 600°F. See Figure 14-7. It misses the nose of the C curve. When the material is between 500°F and 600°F, it is held (soaked) at one temperature for a long period of time. This time period is long enough for structural transformation to take place in the bainite region. After a complete bainite structure evolves, the temperature is held for a short period of time and the steel is then quenched to room temperature. Thus, martensite is never formed.

Austempering is a much more gentle heat-treating process than martempering or general tempering. The resulting structure has less internal stress and is more resistant to cracking and distortion. Austempered steel is not as hard or strong as material treated by martempering or general tempering.

Austempering is generally limited to thin parts that are susceptible to distortion. Springs, lock washers, needles, and other fine metal parts are often austempered.

Austempered metals are usually tougher and more ductile than tempered or

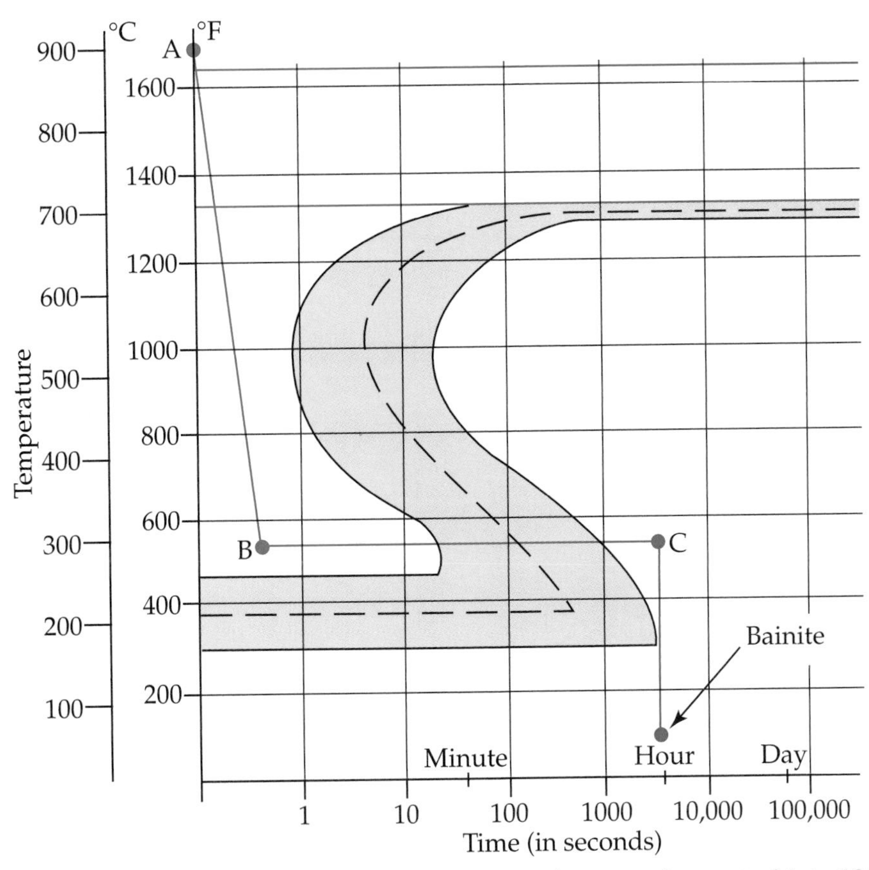

Figure 14-7. *Austempering is a special quenching process that transforms steel into 100% bainite.*

martempered steels. They have better impact resistance since they contain no martensite.

Isothermal Quenching and Tempering

Isothermal quenching and tempering is a heat-treating process that produces a structure consisting of bainite and tempered martensite.

The isothermal quenching and tempering process is illustrated in Figure 14-8. The steel is rapidly quenched until it reaches approximately the midpoint of the martensitic region. When transformation from austenite is about 50% complete, the partially transformed steel is held at constant temperature. The holding temperature is about 300°F to 400°F (150°C to 205°C). Approximately half the steel has been transformed from austenite into martensite; the other half remains austenite.

After the steel is held for a few seconds at the intermediate temperature, it is heated to a higher tempering temperature. The remaining austenite is then transformed into bainite as the steel crosses the bainite region. The steel is soaked at the tempering temperature for a period of time to remove many internal stresses. Finally, it is cooled to room temperature.

Isothermal quenching and tempering can be described as a "happy medium" between martempering and austempering. It produces material that is harder and stronger than austempered metal. It also results in material that is more ductile and stress-free than martempered metal.

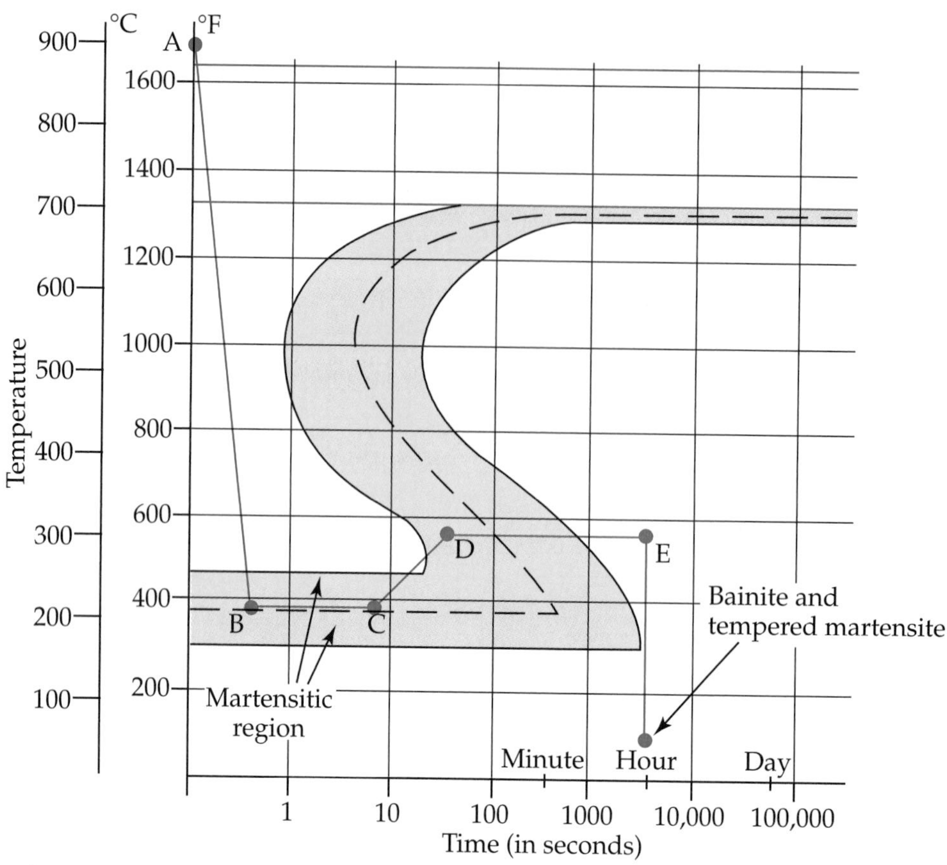

Figure 14-8. *An I-T diagram for isothermal quenching and tempering. A structure of bainite and tempered martensite is produced by this method.*

Comparing Heat-Treating Methods

Tempering uses variable heating and quenching methods to produce different properties in metal. A comparison of the characteristics associated with general tempering, martempering, isothermal quenching and tempering, and austempering is shown in Figure 14-9.

General tempering is the most dramatic quenching method. It produces metal that is higher in hardness and strength but lower in ductility and toughness than the other methods. Austempering is the least dramatic quenching method. It produces metal that is higher in ductility and toughness but lower in hardness and strength than the other methods.

A comparison of the effects produced by tempering and different types of annealing and normalizing is shown in Figure 14-10. Tempering, annealing, and normalizing are all secondary heat-treating operations. Each technique causes quenched steel to become softer.

Tempering, particularly general tempering, has the least effect on hardness; it is the most dramatic secondary heat-treating method. Full annealing (in which heating and cooling occurs in a temperature-controlled furnace) employs the most gentle cooling method and produces the softest steel.

Comparison of Tempering Methods			
General Tempering (Tempered Martensite)	Martempering (Tempered Martensite)	Isothermal Quenching and Tempering (Bainite and Tempered Martensite)	Austempering (Bainite)
Less severity in quenching →			
More ductility →			
More toughness →			
Less internal stress →			
Less hardness and strength →			
Less brittleness →			
Less cracking and distortion →			
More machinability →			
More formability →			

Figure 14-9. *A comparison of tempering methods.*

Effects of Tempering, Annealing, and Normalizing on Steel			
Tempering	Process Annealing	Normalizing	Full Annealing
Less severity in quenching →			
Less hardness →			
More ductility →			
Less internal stress →			

Figure 14-10. *Comparison of tempering, annealing, and normalizing.*

Test Your Knowledge

Write your answers on a separate sheet of paper. Do not write in this book.

1. Does tempering harden metal or soften it?
2. Define *tempering*.
3. Why is tempering also referred to as *drawing?*
4. The purpose of tempering is to _____ ductility and _____ internal stress.
5. Which of the following properties does tempering normally improve?
   a. Strength.
   b. Hardness.
   c. Ductility.
   d. Shock resistance.
   e. Toughness.
6. When is tempering not desirable?
7. What is the standard range in temperature used for tempering metal?
8. How soon after quenching should steel be tempered?
9. What final structure of steel is produced by general quenching and tempering process?
10. What is the difference between *general quenching and tempering* and *martempering?*

11. What special quenching technique produces 100% bainite?
12. What final structure of steel is produced by isothermal quenching and tempering?
13. What is the name given to the special tempering process wherein a metal is quenched to a temperature of about 500°F to 600°F, held there for a few seconds, quenched to martensite, and then tempered?
14. What is the name given to the special tempering process wherein a metal is quenched to a temperature of about 500°F to 600°F and held at that temperature until bainite is formed?
15. Which of the four quenching and tempering processes is considered to be the most drastic?
16. Plot an I-T diagram for tempering 1566 steel. Plot temperature-time lines for four different samples that would be transformed into tempered martensite. Compare the hardness values for each of the samples.
17. Plot an I-T diagram for tempering 52100 steel. Plot lines for four different samples that would be transformed into bainite. Compare the hardness values for each of the samples.
18. Plot an I-T diagram for tempering 1095 steel. Plot four lines that show different quenching rates and holding temperatures before tempering begins. Compare the results and identify the different structures that would occur for each sample.

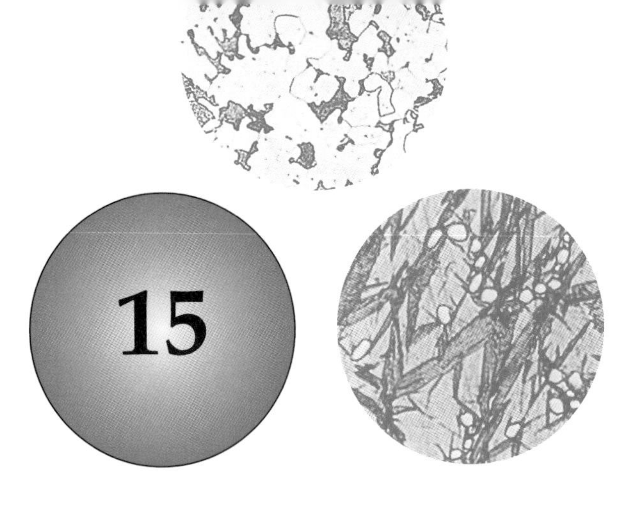

15 Surface Hardening

After studying this chapter, you will be able to:

- Explain the principles of surface hardening.
- Identify metallurgical applications for surface hardening.
- Describe the three basic surface-hardening methods.
- Identify eight different processes used to case-harden materials.
- Compare the advantages and disadvantages of each surface-hardening process.

Introduction to Surface Hardening

Surface hardening is a heat-treating process that creates a thin, hardened, wear-resistant layer on the outer surface of a material while maintaining a soft and ductile inner core. Surface hardening is commonly known as *case hardening*. Although there is a slight technical difference between the two processes, they will be treated as the same in this chapter.

A surface-hardened or case-hardened metal part can be compared to an apple, a loaf of bread, or a sealed plastic bag full of soft wheat grains. See Figure 15-1. In each of these examples, a hard, strong outer skin (or *case*) protects a soft interior. Of course, metal has a much stronger outer case and inner core than those examples. Surface hardening is commonly called "case hardening" because a hard metal case protects an unhardened core.

Surface Hardening in Steel

In steel, the hard case produced by surface hardening is usually martensite, Figure 15-2. The "skin" of the case may be as thick as 0.125" or as thin as 0.001".

The softer, interior material is typically a more ductile form of steel, such as ferrite, pearlite, cementite, or a combination of these structures. This unique mixture in steel is very desirable for many applications. Some typical applications are discussed next.

Figure 15-1. *Common products that have the same characteristics as surface-hardened parts. In each case, a soft core is protected by a hard, outer surface.*

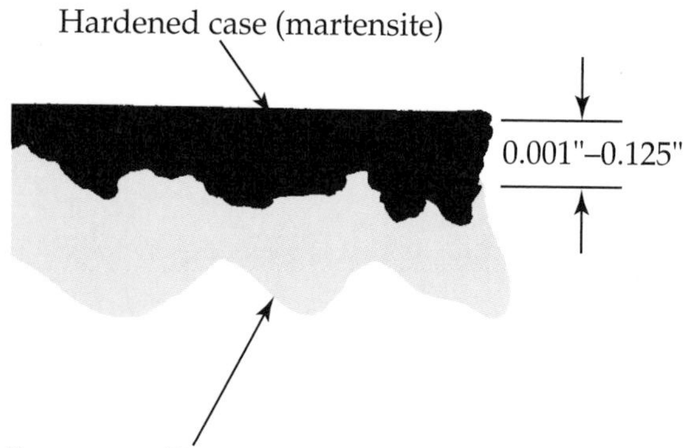

Figure 15-2. *Surface-hardened steel has a hard, outer case and a softer interior.*

Ratchet wheel

The ratchet wheel shown in Figure 15-3A rotates at a constant speed during operation. A cam wheel rides on the teeth of the wheel. A strong spring keeps the cam wheel in contact with the ratchet wheel. After each ratchet tooth lifts the cam wheel, the cam arm actuates a switch. This indirectly measures the rotational speed of the ratchet wheel.

After many hours of operation, the pressure from the cam wheel will wear down the ratchet wheel teeth (as well as parts of the wheel surface). See Figure 15-3B. Unless the ratchet wheel has a hard, wear-resistant surface, it will become worn relatively quickly. Case hardening can correct this problem.

When selecting materials for manufacture of a ratchet wheel, it would *not* be desirable to use a hard material like martensite for the entire part. Martensite would be too brittle to withstand the repeated shock that occurs when the cam wheel drops from the teeth to the wheel surface. The ratchet wheel would probably crack under these forces. Therefore, it would be desirable for the ratchet wheel to have a soft interior (to withstand shock) and a hard outer case (to resist wear).

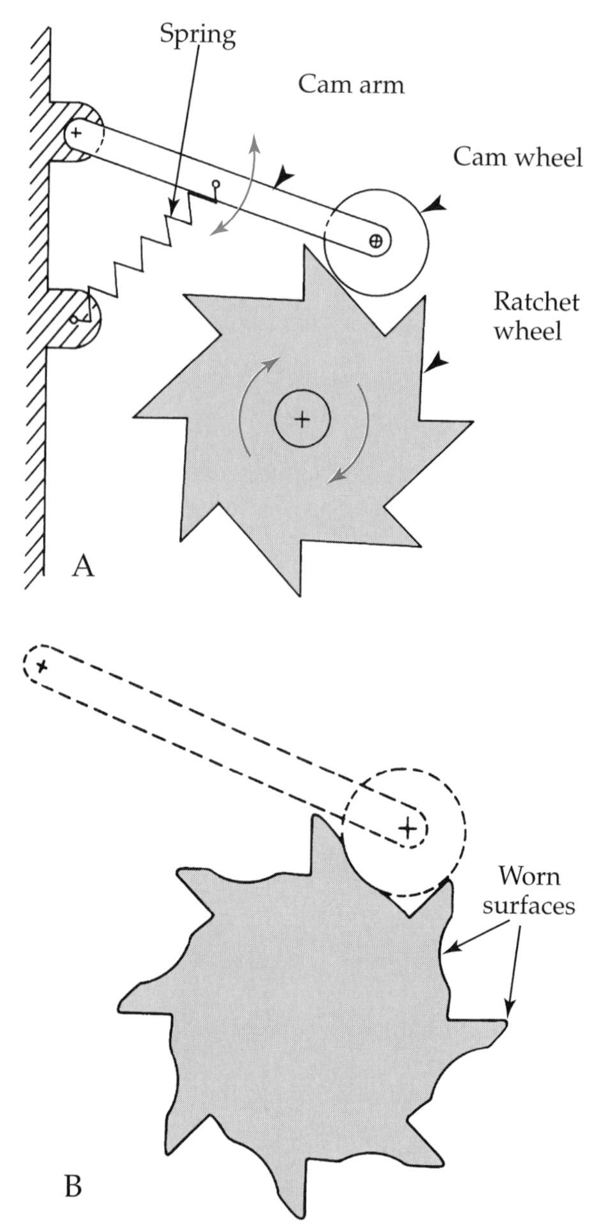

Figure 15-3. *Surface hardening helps this ratchet wheel withstand many hours of surface contact. A—The cam wheel rides on the surface of the teeth during operation. B—Pressure from the cam wheel would wear down an unhardened part quickly.*

Surface Hardening for Tools and Other Applications

Some cutting tools require a hard outside cutting surface, but they will fracture easily if the entire tool is hard and brittle. See Figure 15-4. Many machine parts subjected to high levels of pressure during operation require surface hardening. Bearings, piston pins, crankshafts, and cams must all stand up to

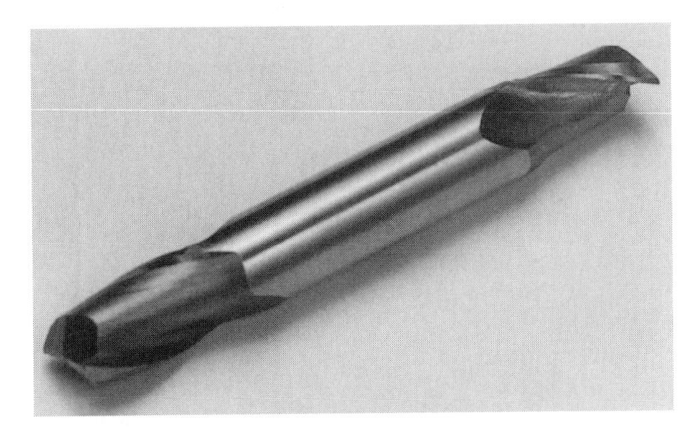

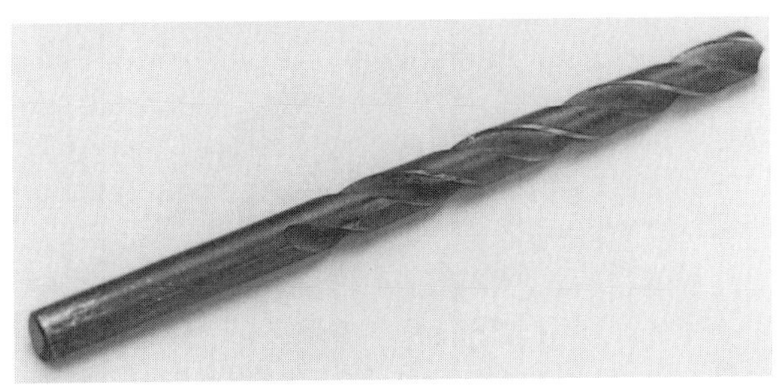

Figure 15-4. *Cutting tools are commonly case-hardened to increase their resistance to wear.*

constant friction and wear while resisting varying shock loads.

One of the first known applications of surface hardening occurred in prehistoric times, when warriors used crude methods to harden the tips of their spears. This allowed them to pierce enemy armor without cracking their spear points.

In all these examples, case hardening is used to produce a hard, protective case and a soft interior. Case hardening is desirable for any application requiring wear resistance and strength under high-pressure conditions.

Basic Surface-Hardening Methods

There are three basic surface-hardening methods used to give metal a hardened case:

- Carburizing.
- Nitriding.
- Localized heating.

Today, there are eight common industrial processes that employ these basic methods. The three methods are discussed first.

Carburizing

Carburizing is a process that impregnates the outer surface of low-carbon steel with considerable amounts of carbon before secondary heat treatment (quenching). Carburizing is the most widely used surface-hardening technique. It is also the least expensive method.

When low-carbon steel is quenched, it does not become much harder. The addition of carbon increases its ability to be hardened. There are two steps in carburizing. First the surface of the part is impregnated carbon and heated above its transformation temperature. Then the part is quenched. The carbon-enriched surface becomes very hard.

It should be emphasized that the addition of carbon to the surface does not cause hardening directly. Rather, hardening is caused by the secondary heat treatment. A major advantage of carburizing is the opportunity to hold the part for quenching after the surface has been carburized. This will be discussed later in this chapter.

Nitriding

In *nitriding*, the surface of the material is hardened by the introduction of nitrogen. This process is similar to carburizing, but secondary heat treatment is not necessary. Atoms of nitrogen combine chemically with neighboring atoms of steel to form extremely hard nitride compounds.

Localized Heating

Localized heating uses heat, rather than chemical elements, to harden selected surface areas of high-carbon steel. In this method, heat is applied to only the surface of the metal that needs hardening. The entire part is then quenched, but only the preheated surface becomes hard. The core of the part remains unchanged, since it was not heated above its transformation temperature. The key to localized heating is being able to heat the outside skin of a metal to red hot without heating the interior as well.

Localized heating is the basic method used by two popular case-hardening processes, flame hardening and induction hardening. These processes, along with other case-hardening processes, are discussed in the following sections.

Surface-Hardening Processes

The principles of carburizing, nitriding, and localized heating are used by various surface-hardening processes. The names of several of the processes suggest the case-hardening method used in the process. The eight processes listed below are among the most widely used in surface hardening:

- Pack carburizing.
- Gas carburizing.
- Liquid carburizing.
- Nitriding.
- Carbonitriding.
- Cyaniding.
- Flame hardening.
- Induction hardening.

Pack Carburizing

Pack carburizing is the oldest and most basic surface-hardening process. In this technique, the parts to be hardened are "packed" into a metal box along with a carbonaceous material and heated. See Figure 15-5. A

Figure 15-5. *In pack carburizing, parts to be case-hardened are loaded into a metal box filled with a carbonaceous material.*

carbonaceous material is a carbon-rich source that can transfer carbon when heated.

The carbonaceous material completely buries the loaded parts. The material can be almost any solid form of carbon. Most commonly, it is either charcoal or coke. Many other materials are frequently used, including coal, shells, peach pits, beans, nuts, leather, and hardwood.

The entire box of parts is heated to a temperature high enough to convert some of the carbon to carbon monoxide. See Figure 15-6. The parts must be heated above the upper transformation temperature so that a transformation to austenite occurs. Heating temperatures vary from 1500°F–1800°F (815°C–980°C), depending on the type of steel.

During heating, carbon monoxide penetrates the surface of the parts. It is absorbed by the austenite, producing a thin case of carbon on the surface.

In the first hours of heating, the case grows at the rate of 0.010"–0.020" (0.25 mm–0.5 mm) per hour. This rate drops to 0.005" (0.13 mm) per hour after five or six hours. After 10 hours, the penetration of carbon is virtually exhausted. Thus, over an eight-hour period, a case depth of 0.060" (1.50 mm) is typically produced, depending on the temperature and type of steel employed. See Figure 15-7.

Figure 15-6. *This large, floor-type furnace is chiefly used for heating temperatures up to 2000°F. It is used for pack carburizing and other forms of heat treatment of low-carbon steel. It is also used for preheating high-speed steel, as well as casting operations. (Charles A. Hones, Inc.)*

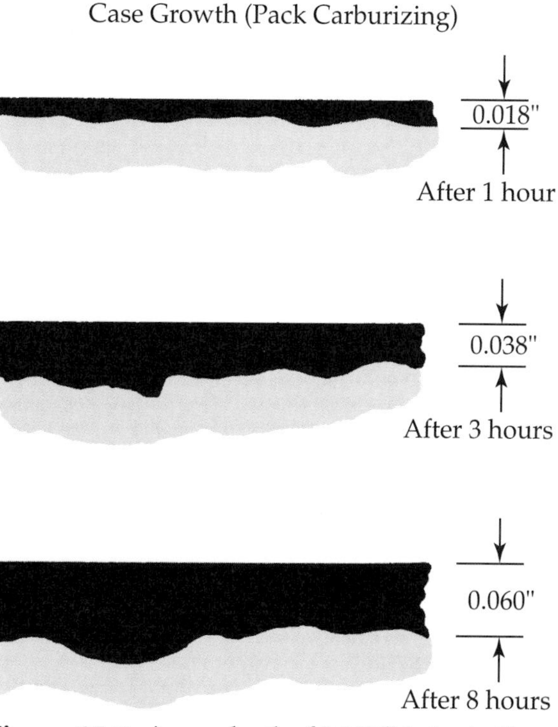

Figure 15-7. *A case depth of 0.060" is typically produced after eight hours in pack carburizing.*

The box used to heat the parts must be able to resist thermal abuse. The box must be able to withstand repeated heating and cooling cycles, which can produce internal stress and deterioration.

Pack carburizing has several advantages over other case-hardening techniques. It involves minimal capital expense, and it is more foolproof than other processes. It is especially practical when only a few small parts require surface hardening at one time. The main disadvantages of pack carburizing are the relative slowness and dirtiness of the process.

Gas Carburizing

In *gas carburizing,* low-carbon steel is heated inside an oven or furnace filled with a carbonaceous gas, Figure 15-8. The gas may be natural gas, ethane, propane, butane, carbon monoxide, or a vaporized fluid hydrocarbon. Carbon atoms from the gas are introduced to the steel parts. As a result, the outer skin of the steel becomes filled with carbon. The longer the steel is left in the gas-filled oven, the thicker the carbon layer becomes.

After the parts have received enough carbon, they may be immediately quenched to produce a hardened case. If machining is required, reheating and quenching can be delayed until the machining is completed.

As is the case with pack carburizing, gas carburizing requires the parts to be heated to an austenitic temperature in order to facilitate full penetration of the carbon. The heating

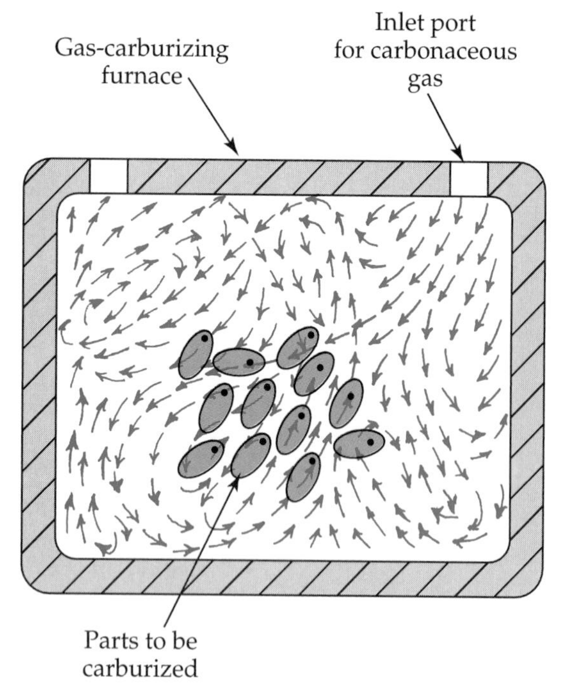

Figure 15-8. *In gas carburizing, parts are heated inside an oven or furnace filled with a carbonaceous gas.*

temperature varies from 1500°F to 1800°F (815°C to 980°C), depending on the type of steel.

Different types of furnaces are used in gas carburizing. In a rotary furnace, the parts are "tumbled" by rotating a retort. Continuous furnaces, Figure 15-9, are widely used today because they are more efficient and less expensive to operate than other furnaces. Many larger furnaces use a fan or similar circulating device to keep the gas in constant contact with the parts. This helps produce a more uniform deposit of carbon.

Comparing gas carburizing and pack carburizing

Gas carburizing presents several advantages over pack carburizing. It is a faster process, and it requires less physical labor and handling. The case depth can also be controlled more accurately. Gas carburizing is more practical than pack carburizing when large quantities of parts must be hardened.

Figure 15-9. *A continuous carburizing furnace. (Seco/Warwick)*

One disadvantage of gas carburizing is the expense of the equipment and materials required. Pack carburizing can be performed without special machinery.

Liquid Carburizing

Liquid carburizing uses a liquid source of carbon to produce surface-hardened steel parts. This process is similar to gas carburizing. The carbon is deposited when steel parts are immersed in a carbon-rich liquid. A molten salt bath is typically used, Figure 15-10. Liquid carburizing can be performed with sodium cyanide (NaCN), barium cyanide ($BaCN_2$), or calcium cyanide ($CaCN_2$).

The salt bath is usually heated by electricity. In most tanks, the solution is stirred to keep it in uniform movement with the parts.

Comparing carburizing techniques

Liquid carburizing has several advantages over both gas carburizing and pack carburizing. Liquid tends to transfer heat rapidly, so carbon penetrates the steel very quickly during the first hour of carburizing. Liquid carburizing is also a more efficient process for producing thin, uniform case depths. The case is more uniform along the surface because liquid tends to flow more evenly than gas. Essentially, liquid covers the immersed parts like a blanket. This reduces the amount of contamination from oxidation.

There are several reasons liquid carburizing is not always the best carburizing technique. In a cyanide salt bath, some nitrogen is absorbed by the steel along with carbon. Nitrogen produces immediate hardening. Therefore, parts carburized in liquid are generally not machined after carburizing. Cyanide salts are also poisonous and present a safety hazard.

Parts must be rinsed after liquid carburizing to prevent rusting. This takes time and requires additional equipment.

Salt baths are usually contained in relatively small chambers or tanks, so it can be impractical to immerse large, odd-shaped parts into the liquid. For this reason, liquid carburizing is usually restricted to the surface hardening of small parts.

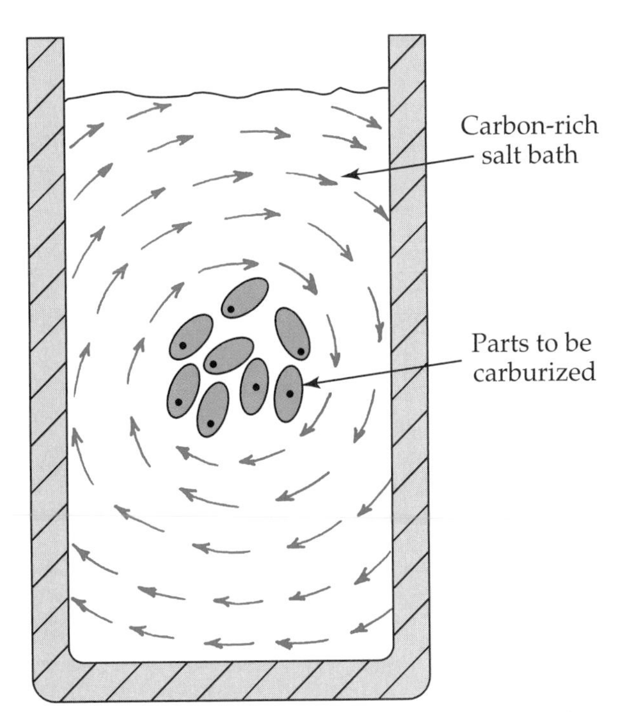

Figure 15-10. *In liquid carburizing, parts are immersed in a carbon-rich salt bath.*

Nitriding

Nitriding is a gaseous surface-hardening process that fills the outer skin of steel with nitrogen. The most common nitrogenous gas used in this process is ammonia (NH_3). When the nitrogen unites with the surface of steel parts, several types of iron nitrides are formed. These compounds are extremely hard, Figure 15-11.

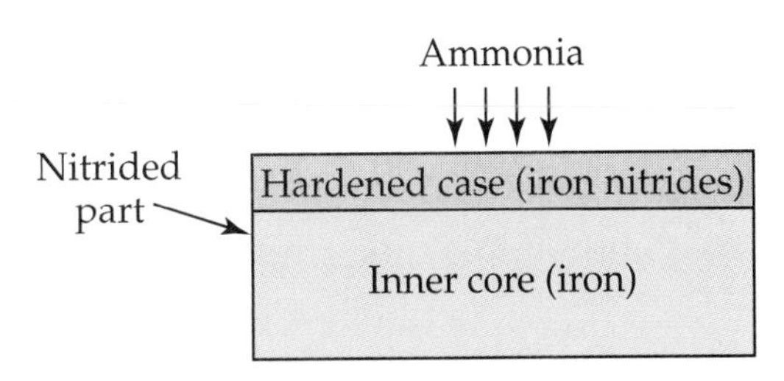

Figure 15-11. *In nitriding, the steel surface is filled with nitrogen to produce iron nitrides.*

Nitriding is performed in a sealed, heated chamber. Ammonia enters the chamber through an inlet. After nitriding, the part is hardened and requires no further quenching.

Nitriding is one of the most widely used processes in surface hardening. The advantages and disadvantages to using this technique are discussed next.

Advantages of nitriding

There are a number of advantages to using nitriding as a surface-hardening process, depending on the application. These advantages are listed and explained below.

- Hardness. Of all the surface-hardening processes, nitriding produces the hardest case. Hardness values measuring over 70 on the Rockwell C scale can be attained through nitriding.
- Immediate hardness. After nitrogen joins the surface, the outer case is hard. No reheating or quenching is required, as is the case with carburizing. This saves time.
- Case-hardening temperature. During nitriding, nitrogen atoms will join the iron surface at a heating temperature below the transformation temperature of steel. Heating temperatures of 900°F–1000°F (480°C–540°C) are normally used. Nitriding is the only surface-hardening process that can produce a hardened case at low temperature. Hardening at a low temperature is more convenient and less expensive than using an elevated temperature.
- Elimination of distortion. Nitriding yields less internal stress and more dimensional stability in steel. The chances of warpage, cracking, and distortion are also reduced. The reasons for this are the lower case-hardening temperature and the elimination of quenching as a secondary heat-treating process. This is of special importance for complex parts with uneven sections that tend to distort easily.
- Corrosion resistance. Nitrided surfaces are more resistant to corrosion than other case-hardened parts. Humidity, salty conditions, water, oil, gasoline, and other corrosive agents are not as harmful to nitrided steel.
- Temperature resistance. Reheating parts to temperatures of 1000°F–1100°F (540°C–595°C) for a short period of time does not affect nitrided cases. These temperatures will soften carburized cases. Prolonged heating at temperatures of 600°F–800°F (315°C–425°C) will not affect a nitrided case, but it will affect a carburized part. Thus, at elevated temperatures, nitrided cases are more dimensionally stable.
- Cleaning. Freshly nitrided parts do not require cleaning. Carburized parts do, to prevent corrosion.

Disadvantages of nitriding

Despite its many advantages, nitriding is not always the best surface-hardening process for steel. Some of the disadvantages of nitriding are listed and explained below.

- Slowness. Nitriding is the slowest of the surface-hardening processes. It can require several days to complete, rather than several hours. For this reason, nitriding is generally limited to producing very thin cases, unless a high level of hardness is extremely important. It might take ten hours to harden a 0.010" case. A 0.030" case might require several days of nitriding.
- Cost. Nitriding is an expensive process. Ammonia gas is much more expensive than some of the most commonly used carburizing gases. The equipment used in nitriding is expensive as well. Inexpensive, low-carbon steel cannot be nitrided. The steel must contain alloys that can react with the nitrogen to form iron nitride compounds. These alloy steels cost more.
- Size growth. As parts take on nitrogen, they begin to swell. This can affect dimen-

sional accuracy. In most cases, however, the amount of growth can be estimated fairly closely. Parts are often made undersized before nitriding to allow for a small change in size.

- Machinability. Little machining can be done to a part after it has been nitrided, because hardening takes place immediately. Only a small amount of grinding is practical for a nitrided part. Greater machinability is one of the advantages of pack carburizing and gas carburizing. Using these processes, parts can be machined after the first step of surface hardening (before quenching takes place). This reduces the effects of distortion and warpage.
- Accuracy of process. Nitriding is a time-consuming process that requires close regulation. The heating chamber must be kept at low temperature and must be sealed with ammonia gas. The percentage of ammonia introduced to the steel must be carefully controlled by an experienced operator.

Carbonitriding

Carbonitriding is a gaseous process that introduces both carbon and nitrogen to steel parts. Ammonia and natural gas are put into contact with the parts, producing a hardened case of iron carbide and iron nitrides. Carbonitriding is generally used for thin cases measuring approximately 0.005"–0.010".

Advantages of carbonitriding

Carbonitriding has some advantages over other surface-hardening processes, such as gas carburizing. The presence of nitrogen alters the austenite structure. This permits the use of a lower transformation temperature and a slower cooling rate.

Carbonitriding can be performed at temperatures of 1400°F–1700°F (760°C–930°C), about 100°F (55°C) lower than the typical heating temperatures used for carburizing. See Figure 15-12. Oil quenching, a less dramatic technique than water quenching, can be used because hardness can occur at a slower

Figure 15-12. *An electrically heated carbonitriding furnace. Heating temperatures range from 1400°F–1700°F, which are lower than the typical temperatures used in carburizing. (The Electric Furnace Company)*

cooling rate. Internal stress is reduced, and there is less chance of warpage, cracking, and distortion.

Cyaniding (Liquid Carbonitriding)

Cyaniding is a form of carbonitriding that uses liquid instead of gas to impregnate the surface of steel with carbon and nitrogen. It is primarily used for the rapid production of hard, thin cases.

Cyaniding is similar to liquid carburizing in that it uses a molten salt bath to heat the steel. See Figure 15-13. The salt bath is made up of a molten cyanide salt, such as sodium cyanide, calcium cyanide, or potassium cyanide. Sodium cyanide melts at 1140°F (615°C), so it is in a liquid state when the steel is heated.

Very thin cases are produced by cyaniding. During the first 30 minutes of hardening, carbon and nitrogen penetrate the steel surface rapidly to a depth of about 0.005", depending on the concentration of the salt bath and the type of material. After the first 30 minutes, penetration slows down considerably. Therefore, cyaniding is seldom used for case depths measuring above 0.010". Since nitrogen is added to the surface, the final case is very hard and can reach a hardness value of 65 R_c.

Figure 15-13. *This gas-fired furnace is used for cyaniding, liquid carburizing, or any other heat-treating process requiring a molten salt bath. (Charles A. Hones, Inc.)*

Three heating tanks are commonly used in cyaniding. The liquid in the first tank is used to preheat the parts. The second tank contains a 30% sodium cyanide salt solution and is used to harden the case. The third tank is for quenching. Normally, parts are quenched immediately after cyaniding. Since part of the hardening effect is due to nitrogen, the quench does not have to be rapid. Therefore, cyanided parts are typically oil quenched, rather than water quenched. This eliminates some of the hazards of distortion and cracking.

Advantages of cyaniding

Cyaniding is a relatively inexpensive surface-hardening process because regular carbon steel can be used. It is a fairly rapid process used in applications requiring a thin, hard case.

Disadvantages of cyaniding

Cyaniding can be a very hazardous process. Cyanide salts are poisonous; the resulting fumes can be fatal if they are inhaled. The area surrounding the furnace must be well ventilated.

Proper care and safety must be practiced when handling materials to be cyanided. If liquid from the salt bath makes contact with an open cut or wound, the results can be very serious.

Flame Hardening

Flame hardening is a surface-hardening process that heats selected areas of steel with a direct flame before quenching. It is different from any of the case-hardening processes previously discussed. No carbon or nitrogen is introduced to the surface, only heat is added.

In flame hardening, a direct flame from an oxyacetylene torch is brought into contact with the surface to be hardened. As soon as the surface is heated to the elevated temperature, quenching follows. Heating and quenching must occur very rapidly. If there is enough time for heat to penetrate deeply into the part, not only will the surface be hardened, the core will be hardened as well. If enough heat is applied to the surface to reach the upper transformation temperature before quenching, martensite will be formed and a hard outer shell will be produced.

Flame hardening is useful for such components as cylindrical pins, lathe beds, cams, engine push rods, and pulleys. In these applications, a thin case depth is not critical, and only certain portions of the part require hardening.

Figure 15-14. *Flame-hardening equipment is used to harden the jaws of the wrenches shown. The jaws are selectively heated while the rest of each tool remains relatively soft and impact resistant. After a hardened case is produced, the wrenches are quenched in oil. (Tocco Division, Park-Ohio Industries, Inc.)*

Advantages of flame hardening

Flame hardening is suitable for certain case-hardening applications. The advantages of this process are listed and explained below.

- Case depth. Flame hardening is a very rapid and efficient method for producing cases as deep as approximately 1/4".
- Localized heating. Selected areas of steel can be heated using the flame-hardening process. This technique is very useful when only certain portions of a part require hardening, such as the jaws of a wrench. See Figure 15-14. When a part is quenched after flame hardening, the chance of distortion is reduced since only a small portion of the part has been heated. Tempering is still recommended after quenching in order to reduce localized stress, but distortion of the entire part is minimized.
- Cost. Flame hardening is a relatively inexpensive process. No expensive ovens are required, and no special chemicals or gases are used.
- Quantity of parts. Small quantities of parts are very suitable for flame hardening. It is usually not practical to heat a large oven or salt bath when only a few parts need hardening.
- Size of parts. Flame hardening is useful for large, bulky parts that cannot fit into a furnace or tank. It is also more convenient to use a torch for heavy materials that cannot be easily transported from one location to another.
- Automation. The equipment required for flame hardening can be as simple as a welding torch and a water hose. At the

other extreme, highly mechanized, automated systems are used to produce flame-hardened parts. Automated systems are typically designed with combination tools that serve as both a torch and cooling source. After selected areas of the part are hardened, a spray of water quenches the part. This produces a fairly uniform case depth. In some systems, the parts may be immersed in a quenching medium, rather than sprayed. An automated system used to harden gear teeth is shown in Figure 15-15.

Disadvantages of flame hardening

Flame hardening has several disadvantages as compared to other case-hardening processes.

- Case depth.When using a torch, it is very difficult to accurately control the penetration of heat into the metal. As a result, flame hardening is usually not used for thin cases.
- Type of steel. Flame hardening can only be used for certain types of steel. No additional carbon or nitrogen is added to the surface, so the source of the hardening must come from the metal itself. Flame hardening cannot be used with low-carbon steel. Medium-carbon steels with 0.35% to 0.60% carbon are most commonly used. This type of steel is slightly more expensive than low-carbon steel.
- Maximum hardness. The maximum hardness that can be obtained from flame hardening is usually less than that attained by other surface-hardening methods. In this process, the attained hardness depends on the carbon content of the steel. Flame hardening produces hardness values of 50–60 R_C.

Figure 15-15. *This automated flame-hardening system is used to harden gear teeth. As the gear revolves, stationary flaming heads around the outer periphery heat the teeth. Flames shown here are coming from pilot jets before flame-hardening operaton is started. A quench spray is applied after each tooth is hardened. (Tocco Division, Park-Ohio Industries, Inc.)*

Induction Hardening

Induction hardening could be called "high class flame hardening." Like flame hardening, only the outer surface is heated above the transformation temperature. No carbon or nitrogen is added.

The big difference between flame hardening and induction hardening is the fascinating way in which the heating is accomplished. In induction hardening, the part to be hardened is surrounded by a coil of wire. The coil acts like the primary winding of a transformer, Figure 15-16. High-frequency electrical current (3000 Hz–1,000,000 Hz) passes through the coil, producing a magnetic field around the part. This induces eddy currents in the part. The electrical resistance of the part generates heat. Due to a phenomenon called "skin effect," the current and the heat stay only on the outer surface, or skin, of the steel. When the surface of the part has received sufficient heat, it is quenched in water or oil. See Figure 15-17.

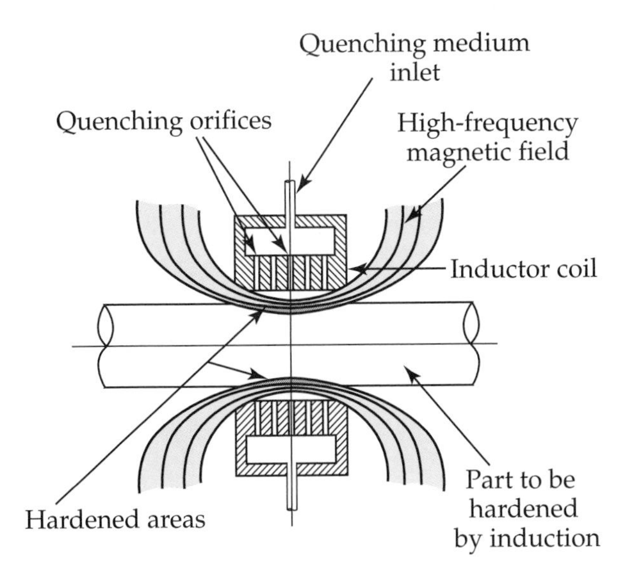

Figure 15-16. *The induction hardening process. Magnetic field emitted from an inductor coil cut through the surface of the part. The darkened areas represent the sections to be hardened. Long, narrow holes in the inductor transfer the quenching medium to the part after it is heated. (Tocco Division, Park-Ohio Industries, Inc.)*

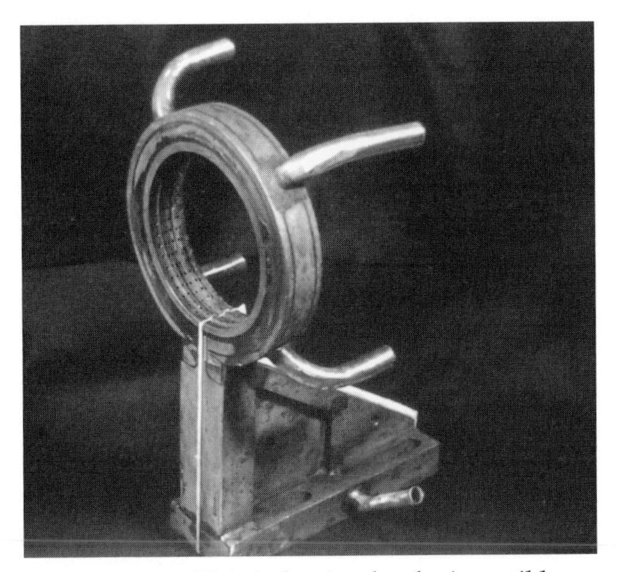

Figure 15-17. *This induction-hardening coil has four quenching orifices. The part to be hardened passes through the center of the coil. After heating, a liquid quenching medium flows through the four rows of inlets around the inner periphery of the coil. (Tocco Division, Park-Ohio Industries, Inc.)*

Induction hardening is used for a wide variety of applications. Induction hardening is commonly used for such components as sprockets. Other practical applications include cams, crankshafts, piston rods, axles, ball and roller bearings, chain links, and parts for firearms.

Advantages of induction hardening

Induction hardening has many advantages over other case-hardening processes. Of all the processes, it is the most efficient. No heating time is required prior to the induction of current, and parts can be heated in five seconds or less. When parts are hardened in large quantities, the unit cost is very low.

Irregular shapes can be handled quite readily with induction hardening. Skin current can penetrate crevices and holes, as well as interior surfaces.

The case depth can be controlled more accurately in induction hardening than it can in any other process. The depth can be controlled by varying the frequency, the current, and the amount of time the coil is in contact with the part. The higher the frequency, the more the current tends to flow over the outer surface only.

Induction hardening is generally used to produce *thin* cases. Larger depths such as 1/8" can be attained by leaving the current in contact with the surface for a longer period of time and operating at lower frequencies.

Induction hardening provides outstanding resistance to warpage, distortion, oxidation, and scale. This is due to the short period of heating time and the fact that only a small portion of the part requires heating.

Disadvantages of induction hardening

The main disadvantage of induction hardening is the cost of machinery and materials required for the process. More expensive carbon steel is required, since no carbon or nitrogen is introduced to the parts. Medium-carbon steel with 0.35% to 0.60% carbon is most commonly used.

The level of hardness attained by induction hardening is dependent on the original carbon content of the steel. Therefore, high levels of hardness cannot be produced unless special high-alloy steel is used.

Selecting a Surface-Hardening Process

Each surface-hardening process has a number of advantages and disadvantages in comparison to other processes. The selection of a process depends on the quantity of parts involved, the accuracy and depth requirements for the case, the shape and size of the parts, and the equipment and facilities available.

The table in Figure 15-18 summarizes the comparative advantages and disadvantages of each process. Keep in mind that the table is based on average typical case. With slight adjustments, a case-hardening process can be applied beyond its typical applications.

Test Your Knowledge

Write your answers on a separate sheet of paper. Do not write in this book.

1. What is the purpose of *surface hardening?*
2. Why is surface hardening also known as *case hardening?*
3. Name four different applications or components that require surface hardening.
4. Explain two differences between carburizing and nitriding.
5. What two surface-hardening processes use the principles of *localized heating?*
6. Explain the process of *pack carburizing.*
7. Name two advantages gas carburizing has over pack carburizing.
8. What type of carbon-rich source is typically used in liquid carburizing?
9. Of all the surface-hardening processes, which one is capable of producing the hardest case?
10. What range of heating temperatures is typically used in nitriding?
11. What is *carbonitriding?*
12. Which surface-hardening process uses liquid to impregnate the surface of steel with carbon and nitrogen?
13. What two surface-hardening processes do not involve the addition of carbon or nitrogen to steel?
14. Explain the advantages of flame hardening over other surface-hardening processes.
15. What is the main disadvantage of cyaniding?
16. In induction hardening, what is the source of electromagnetic induction?
17. Which surface-hardening process is considered to be the most efficient and most accurate?
18. A part used in the passenger seat mechanism in a commercial aircraft requires surface hardening. Discuss the factors that would help determine the best case-hardening process to use for the part.
19. A tiny gear used in a clock mechanism requires surface hardening. Which case-hardening techniques would you recommend for this application?
20. A part used in a business machine comes into contact with the sharp edges of coins during operation. The part needs to be case hardened to ensure greater life of the part. The part measures 8" × 8" but is only 0.060" thick. Which surface-hardening processes would you consider for this part and why? Which processes would you not consider?

Comparison of Surface-hardening Processes												
Surface Hardening Process	**What is Added to Surface**	**Order of Hardness Attainable**	**Heating Temperature**	**Hazard of Process**	**Equipment and Tooling Cost**	**Cost per Unit (High Quantities)**	**Cost per Unit (Low Quantities)**	**Quenching Required after Hardening**	**Tendency for Distortion**	**Machinability after Hardening**	**Control and Accuracy of Process**	**Average Case Depth over Time (in inches)**
Pack Carburizing	Carbon	4 (tie)	High (Austenitic) 1500°F–1800°F	Little hazard	Very low	High	Very low	Yes	Some	Yes	Poor	0.010 (first hour)
Gas Carburizing	Carbon	4 (tie)	High (Austenitic) 1500°F–1800°F	Fairly poisonous gas	High (furnace required)	Low	High	Yes	Some	Yes	Good	0.013 (first hour)
Liquid Carburizing	Carbon	4 (tie)	High (Austenitic) 1500°F–1800°F	Fairly poisonous	Average	Average	Average	Yes	Some	Some machinability	Good	0.018 (first hour)
Nitriding	Nitrogen	1 Hardest	Lowest (900°F–1000°F)	Proper safety required	High (furnace required)	Average	High	No	Very little	Difficult	Very good	0.010 (per first 10 hours)
Carbonitriding	Carbon and Nitrogen	2 (tie)	Low (1400°F–1700°F)	Fairly poisonous gas	High (furnace required)	Low	High	Yes	Little	Difficult	Good	0.008 (first hour)
Cyaniding	Carbon and Nitrogen	2 (tie)	Low (1400°F–1700°F)	Very poisonous fumes	Average	Average	Low	Yes	Little	Difficult	Good	0.010 (first hour)
Flame Hardening	Heat	8 Hardness depends on material	Very high (over upper transformation temperature)	Caution with flame	Low	High (without automation)	Low	Yes	Yes	Difficult	Poor	Instant depth (for thick cases up to 1/4")
Induction Hardening	Heat	7 Hardness depends on material	Very high (over upper transformation temperature)	Caution with equipment	High (electrical equipment required)	Lowest cost	Variable	Yes	Little	Difficult	Most accurate process	Instant depth (for very thin cases)

Figure 15-18. *This table compares the various characteristics of each surface-hardening process.*

Section Four

Nonferrous Metallurgy

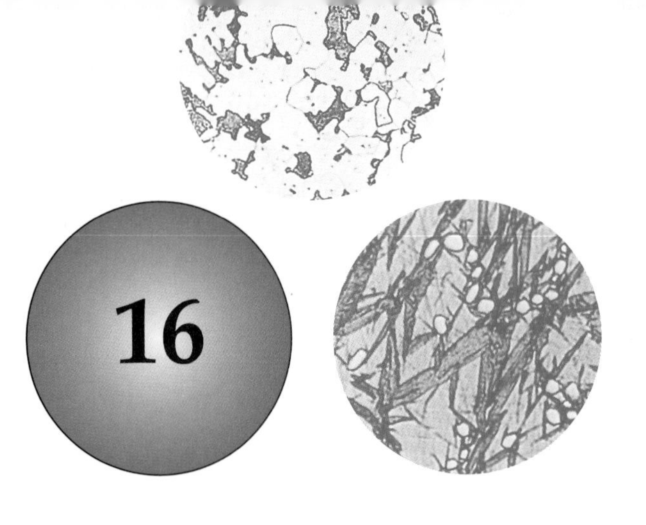

Processing Nonferrous Metals

After studying this chapter, you will be able to:

- Identify and explain the basic processing methods used for nonferrous metals.
- Describe the different types of atomic structures in metals.
- Explain the principles of alloying and the structural effects produced by alloying elements.
- Summarize the effects of cold working on metal.
- Recognize how precipitation hardening is applied to improve the properties of nonferrous alloys.
- Use phase diagrams to identify the various structures that occur in alloy systems.

Introduction to Nonferrous Metallurgy

Iron and steel have mechanical properties useful for many applications. When steel is manufactured from iron, its composition can be changed through a number of metallurgical processes to attain different characteristics. *Ferrous metallurgy,* the science of developing metals that use iron as their major alloying element, has been the main focus of this text.

As versatile as iron and steel are, however, there are many types of *nonferrous* metals and alloys that are widely used by metallurgists. *Nonferrous metallurgy* is the study of metals that do not use iron as their principal alloying element. The basic properties of nonferrous metals (such as copper, aluminum, and titanium) make them well suitable for many applications.

Can you imagine an airplane made entirely of steel? It would be impossible for the aircraft to carry much payload, even if it could take off! Such an application would require lighter metals for manufacture. Light weight and other special properties can be attained through the application of nonferrous metallurgy and alloys.

Often, the basic methods for improving the properties of nonferrous metals are similar to the metallurgical processes used for steel. This chapter will introduce some of these methods and discuss how they are applied to make useful products from aluminum, copper, magnesium, and other nonferrous metals.

Basic Processing Methods Used in Nonferrous Metallurgy

Most pure metals are quite soft by nature. They will bend easily and they will wear out if rubbed against other hard materials for a period of time. If you grind a sharp edge into a piece of pure metal to make a knife, for example, the edge becomes dull very quickly after use.

Recall from earlier chapters that in ferrous metallurgy, the use of alloys and different heat treatment processes can improve the hardness and strength of iron. It is often necessary to improve similar characteristics in nonferrous metals. The mechanical properties of pure nonferrous metals can be changed using three basic processing methods:

- Alloying.
- Cold working and annealing.
- Precipitation hardening.

Each of these methods will be discussed in this chapter. The metallurgical processes used for nonferrous alloys are very similar to those used for some types of alloy steel, especially stainless steel.

Nonferrous metallurgy can be used to produce other desirable properties besides hardness and strength. Some applications require not only tensile strength or hardness, but also ductility and formability. Other special properties offered by nonferrous metals include ease of processing, lighter weight, corrosion resistance, and electrical conductivity.

In metallurgy, different structural forms of metal are associated with specific mechanical properties. This was covered in Chapter 7. When a sample of metal is heat treated, the arrangement (or "pattern") of atoms in the metal undergoes physical change. The most common atomic structures in metals and the properties associated with these structures are discussed next.

Atomic Structures in Metal

Internal structures of metal are made up of large numbers of atoms. The atoms occur in regular formations called crystal structures; each atom is carefully lined up in a basic arrangement called a unit cell. A crystal structure is a group of unit cells. This was covered in Chapter 7. The atoms are arranged so that they have minimum chemical energy. If a sample of metal is solidified (or cast) slowly, the resulting crystal structure is very uniform. In a sample of pure metal containing all like atoms, the crystal structure is repeated in every direction. See Figure 16-1.

There are several different types of atomic unit cells that occur in metal. The most common are body-centered cubic, face-centered cubic, close-packed hexagonal, and body-centered tetragonal. These structures are found in specific types of ferrous and nonferrous metals. Each one is discussed below.

A body-centered cubic structure, Figure 16-2, commonly occurs in ferritic iron at room temperature. Metals with this structure include molybdenum, niobium, potassium, tungsten, and vanadium. These metallic elements have the same structure at all temperatures below their melting points. Titanium at high temperature also has a body-centered cubic structure.

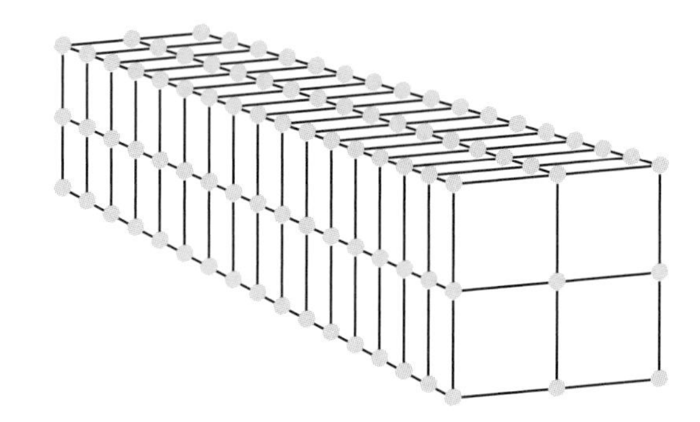

Figure 16-1. *A group of unit cells makes up a crystal structure. Only the atoms located at the corners of each cell are shown for clarity.*

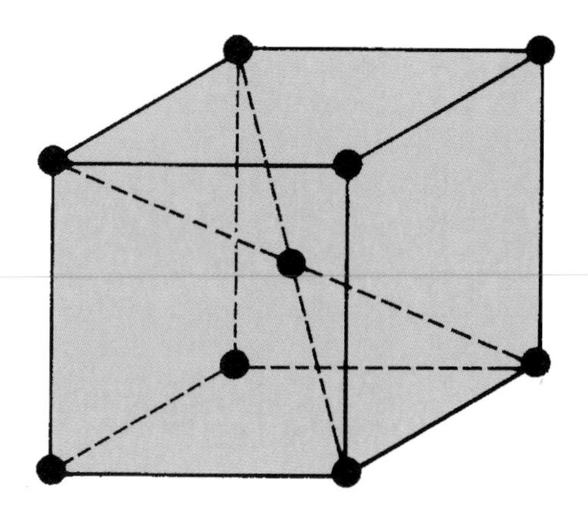

Figure 16-2. *The arrangement of atoms in a body-centered cubic unit cell.*

A face-centered cubic structure, Figure 16-3, commonly occurs in austenitic iron at high temperature. Aluminum, copper, gold, silver, lead, nickel, and platinum all have a face-centered cubic structure.

A close-packed hexagonal structure, Figure 16-4, is not typical of any form of iron. Metals such as beryllium, cobalt, magnesium, and zinc take this structure. Titanium and zirconium also have a close-packed hexagonal structure at room temperature.

The arrangement of atoms in a body-centered tetragonal structure, Figure 16-5, is similar to that of a body-centered cubic structure. However, the body-centered tetragonal structure is not a perfect cube. One edge length is different than the other two. Tin at room temperature takes on a body-centered tetragonal structure.

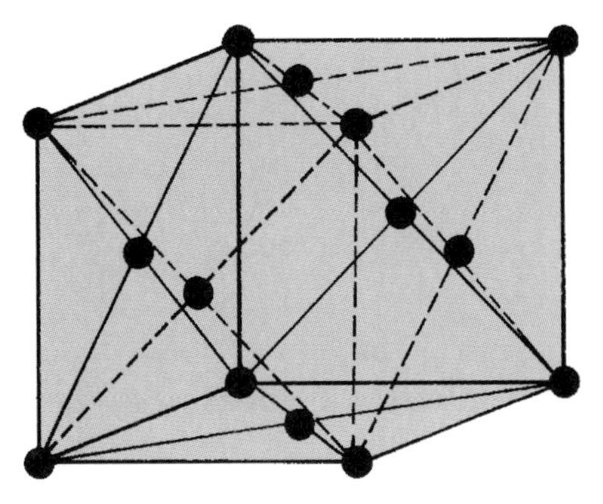

Figure 16-3. *The arrangement of atoms in a face-centered cubic structure.*

The atomic structure of a metal has a strong influence on the mechanical properties of the metal. For example, metals with a close-packed hexagonal structure do not bend as easily as other structures. For this reason, wheel rims made from magnesium can be chipped easily (magnesium has a close-packed hexagonal structure).

Metals with body-centered and face-centered cubic structures are more ductile than other structures. Assume a metal sample with a body-centered cubic structure measures 1/8" square and 6" long. When a large bending force is applied to this sample, it will bend and deform. It will remain in the deformed position when the force is removed.

When metal deforms, the atoms in a crystal structure slide past one another along slip planes, Figure 16-6. This was discussed in Chapter 8. Since pure metal bends easily, a relatively small force will cause some slippage. For even a large amount of atomic slip to occur (relative to the size of the atoms), the force required is not great.

It is impossible to see the sliding movement of the atoms because they are so small. It takes over a million individual atomic

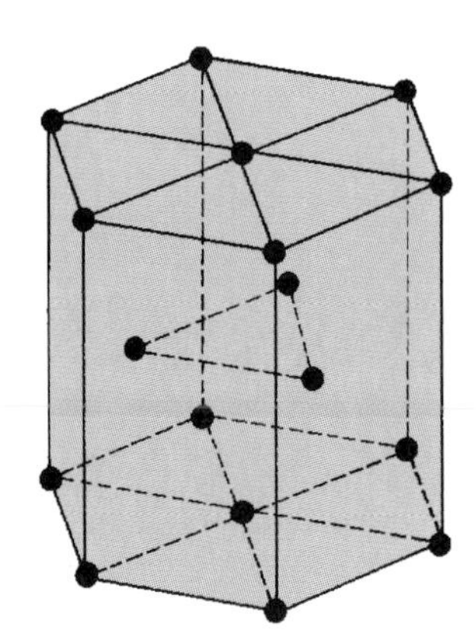

Figure 16-4. *Close-packed hexagonal structures occur in nonferrous metals such as beryllium, titanium, and zirconium.*

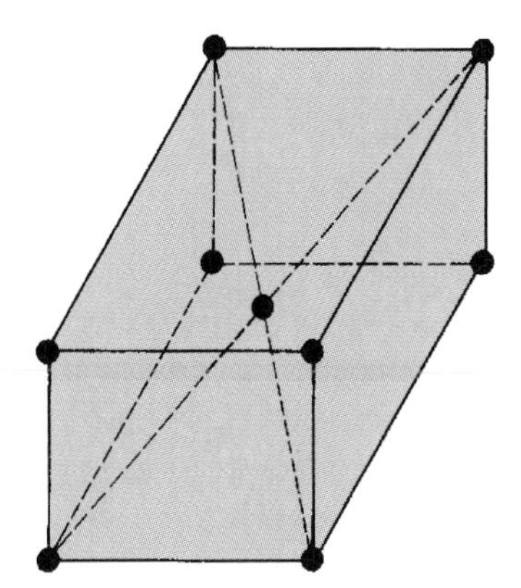

Figure 16-5. *A body-centered tetragonal structure has a similar atomic arrangement to that of a body-centered cubic structure.*

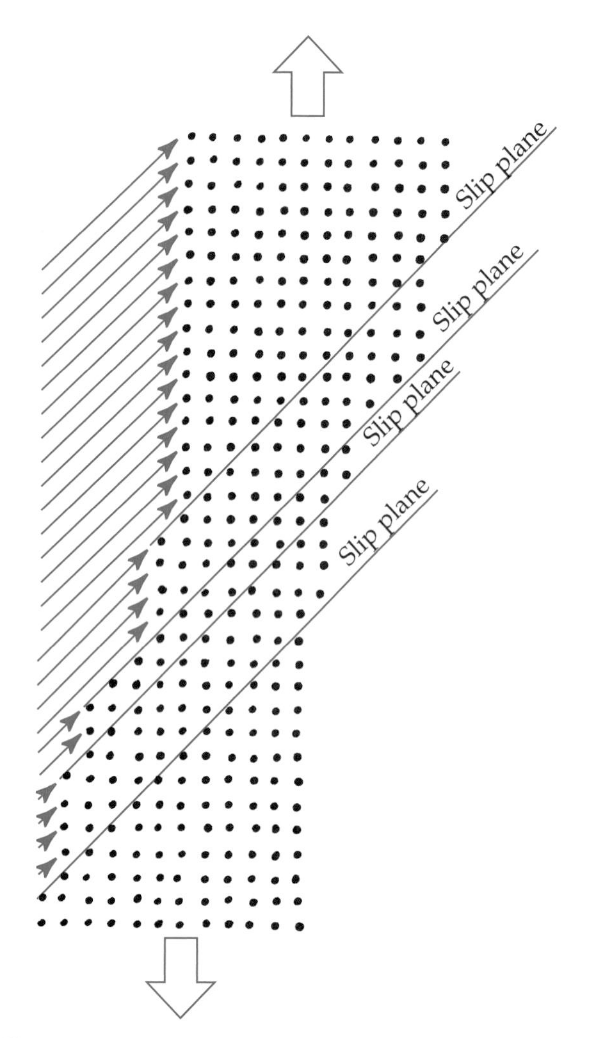

Figure 16-6. *Atoms slide past one another along slip planes during metal deformation.*

displacements before slip planes can be seen in a typical sample.

Alloying

Alloying is the addition of a second metal or several metals to produce a metal alloy with desirable properties. An *alloy* is a metal composed of two or more elements, at least one of which is a metal.

Alloys may be one of two types. When only a small amount of an element is added to a metal, the metal's crystal structure remains the same. This type of alloy is called a *solid solution alloy.* When a new compound is formed in metal, such as cementite (Fe_3C) in steel, the alloy is technically a *mixture.* Different alloying techniques produce solid solution alloys and mixtures by altering the atomic structure of metal.

Assume small amounts of an alloying element are added to a sample of pure metal during melting and casting. The atoms of the alloying element are a slightly different size than those of the original metal. Alloying causes the uniform arrangement of atoms in the original metal to become distorted, as shown in Figure 16-7.

When oversized atoms from the alloying element are introduced, they displace atoms in the original structure. Referring to Figure 16-7, the large atom shown has changed the arrangement of the surrounding atoms. The atom identified at point A has been pushed slightly to the left. The atom at point B has been pushed forward. The atom at point C has been pushed right, and the atom at point D has been pushed back slightly.

This alloyed sample is a solid solution alloy. Although the crystal structure remains the same, the sample has become stronger. When a bending force is applied to the new material, more force (or load) is required to deform the metal than the force that was previously needed to bend the pure metal.

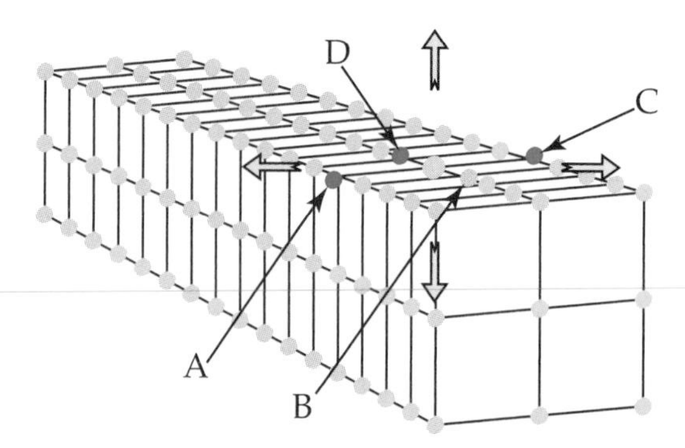

Figure 16-7. *Atomic displacement occurs when an alloying element is added to a sample of pure metal, causing distortion.*

This is due to the displacement of atoms after alloying.

The new atoms produce additional forces on the surrounding atoms and structural distortion due to their different size. The displaced atoms will not slide along the slip planes as easily as they would have before alloying. Therefore, the individual atoms must be subjected to more force to make them slide. Interestingly enough, in many cases, the new alloy is not only stronger than the pure metal, it also has more formability.

This method of alloying is referred to as *alloy strengthening* because it increases the strength of metal. Alloying is also called *solid solution hardening* when a second element (or additional element) is completely dissolved into the original metal.

Cold Working

Cold working is the repeated application of force to deform and strengthen metal. This process is also known as *work hardening.* It was briefly discussed in Chapter 8. Cold working is commonly used to process and strengthen nonferrous metal.

In cold working, metal is strengthened when its shape is changed through hammering or bending. To apply this technique, place a small piece of soft copper tubing in a vise and strike it with a hammer. On the first blow, the tubing will flatten considerably. As you strike it repeatedly, however, the shape will change less and less with each blow. Eventually, you will have to strike the metal much harder to flatten it any further.

The repeated application of flattening force has caused the metal to become harder and stronger. The metal has also become less soft and ductile. This is because the atomic structure of the metal has been changed into a stronger formation, and it is more resistant to stress. Work-hardened metal can withstand more force than metal in its original shape. Work hardening is called "cold working" because the metal is not heated, so strengthening occurs at normal temperature.

Rolling

Rolling is a cold-working technique in which metal is compressed as it is passed through forming rolls. In this process, a section of heavy sheet is reduced to a final thickness after it is "cold rolled" in a rolling mill. See Figure 16-8. The metal is hardened and strengthened as it is rolled from its original size.

When force is applied to metal, the amount of cold working that takes place depends on the ductility of the metal. Metal can tolerate a certain amount of deformation (or cold work) before failure occurs. The amount of cold work that occurs during rolling can be determined using the following equation:

$$\%CW = \frac{t_{orig} - t_{exit}}{t_{orig}} \times 100\%$$

where

$\%CW$ = percent cold work

t_{orig} = original thickness of sheet

t_{exit} = exit thickness of sheet

The original sheet thickness and the thickness after rolling are shown in Figure 16-8. Assume the sheet enters the rolling mill at 0.250" and is reduced to 0.150" after rolling. The

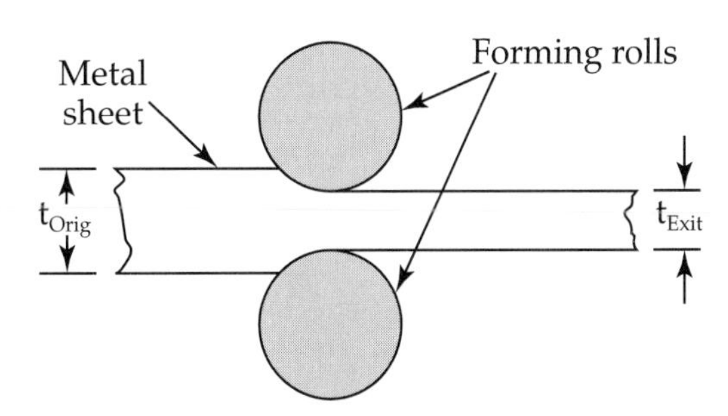

Figure 16-8. *Rolling reduces the thickness of a sheet of metal to a final size.*

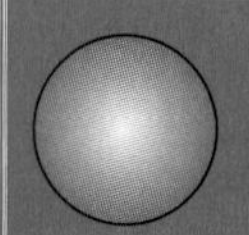

Self-Demonstration
Effects of Cold Working on Copper

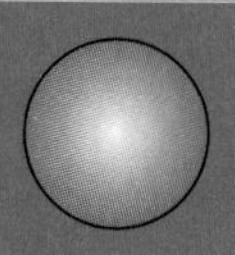

Obtain several pieces of soft copper tubing measuring approximately 3/8" in diameter and 8" long. Soft copper tubing is a form of highly pure copper that will bend easily. It can be commonly found at hardware stores.

Hold the tubing in front of you, with one end in each hand. Try to bend it in the center of the piece. The ends should make an angle of about 90°. As you bend the tubing, try to *feel* the force required to make the bend.

Next, straighten out the tubing. Does this require the same force, more force, or less force?

Now, bend the tubing again. How much force is needed? Is it possible to straighten it out once more?

Look closely at the metal after each bend. Look at the area where the inside of the bend was. After the second bend, examine the outside of the bend. Do you see small cracks in the surface? You may need to bend the tubing back and forth two or three times to see the cracks.

Take a new piece of tubing and bend it once, but do not straighten it. Try to make the bend radius as small as possible while holding the metal in your hands. Look at the outside surface. Do you see any small cracks similar to the ones you saw before?

Cracks are unlikely to appear after the first bend. Soft copper can be worked once, but not twice, before the ductility of the metal is lost through deformation. When the metal has low ductility, but deformation continues, the atoms can no longer be displaced along the slip planes. Instead, very small cracks are formed in the surface of the metal. This is one way in which cracks form in metal.

To understand how the atomic structure of the copper has been changed, refer to Figure 16-6. In its original form, the tubing has a nearly perfect arrangement of crystals. The atoms are arranged in a face-centered cubic structure. During deformation, the atoms of each crystal can be displaced along many different slip planes. When you worked the metal, some of these planes intersected and slip occurred across different planes. The intersection of planes made it harder for the atoms to slip. Therefore, the metal became stronger and the force required to bend it increased. The intersection of slip planes also caused the metal to lose ductility. The copper could bend once, but not twice, before cracking.

percentage of cold work can be calculated as follows:

$$\%CW = \frac{t_{orig} - t_{exit}}{t_{orig}} \times 100\%$$

$$= \frac{0.250" - 0.150"}{0.250"} \times 100\%$$

$$= 40\%$$

Thus, the thickness of the sheet has been reduced by 40% after one pass through the rolling mill. Most metals can be cold rolled no more than 75% before they lose ductility and cracking begins to develop.

Drawing and Extrusion

Drawing and extrusion are similar finishing processes. *Drawing* is a shaping method in which metal is pulled (drawn) through a die to produce a smaller size. In *extrusion,* the metal is reduced to a final size when it is forced through a die by a ram. Metal can be extruded hot or cold. Hot extrusion occurs when heated metal is extruded to a final size.

Metal rod, tube, and wire are commonly produced through drawing. A drawn rod reduced from its original size to a smaller diameter is shown in Figure 16-9. In drawing, the amount of cold work can be determined by first computing the original and final area of the part and then using these values to find the percentage of cold work. The following equations are used to determine cold work for drawn metal:

$$A_{orig} = \frac{\pi \times d^2_{orig}}{4}$$

$$A_{exit} = \frac{\pi \times d^2_{exit}}{4}$$

$$\%CW = \frac{A_{orig} - A_{exit}}{A_{orig}} \times 100\%$$

where

d_{orig} = original diameter
d_{exit} = final diameter
A_{orig} = original area
A_{exit} = exit area
$\pi = 3.14159$

Assume a piece of metal stock is drawn cold from an original size of 0.350" to a final diameter of 0.325". The percentage of cold work can be calculated as follows:

$$A_{orig} = \frac{\pi \times d^2_{orig}}{4}$$

$$= \frac{\pi \times (0.350")^2}{4}$$

$$= 0.09621 \text{ in}^2$$

$$A_{exit} = \frac{\pi \times d^2_{exit}}{4}$$

$$= \frac{\pi \times (0.325")^2}{4}$$

$$= 0.08296 \text{ in}^2$$

$$\%CW = \frac{A_{orig} \times A_{exit}}{A_{orig}} \times 100\%$$

$$= \frac{0.09621 \text{ in}^2 - 0.08296 \text{ in}^2}{0.09621 \text{ in}^2} \times 100\%$$

$$= 13.8\%$$

Thus, the percentage of cold work in the drawn rod is 13.8%.

When metal is extruded hot, the percentage of size reduction is calculated using the same method. However, the change in size is calculated as percent hot work instead of percent cold work.

Annealing Metal after Cold Working

It is not always possible to roll a sheet of metal to final size through cold working only. When a relatively thick sheet must be reduced to a small thickness, other processes are required. For example, sheet metal used for the body of an automobile measures 0.040" or less thick, and the thickness of fin stock for an automobile radiator is normally less than 0.005".

To produce these sizes, aluminum ingots are rolled hot to a thickness of about 0.250" and then reduced further by cold rolling. However, it is not possible to cold roll the metal directly from 0.250" to 0.040" or 0.005". Even if it were possible, the ductility of the finished thin strip would be very low. This hard, strong strip would crack from a small amount of bending and it would not be possible to form the sheet into parts.

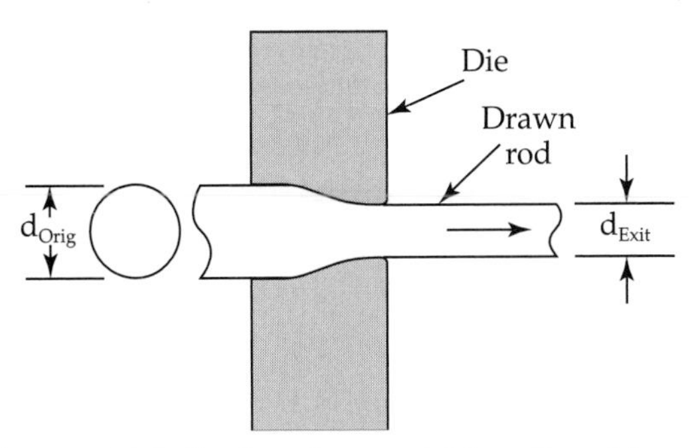

Figure 16-9. *In drawing, metal is reduced in thickness as it passes through a die.*

When further cold working is required to produce a final size, metal can be annealed to regain softness and ductility. Annealing was discussed in Chapter 12. In this process, metal is heated and slowly cooled in a furnace. In nonferrous metallurgy, annealing consists of heating cold-worked metal to a moderately high temperature (below the melting temperature) and then cooling it in air to room temperature. When the metal is fully cooled, it will be as soft and ductile as it was before cold working.

The time required in annealing varies with the annealing temperature. Heating time, annealing temperature, and cooling time must be specified to ensure the metal undergoes the proper heat treatment. At moderately high temperatures, heating will often require an hour or longer in the annealing furnace. At higher temperatures (near the melting point), the annealing time is much shorter. For a copper alloy at a heating temperature approximately 200°F below the melting point, annealing takes less than two minutes.

Cold working and annealing both change the atomic structure of metal. Before cold working, the metal has a nearly perfect crystal structure. Cold working forces slip to occur on many planes at once; slip becomes more difficult as the slip planes intersect. This causes structural distortion on the submicroscopic scale. When the metal is annealed, the atoms return to the original, near-perfect crystal structure.

In annealed material, the areas made of one crystal are called grains, and grain boundaries can be clearly seen under a microscope at 100X magnification. Cold working stretches and bends the crystals. It is difficult to recognize the grain boundaries in a photomicrograph. During annealing, new grains form and the final structure looks virtually identical to the original, clean grain structure.

Precipitation Hardening

Precipitation hardening is a heat treatment process that strengthens alloys by causing *phases* (regions in the metal with different crystal structures) to precipitate at various temperatures and cooling rates. The metal is changed from a solution to a mixture. Suppose additional elements are added to a pure metal to form an alloy. When this alloy is heated to a high temperature, the alloying elements form a solution in the metal. Upon rapid quenching, a new phase will precipitate in a few minutes or weeks. The new phase strengthens the metal as it grows, in this process called *age hardening*. It is used for many types of nonferrous alloys, such as copper-beryllium, copper-aluminum, and some stainless steel.

Precipitation hardening consists of two separate heating and cooling cycles. First, the alloy is heated to a *solutionizing* temperature (an elevated temperature below the melting point). In this condition, the alloy is a single-phase (or single-crystal structure) solid solution. The alloy is held at this temperature for a time before rapid quenching. Quenching prevents a second phase (or additional phases) from precipitating in the alloy. It also strengthens the metal somewhat.

In the second stage of precipitation hardening, the alloy is reheated to a moderate temperature (well below the solutionizing temperature), causing the second phase to precipitate. This allows the second phase component to be uniformly distributed in the grains of the original solution. The resulting structure is very strong and has the properties that are essential to the alloy.

The two different stages of heat treatment in precipitation hardening allow nonferrous alloys to attain maximum hardness and other desirable characteristics, such as toughness.

Using Precipitation Hardening to Strengthen Alloys

Precipitation hardening is commonly used to process copper alloys and other nonferrous metals for commercial use. The following is an example of using precipitation hardening to

Self-Demonstration
Cold Working and Annealing Copper

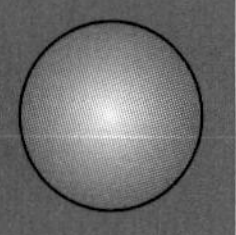

Obtain a piece of soft copper tubing measuring 3/8" in diameter and about 1" in length.

Carefully flatten the tubing by applying just enough pressure to join the two sides together. Use a hand vise if one is available. A piece of fully annealed brass bar stock, measuring 0.5" square and 1" long, will work even better than copper tubing for this exercise.

Using the Rockwell F scale, measure the hardness of the metal.

Next, hammer the piece on one side of the metal only. Reduce the thickness by 50%. If you are using a sample of brass bar stock, you may reduce the thickness from the original 0.5" size to 1/8" (a 75% reduction) to produce a more dramatic effect. Try to strike the metal so that the center of the piece becomes flat and an accurate hardness measurement can be made. Measure the hardness, using the Rockwell F scale.

Heat the sample in a furnace at 800°F for 1 hour. Cool in air, then measure the hardness again and compare the new value to the hardness prior to annealing. The hardness should return to its original value (or change to an even lower value). After annealing, the crystal structure should also be back in its original, neat pattern.

process C17200, a copper alloy with 1.9% beryllium, for an electrical application.

Assume a thin strip of copper-beryllium measuring 0.020" thick and 0.3" wide is used to make the electrical contact components in a computer. The metal is formed into contact arms by a stamping machine. In many cases, these components must perform two functions. The arms put force on a pin to make the electrical connection while also holding the pin in place. If you look closely at the socket of a ribbon connector in a personal computer, you may be able to see a similar application. The sockets for edge card connections in industrial computers are typically made in the same fashion.

In its original form, C17200 is fully annealed and very soft. It also has high formability. This allows the stamping machine to make sharp bends in the metal and produce the required shape without cracking. However, the strength of this soft metal is not suitable for the computer application. The metal cannot press firmly enough against the electrical connection points.

Precipitation hardening can be used to produce the required strength after the metal is formed to the desired shape. In the first stage of heat treatment, the copper-beryllium parts are heated to a solutionizing temperature and held at that temperature for about 30 minutes. The parts are then rapidly quenched in water. This makes the alloy harder and stronger than it was originally. However, the parts are not yet at full strength.

In some cases, the quenched parts are *naturally aged* at room temperature. In this step, the metal gains strength slowly over a period of time, sometimes years. For some commercial applications, however, processing must occur rapidly and the metal must have the absolute maximum strength.

To acquire the maximum strength in precipitation hardening, the solutionized and quenched parts are *artificially aged* (strengthened) at moderate temperature. This form of secondary heat treatment takes one to four hours. After the parts are cooled, they are at maximum strength. In some nonferrous alloys, such as aluminum A96061, some ductility is retained after precipitation hardening. By contrast, C17200 has very high strength but little ductility.

After processing, the copper-beryllium parts are strong enough to perform as electrical connectors. Solutionizing (the initial heat treatment) gave the metal excellent form ability. Precipitation hardening strengthened the metal and made it suitable for commercial use.

Heating temperatures in precipitation hardening must be carefully controlled to produce the desired results. During heat treatment, different atomic structures, or phases, are present in the alloy at various temperatures. These phases have a direct effect on the properties of the metal and can be identified at each stage of heating and cooling. Through the use of phase diagrams, metallurgists can predict the phases that occur during precipitation hardening. Phase diagrams are discussed next.

Phase Diagrams

Phase diagrams identify alloy phases that occur at various temperatures and percentages of alloying elements. Phase diagrams are commonly used for *binary alloy systems* (alloys with two metallic elements). Iron-carbon phase diagrams were covered in Chapter 9. The iron-carbon phase diagrams are graphic representations of steel structures at different temperatures and percentages of carbon.

Phase diagrams are also called *equilibrium diagrams* because they identify structures that occur in equilibrium conditions (as different compositions of the alloy *equalize* over time). In other words, the structure represented at any given composition and temperature is that with the lowest chemical energy. Structures are produced after the alloy is held at an indicated temperature for a period of time. At temperatures close to the melting point, equilibrium is typically reached in a few minutes. At room temperature, most structural metals will never reach equilibrium, in practical terms.

Often, the most desirable material properties in an alloy system are produced when the atoms are not at equilibrium. By referring to a phase diagram and varying heating temperatures and cooling rates, metallurgists can determine the proper heat treatment and develop the desired properties for an alloy.

Identifying Alloy Phases in Precipitation Hardening

The different alloy phases that occur during precipitation hardening can be identified using a phase diagram. A diagram for copper-beryllium is shown in Figure 16-10. The various structures and compositions of copper (with small percentages of beryllium) are graphed at different temperatures. The copper alloy in its original form, C17200, contains about 1.9% beryllium. This alloy is represented by point A on the dashed line in Figure 16-10.

The original alloy consists of copper (Cu) and relatively large particles of CuBe, a copper-beryllium compound. A sketch of the microstructure of this alloy system is shown in Figure 16-11. The portion of copper in the alloy has a face-centered cubic structure. The CuBe particles have a crystal structure that resembles a body-centered cubic structure, with the exception that the center atoms are always beryllium. These particles have little effect on the strength of the metal.

During solutionizing (the first stage of precipitation hardening), when the alloy is heated to an annealing temperature, the alloy moves from the Cu plus CuBe region to the Cu region (point B in Figure 16-10). The CuBe particles are dissolved into the solution, Figure 16-12.

When the alloy is rapidly quenched to room temperature (point A or below in Figure 16-10), the rate of cooling prevents development of a second phase. The beryllium parti-

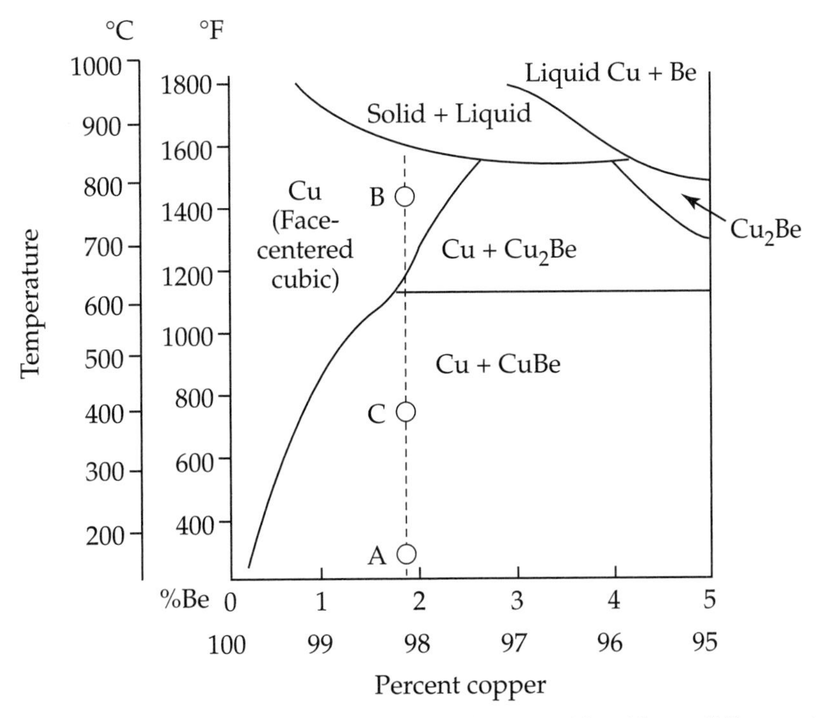

Figure 16-10. *A phase diagram for copper-beryllium. Each region identifies a different phase of the alloy during heat treatment. (Adapted from Metals Handbook, Eighth Edition, Volume 8; ASM International)*

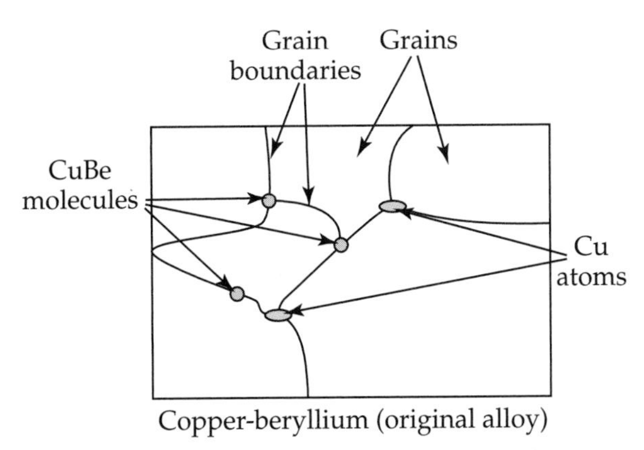

Figure 16-11. *A microstructural sketch of a copper-beryllium sample before heat treatment.*

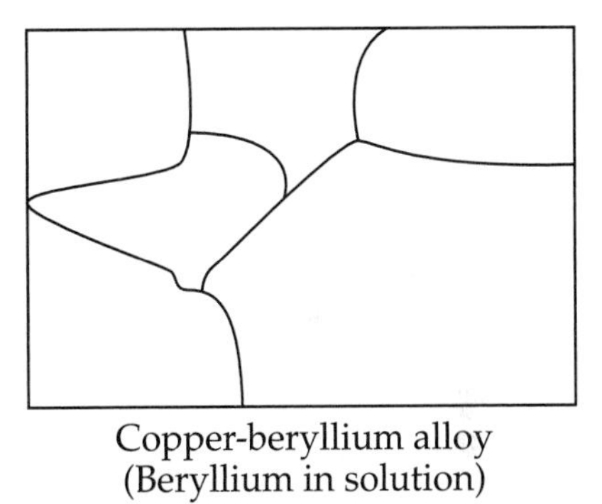

Figure 16-12. *Particles of beryllium are dissolved into solution when the alloy reaches the solutionizing temperature.*

cles do not have enough time to precipitate. The beryllium remains in solution. At this point, the single-phase alloy is a *supersaturated solution.* The alloy is stronger as a result of quenching. Any small particles of CuBe are so small they cannot be seen even at 20,000X magnification. The beryllium has strengthened the alloy through the formation of a few CuBe particles that are nearly on the atomic scale in size.

Additional strength can be attained by secondary heat treatment. When the alloy is reheated to a moderate temperature for approximately one hour (at point C in Figure 16-10), beryllium reacts with the copper to form CuBe particles. When the particles precipitate, they are no longer in solution, and a true mixture is produced.

At moderate temperature, the beryllium atoms can move inside the metal, but they cannot move very far or fast. These atoms tend to move toward the first particles that form. Since their movement is limited, however, additional particles are formed and more precipitation occurs. The particles (or *precipitates*) are very small and evenly spread out in the grains of the alloy, unlike the original structure. See Figure 16-13. This makes it more difficult for slip to occur. Therefore, the metal is much stronger than before.

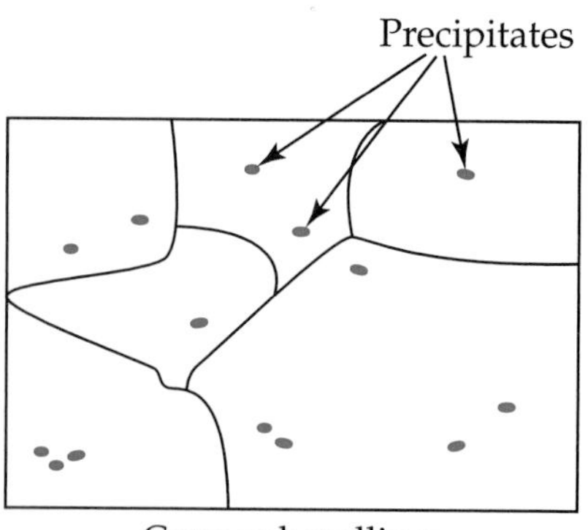

Figure 16-13. *A microstructural sketch of copper-beryllium shows the development of precipitates in the alloy after precipitation hardening.*

Precipitates are so small they cannot be seen with an optical microscope (at 50X to 2000X magnification). A very high magnification is needed to view the precipitates in an alloy of aluminum and silver, Figure 16-14.

Interestingly enough, copper-beryllium processed through precipitation hardening does not lose as much ductility as it would lose during cold working or other strengthening methods. Thus, precipitation hardening produces maximum strength in nonferrous alloys with only a moderate loss of ductility. A comparison of hardening methods is shown in the table in Figure 16-15.

Figure 16-14. *The Ag_2Al precipitates in this aluminum alloy are visible at 16,000X magnification after precipitation hardening. The Ag_2Al precipitates are the thin dark lines (particles seen on edge) and dark platelets (particles seen on a broad flat side). The alloy has a silver content of 4.2%. (J.M. Howe/Department of Materials Science and Engineering, University of Virginia)*

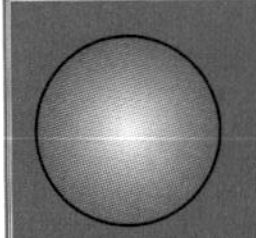

Self-Demonstration
Solutionizing and Precipitation Hardening

Pour about one cup of water into a beaker. Add one teaspoon of sugar. Stir until the sugar is not visible. This is a solution of sugar in water.

Add two tablespoons more sugar into the solution and stir. Continue adding sugar until some of it remains on the bottom of the beaker even after stirring for a minute or more. This is a mixture of solid sugar in water. It is a two-phase structure, similar to the composition of copper-beryllium before solutionizing. The clear liquid (a solution of water and sugar) is one phase. The solid sugar grains at the bottom of the beaker constitute a second phase.

Next, heat the solution and bring it to a boil. Add two more tablespoons of sugar. Stir and heat the solution until you can no longer see the sugar on the bottom (that is, when it has been completely dissolved into the solution).

Be careful not to let the solution boil over the sides of the beaker. Reduce the heat if necessary. If the water appears thin, add more sugar, two tablespoons at a time.

When the solution is still boiling, but has become much thicker than pure water, check to make sure that all of the sugar is in solution. There should be no solid grains at the bottom of the beaker. At this point, the solution is in a single phase, similar to the composition of copper-beryllium after solutionizing.

Reduce the heat and let the beaker cool. Handle the hot beaker carefully. Cover it to keep the solution free of dust.

After cooling, the solution should remain in liquid form. No grains of sugar should be visible. This is a *supersaturated solution.* It is not at equilibrium.

Store the cooled beaker on a shelf for several days. Each day, check the solution for solid crystals of sugar. If you added enough sugar while the solution was boiling, it will take a week or so for a new crystal to form.

The new formation represents precipitation of the sugar. The second phase has formed and the solution has returned toward the same composition it was in before solutionizing. At this point, the solution is moving toward equilibrium.

Comparison of Processing Methods for Nonferrous Metals			
Process	**Equipment used (after casting)**	**Strength (relative to a pure metal at low strength)**	**Ductility (relative to a pure metal at high ductility)**
Alloying	None	Moderate	Fairly high
Cold working	Rolling, forging, and drawing mills; hammers	High	Low
Precipitation hardening	Furnaces, quench tank	Very high	Moderate

Figure 16-15. *A comparison of the material properties associated with processing methods for nonferrous metals.*

Test Your Knowledge

Write your answers on a separate sheet of paper. Do not write in this book.

1. Define *nonferrous metallurgy.*
2. Name four types of atomic unit cells that occur in metals.
3. What type of atomic structure occurs in austenitic iron and metals such as aluminum, copper, and silver?
4. Explain what happens inside a crystal structure when metal deforms.
5. What is the difference between a *solid solution alloy* and a *mixture?*
6. What two metallic elements are alloyed to produce bronze?
7. What is *cold working?*
8. A section of heavy sheet 0.100" thick is reduced to a final thickness of 0.040" by passing through a rolling mill. What is the percentage of cold work?
9. Explain the difference between *drawing* and *extrusion.*
10. How can annealing be used to regain softness and ductility in a nonferrous metal after cold working?
11. Describe the two stages of heat treatment that alloys undergo in precipitation hardening.
12. What are the two functions of quenching in precipitation hardening?
13. What is a *phase diagram?*
14. Why is a phase diagram also called an *equilibrium diagram?*
15. What are *precipitates?*
16. Of the basic processing methods used for nonferrous metals, which one produces the most hardness?

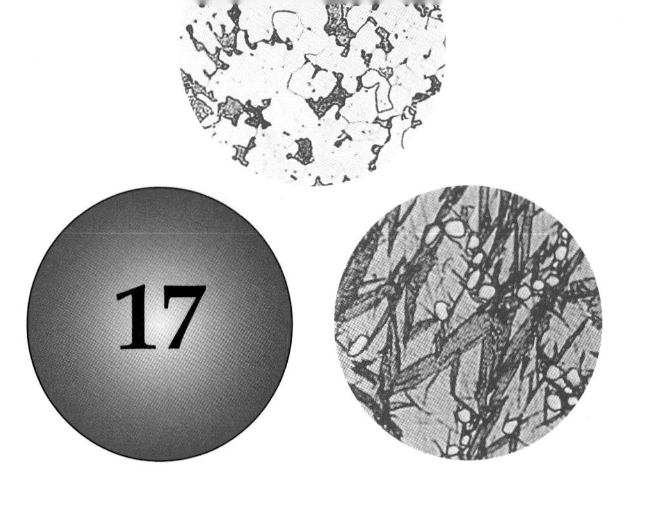

17 Aluminum and Aluminum Alloys

After studying this chapter, you will be able to:

- Describe the desirable properties of aluminum and aluminum alloys.
- Explain how aluminum is refined.
- List common applications of aluminum and aluminum alloys.
- Explain how alloying, cold working, and precipitation hardening can be used to change the properties of aluminum.

Introduction

Aluminum and aluminum alloys are the most widely used nonferrous metals. They have many desirable properties, including high strength, good corrosion resistance, low density, and high electrical and thermal conductivity. Aluminum is nonmagnetic and easily recyclable.

Aluminum can be easily formed by rolling and forging. Aluminum parts formed by stamping, bending, and welding are often found in applications where low weight is important, such as in airplanes and automobiles.

Manufacturing Aluminum

Aluminum is made from *bauxite,* an abundant ore that contains aluminum hydroxide (alumina), water, iron oxide, silica, and titanium oxide. Many large deposits of bauxite are found near the earth's equator, but bauxite is also found in other areas of the world. After bauxite is mined, it is crushed, washed, dried, and ground into a powder before being transported to a refining plant.

At the refining plant, the powered bauxite is mixed with sodium hydroxide and heated under pressure in a large vessel called a digester. The digester causes the alumina in the bauxite to dissolve in the sodium hydroxide. This forms a solution of sodium aluminate. The other materials in the bauxite remain solid. These solid materials are called *red mud.*

Filtering separates the sodium aluminate solution from the red mud. The sodium aluminate solution is then sent to a precipitator. In the precipitator, aluminum hydroxide crystals are added to the sodium aluminate and the entire solution is agitated. This causes the alumina in the solution to collect on the aluminum hydroxide crystals.

After the precipitation process, the solution passes through filters that separate the aluminum hydroxide crystals from the solution. The crystals are washed and then heated in a kiln to remove any water. The result is a fine white alumina (aluminum and oxygen) powder.

To separate the aluminum from the oxygen, the alumina is dissolved in a high-temperature bath of molten cryolite salt. The molten cryolite salt is contained in a carbon-lined steel pot. Carbon blocks suspended in the pot send electric current through the salt bath, causing the alumina to break apart. Oxygen from the alumina combines with carbon from the blocks and is released as carbon dioxide gas. The molten aluminum metal settles to the bottom of the pot. A pipe is used to transfer the molten aluminum from the bottom of the pot to a ladle, or crucible. See Figure 17-1.

Figure 17-2 shows a typical aluminum "pot line." The heavy bus bars that carry electric current to the carbon blocks can be seen extending over the sides of the pots. In the foreground, an operator handles the pipe used to transfer the liquid metal to the crucible.

The metal in the crucible is cast into pigs (crude castings). Each pig weighs about 50 lb. (23 kg). The pigs are remelted in a large holding oven. Selected alloying elements and, in some cases, scrap aluminum are added to the melt. After the alloying elements and scrap have been added, the melt is poured through filters and cast into a large ingot. These ingots are about 2' thick, 5' wide, and up to 20' long.

Unlike steel ingots, most aluminum ingots are not poured in a single step. Instead, only enough liquid metal is poured in to cover a chilled copper plate that supports the ingot in the mold. Then the plate is lowered, and more metal is poured in. Water jets spray the bottom and sides of the ingot to make it solidify faster. This process is called *drop casting*, or *semi-continuous casting*. Today, some aluminum is poured by continuous casting.

Figure 17-2. *Aluminum oxide is reduced to aluminum metal in a pot line. (Kaiser Aluminum)*

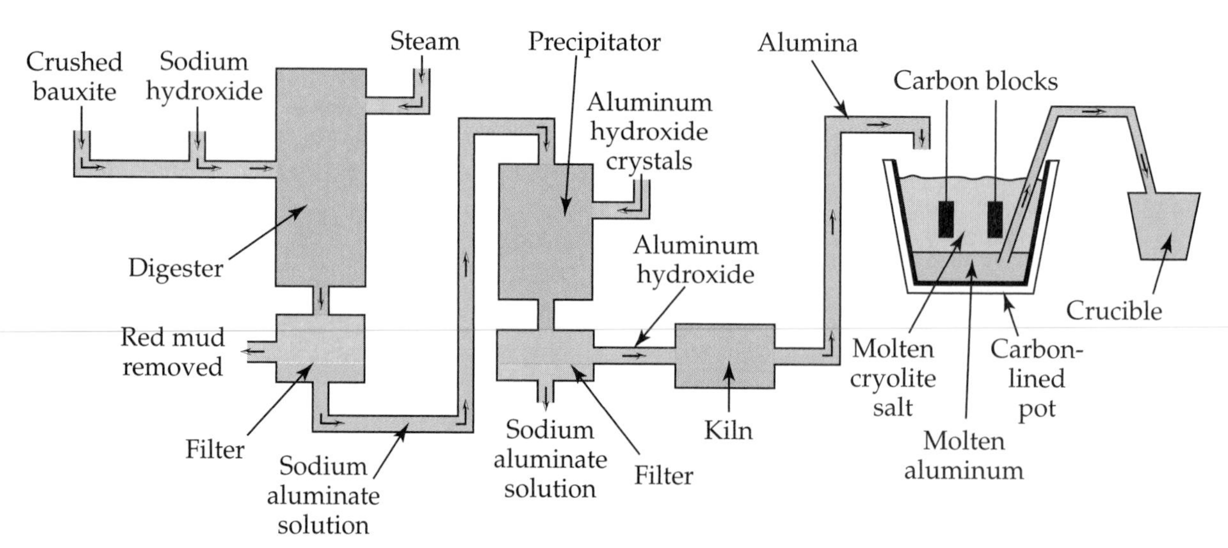

Figure 17-1. *This schematic shows how bauxite ore is refined.*

After the ingot has cooled, the sides are milled to remove the rough cast surface. The ingot then goes to a soaking pit, where it is reheated to a uniform temperature. The hot ingot is placed on a runout table. While the ingot is still hot, large rolls press it down from 24" (60 cm) thick to 1" (2.5 cm) thick or less. Figure 17-3 shows the "hot line" that Alcoa uses to produce aerospace plate and sheet.

Most wrought aluminum is rolled into plate, sheet, or strip. Some aluminum is cast into ingots about 10" (25 cm) in diameter and five or more feet long. These ingots are heated to a high temperature and placed into an *extrusion press.* Then the press pushes the metal out through a die, forming it into the desired shape, Figure 17-4.

Figure 17-4. *In this extrusion press, hot metal is pressed through a die to form seamless pipe. The seamless pipe can be seen between the left and center operators. The hydraulic press is visible on the right. (Kaiser Aluminum)*

Aluminum and Aluminum Alloy Classifications

Aluminum and aluminum alloys are identified by number designations developed by the Aluminum Association (AA).

Figure 17-3. *Aluminum ingots are rolled to the desired thickness on a hot line. (Alcoa)*

Wrought aluminum and wrought aluminum alloys are identified by a simple four-digit number. Each digit in the number has a specific meaning. The first digit identifies the major alloying element, Figure 17-5. The second digit indicates a modification of impurity limits for group 1xxx and a modification of the original alloy for groups 2xxx–8xxx. The last two digits indicate purity for group 1xxx.

Wrought Aluminum and Aluminum Alloy Designations	
Material	**Number**
Aluminum (99.00% minimum and greater)	1xxx
Aluminum alloys are grouped by major alloying elements:	
Copper	2xxx
Manganese	3xxx
Silicon	4xxx
Magnesium	5xxx
Magnesium and Silicon	6xxx

Figure 17-5. *Aluminum Association designations for wrought aluminum.*

Cast Aluminum and Aluminum Alloy Designations	
Material	**Number**
Aluminum (99.00% minimum and greater)	1xx.x
Aluminum alloys are grouped by major alloying elements:	
Copper	2xx.x
Silicon with added copper and/or magnesium	3xx.x
Silicon	4xx.x
Magnesium	5xx.x
Zinc	7xx.x
Tin	8xx.x
Other elements	9xx.x

Figure 17-6. *Aluminum Association designations for cast aluminum.*

For groups 2xxx–8xxx, the last two digits identify the specific aluminum alloys.

Cast aluminum and cast aluminum alloys are also identified by a four-digit number. However, the numbers used to represent cast materials contain a decimal point between the third and fourth digits. See Figure 17-6. The first digit represents the alloy group. The second two digits identify the specific aluminum alloy (groups 2xxx–8xxx) or, in the case of aluminum (group 1xx.x), specify purity. The final digit indicates the form of the product: 0 indicates a casting and 1 indicates an ingot.

The Unified Numbering System (UNS) has established designations similar to those developed by the Aluminum Association. Figure 17-7 compares UNS designations to AA designations.

Applications of Aluminum

There are many applications that take advantage of aluminum's desirable properties, especially the properties of low density, high strength, toughness, and weldability. Some truck body frames are made of aluminum. Aluminum parts reduce weight, improving fuel economy and performance. Using lightweight materials allows trucks to carry more payload without exceeding highway weight restrictions.

Aluminum is also a good conductor of electricity, and it costs less than copper. The cables in electrical transmission lines are made of aluminum. Aluminum reduces the weight of the transmission lines, Figure 17-8. Less weight means there is less load on the cables and towers. Consequently, the towers

UNS Designations and AA Designations			
Wrought Aluminum and Aluminum Alloys		**Cast Aluminum and Aluminum Alloys**	
UNS Designation	AA Designation	UNS Designation	AA Designation
A91xxx	1xxx	A01xxx	1xx.x
A92xxx	2xxx	A02xxx	2xx.x
A93xxx	3xxx	A03xxx	3xx.x
A94xxx	4xxx	A04xxx	4xx.x
A95xxx	5xxx	A05xxx	5xx.x
A96xxx	6xxx	A07xxx	7xx.x
A96xxx	7xxx	A08xxx	8xx.x
A98xxx	8xxx	A09xxx	9xx.x

Figure 17-7. *Unified Numbering System designations for both wrought and cast aluminums.*

Figure 17-8. *Lightweight aluminum has very good electrical conductivity. It is a good material for conductor cables. (The Aluminum Association.)*

can be farther apart. This translates to lower costs and less environmental impact.

Due to its good ductility and high strength, aluminum alloys are often used for aircraft and aerospace applications. As discussed in Chapter 7, a part that is not ductile is brittle, and a sudden blow will destroy it. Clearly, we do not want to build an aircraft from brittle material. Instead, we want a material that is very strong *and* ductile. The combination of high strength and high ductility is called *toughness.* Most technical developments of aerospace alloys aim to develop extreme toughness. The aluminum alloys in the 7xxx series have high toughness. These aluminum alloys contain zinc, magnesium, and copper. Processing is critical to attain successful properties.

Figure 17-9 shows wing frame sections made from 7075 alloy. This aluminum alloy contains 5.6% zinc, 2.5% magnesium, and 1.6% copper. The way this material is processed is as important as its alloy content. The cast ingot is heated to 950°F (510°C); then it is hot rolled from a 16" thick ingot to a 4.5" thick slab. The

Figure 17-9. *The frame sections can be seen in this wing for a Boeing 777-200 under construction. (The Boeing Company)*

slab is reheated to 950°F and then hot rolled into plate (over 0.25") or coil (0.25" thick) form. The coil may be subsequently cold rolled to the desired thickness, with intermediate annealing. When the coil reaches the required thickness, it is solution heat treated and quickly quenched to room temperature. After cutting and forming, the material is aged in a furnace at 250°F (120°C) for 24 hours. This produces the fine particles that give maximum strength and high ductility.

Certain aluminum alloys are used in ocean-going applications because they have good corrosion resistance and are, of course, light. Aluminum was used extensively in the manufacture of the yacht shown in Figure 17-10.

Aluminum can be readily welded and brazed. Complex shapes can be brazed in a single furnace operation to make high-efficiency radiators and other heat-transfer devices. These applications take advantage of the high thermal conductivity of aluminum.

Aluminum products are also easy to recycle. Over 60% of all aluminum cans are now returned to the casting plants. The scrap metal is melted, mixed with new aluminum, and then poured into ingots for new cans. Recycling aluminum saves 95% of the electricity used to refine the aluminum in the first place. Considering the overall aluminum production process, making aluminum ingots from scrap costs much less than refining aluminum from bauxite. This is why aluminum companies happily pay between 25 and 45 cents per pound for scrap aluminum cans. Recycled aluminum is also used for aluminum castings.

Figure 17-10. *This luxury yacht has an aluminum hull. (Bill Prince Yacht Design)*

Changing the Properties of Aluminum

Pure aluminum is soft and easily dented or bent. To make it a more useful material, we must strengthen it using one or more of the metallurgical processes discussed in the previous chapter: alloying, cold working, or precipitation hardening. These processing methods can also affect other properties, such as toughness and ductility.

Alloying

Adding 0.5 to 5.0% magnesium to aluminum increases the strength through solid solution hardening. Adding magnesium also makes aluminum work harden well while maintaining good formability.

Aluminum-magnesium alloy designations begin with the number 5, such as 5054 (used for tank truck bodies) or 5152 (used in the manufacture of air cleaners and inner strengthening members of automobile hoods). Many applications that require great formability and reasonable strength use these aluminum-magnesium alloys. The covers of aluminum beverage cans are made from this type of alloy.

Cold Working

The second method for improving strength is work hardening. Cold rolled aluminum sheet is stronger than fully annealed sheet. Using a stronger material means we can build devices with less metal. The oil cooler shown in Figure 17-11 uses aluminum tubing and thin sheet fins. For the cooler to work well, the thin fin material must press tightly against the tubing wall. Thermal energy must transfer easily from the oil, through the tubing, to the fins, and into the air. Both the tubing and the fins have been cold rolled to increase their

Figure 17-11. *An aluminum oil cooler. (Modine Manufacturing Company)*

strength. The increased strength is needed so they can be pressed tightly together, allowing the efficient transfer of thermal energy.

Solution Heat Treating and Precipitation Hardening

The first aluminum alloys strengthened by solution heat treatment were the 2xxx alloys. These alloys are composed of aluminum and copper (2–6%). Additional elements are sometimes added to achieve specific properties. The alloy is cast, hot rolled, and then cold rolled to the desired thickness. The final sheet may be 0.062" (1.6 mm) thick or less. After cold rolling, the sheet passes through a long solutionizing oven. The oven quickly heats the sheet to a temperature slightly below its melting point. The speed of the metal sheet, or *web,* and the length of the oven must be set so the metal is heated to the desired temperature for a minute or more. Then, the sheet passes directly into a water spray, where it is quenched to room temperature in less than a second.

Usually, the metal is coiled and shipped to the customer in the *as-quenched* condition. The customer can form the as-quenched metal, as it still has high ductility. After forming, the material can be artificially aged, or it can simply be allowed to naturally age at room temperature. The artificial aging will develop higher strength than natural aging. However, artificial aging is sometimes impossible because the complete part will not fit into a low-temperature oven. Aluminum alloy 2024 applications include aircraft skin sheet, the outer surface of passenger airplanes, Figure 17-12.

Certain aircraft rivets are solution heat treated and quenched directly into a freezer, where they are held at –40°F (–40°C) until they are used. The rivets are still in the as-quenched condition, so they are quite ductile. Rivets work well when they can expand to fit the hole in which they are placed, so ductility is needed. The rivets are held at such a low temperature because they naturally age very quickly at room temperature. Within two days, the rivets have reached full strength. This application makes use of the high ductility of the metal in the as-quenched condition and the high strength of the metal in the aged condition.

All aluminum alloys with an alloy number that begins with a 6, such as 6061 or 6063, can be heat treated to increase strength. They all contain magnesium and silicon as their primary alloying elements.

One "workhorse" aluminum alloy, 6061, can be heat treated to provide high ductility in the as-quenched condition and high strength in the final condition. This alloy is mostly aluminum, with 1% magnesium and

Figure 17-12. *The outer surface of this Boeing 747-400 airplane is made of aluminum skin sheet. (The Boeing Company)*

0.6% silicon added. When the metal is heated to 950°F (510°C), these elements go into solution easily and stay in solution upon rapid quenching. In the as-quenched condition, the metal bends and forms easily. After forming, the metal is reheated to 350°F (175°C) for an hour. This causes fine particles of magnesium silicide (Mg_2Si) to precipitate in the metal. The strength more than doubles.

Today, the 6061 alloy is used in truck frames, where the high strength of the heat-treated metal is needed. Careful design and expert welding during assembly allow single-piece construction for added strength and a larger payload. Figure 17-13 shows a dump trailer body with welded stiffening ribs visible along the side.

An extruded window frame made from 6063 aluminum is an excellent example of the solutionizing, quenching, and aging sequence. The frame segments surrounding the glass have a complex cross section and are relatively long. The complex cross section and long length make ideal extrusions, and aluminum is an ideal metal for extruding. The aluminum ingot is extruded at a high temperature, when the metal is soft and ductile. In some cases, it is quenched directly as it exits the extruder. In other instances, it is solution heat treated in a separate furnace and then quenched. In the as-quenched condition, the metal can be bent and drilled as needed. It can then be assembled and welded. After final assembly, the entire frame can be placed in an oven at 350°F (175°C) for an hour to further increase the strength.

Alloy 6010 is another Al-Mg-Si alloy that can be strengthened by precipitation hardening. Aluminum canoes made from 6010 take advantage of the increased strength gained from a final furnace aging treatment. Figure 17-14 shows a canoe made from this alloy.

Designers often specify 6010 for auto body parts because it can be formed easily in the as-quenched condition. The aluminum supplier ships the metal to the automotive manufacturer immediately after the solutionizing and quenching steps. After forming and fabrication, the strength is increased during the paint-baking operation, which all automobile frame and body parts go through. The moderate temperatures (slightly higher than boiling

Figure 17-13. *Aluminum frame rails reduce the weight of trucks, so they can carry more payload. (Ravens Metal Products, Inc.)*

Figure 17-14. *This canoe is made from 6010 alloy. (Marathon Canoe, successor to Grumman Canoe)*

Figure 17-15. *The frame of this bicycle is made entirely of heat-treated aluminum. (Trek Bicycle)*

water) reached during the paint-baking operation produce the artificial aging.

After welding or brazing, all 6xxx alloys can be heat treated in the same way as unwelded metal. The frame of the bicycle in Figure 17-15, for example, was first assembled, and then the tubes were brazed in place. After brazing, the entire frame was placed in a heat treatment oven for the solutionizing, quenching, and aging process.

As discussed, heat treatment produces maximum strength in the *heat treatable alloys*—the 2xxx, 6xxx, and 7xxx groups. However, if these alloys are heated to the aging temperature again after heat treatment, they will lose their strength in a few hours or weeks. Never weld or braze heat treatable alloys after heat treatment.

Properties of selected aluminum alloys are compared in Figure 17-16.

Test Your Knowledge

Write your answers on a separate sheet of paper. Do not write in this book.

1. List five properties of aluminum.
2. Briefly explain the procedure for refining aluminum.
3. An aluminum alloy with "4" as the first digit of its alloy designation number contains what metal as its major alloying element?
4. List two reasons why aluminum is used in electrical transmission lines.
5. Name three alloying elements that improve toughness in aluminum alloys.
6. In addition to low density, what other property of aluminum makes it attractive for use in marine applications?
7. How are auto body parts made from 6010 alloy artificially aged?
8. Why must you never weld parts that have been heat treated?

Selected Properties of Aluminum and Aluminum Alloys									
AA Designation	**Condition**	**Tensile strength (ksi)**	**Yield strength (ksi)**	**Elongation (% in 2")**	**Shear Strength (ksi)**	**Modulus of elasticity (thousands of ksi)**	**Specific Gravity**	**Melting point (°C)**	**Melting point (°F)**
1100	annealed	13	5	40	9	10	2.71	643–655	1190–1215
2024	precipitation hardened	68	47	20	41	10.6	2.78	500–638	935–1180
3003	annealed	16	6	35	11	10	2.73	643–655	1190–1210
5052	cold worked	38	31	10	21	10.2	2.68	607–650	1125–1200
6010	precipitation hardened	42	25	24	N/A	10	2.7	585–650	1085–1200
6061	annealed	18	8	25	12	10	2.7	580–650	1080–1200
6061	precipitation hardened	45	40	12	30	10	2.7	580–651	1080–1206
7075	precipitation hardened	83	73	11	48	10.4	2.81	475–635	890–1175

Figure 17-16. *This table contains properties of aluminum alloys.*

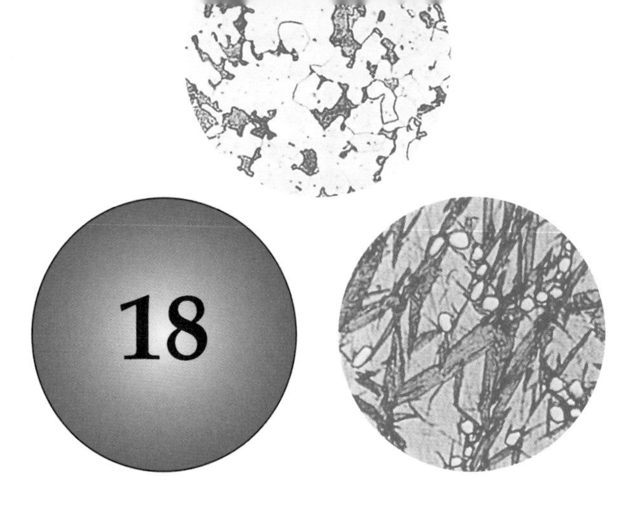

18 Copper, Bronze, and Brass

After studying this chapter, you will be able to:

- List and explain the desirable properties of copper and its alloys.
- Explain how copper is refined from copper ore.
- Describe the favorable properties of bronze and brass.
- Explain the effects of work hardening and precipitation hardening on copper and its alloys.

Copper was one of the first metals used by man. Lumps of pure copper, or native copper, were sometimes found exposed on the surface of the ground. Early inhabitants of the Euphrates Valley, part of what is known today as Iraq, made spear heads of both native and smelted copper. Ornaments of native copper were displayed by the North American Indians as signs of wealth.

The most useful properties of pure copper include excellent electrical and thermal conductivity, as well as high ductility and malleability. The electrical conductivity of copper is second only to that of silver. Copper and copper alloys can be easily soldered and brazed. Copper wire and other electrical applications account for 60% of all copper produced. Copper also has excellent corrosion resistance.

A typical application for copper and copper alloys is the radiators used on over-the-road trucks, Figure 18-1. The large section at the top of the radiator is the tank, and the many tubes in the radiator are held in place by a header. The tank and header distribute water to the tubes for uniform cooling. Figure 18-2 shows the header section and tubes of a truck radiator. Forming and welding are both used to make a strong, long-lasting unit. Soldering joints ensures a leak-free unit.

Manufacture of Copper

Native copper rarely occurs in nature, and the majority of native copper reserves have been exhausted. Today, most copper comes from copper ore mined in the western United

Figure 18-1. *The radiator for an over-the-road truck must withstand a great deal of vibration and temperature variation. (Modine Manufacturing Company.)*

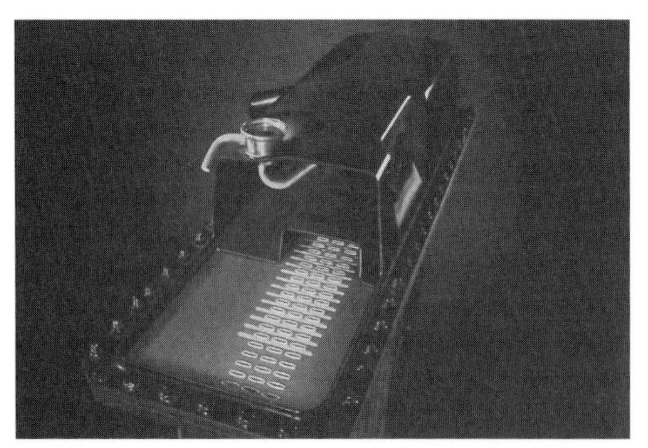

Figure 18-2. *The ribs formed in the header of a Beta-Weld© radiator add strength to the design. (Modine Manufacturing Company.)*

Figure 18-3. *Most copper ore is obtained from large open-pit mines, such as this mine in Southeastern Arizona. (Phelps Dodge Corporation.)*

States, the Andes Mountains of South America, Central Africa, and Russia, Figure 18-3. Copper ore generally contains between 0.2 and 3.0% copper, with an average of 0.7%.

To extract usable copper from copper ore, the ore must be refined through a long and complex process. The ore is first crushed into very fine powder. The powder is mixed with water and chemicals so the part of the ore containing copper (copper concentrate) floats to the surface. The copper concentrate is skimmed off the top of the solution. This process is called *ore benefaction.*

The copper concentrate is heated in a reverberatory furnace, where some of the impurities are driven off. The product that emerges from the furnace is called *copper matte.* The matte is converted to *blister copper* in another furnace, Figure 18-4. Blister copper is 97–99% pure.

If the blister copper does not contain gold or silver and is not going to be used for electrical applications, it can be heated in a fire-refining furnace to burn off most remaining impurities.

If, however, the copper is to be used as an electrical conductor, it must be refined using an electrolytic refining stage called *solution extraction,* or *electrowinning.* Large flat plates are cast from molten blister copper. These plates, called anodes, are suspended in an electrolytic cell that contains a solution of copper sulfate and sulfuric acid. Thin sheets of pure copper are placed between the anodes. These sheets are called the cathodes, or starter plates. See Figure 18-5.

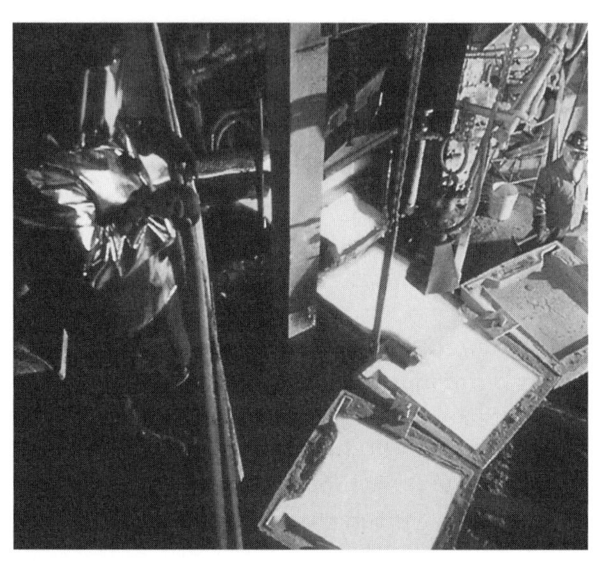

Figure 18-4. *Blister copper is being poured into flat plates for anodes. (Phelps Dodge Corporation.)*

Electrons pass through the solution from the cathode plates to the anode plates. At the same time, the copper atoms from the anodes dissolve into the solution and are deposited on the cathodes. At the end of the electrolytic process, the cathodes are nearly pure copper.

During this electrolytic refining stage, small amounts of silver and gold in the cop-

Figure 18-5. *Copper is refined to pure metal in an electrolytic process. (Phelps Dodge Corporation.)*

per drop down to form a sludge in the electrolytic tank. This sludge is a major source of gold and silver metal. The concentration of these metals in the sludge is much higher than in their natural ores.

Recycled scrap is a major source of copper metal. Over 50% of the copper in products made today has been recycled from scrap.

To make sheet, wire, tubing, and other useful semifinished products, the copper from the electrolytic cathode is charged in a melting furnace. Suitable clean scrap may also be added. For metal to be used in nonelectrical applications, fire-refined metal may be used. Selected alloying elements, such as zinc, aluminum, or phosphorous, may be added as needed to achieve a desired alloy.

The liquid copper or copper alloy is then sent to a holding furnace. From the holding furnace, the metal is poured into a mold. Modern molds for casting copper are semicontinuous, like the molds for aluminum. Typical ingots are 5" thick, 32" wide, and up to 25' long. Some casting operations produce a continuous strip in a horizontal caster.

The finished ingot is ready for reheating and hot rolling or extrusion. All copper alloys intended for sheet or strip are milled after the first hot roll reduction. This guarantees that the final surface finish will be uniform and clean.

Copper and Copper Alloy Designations

Copper and copper alloys are identified by UNS (Unified Number System) designations, which are maintained by the American Society for Testing Materials (ASTM) and the Society of Automotive Engineers (SAE). Each designation contains a five-digit number and is proceeded by the letter "C." The number designations are based on the alloying elements added to copper. See the table in Figure 18-6.

Changing the Properties of Copper

In many cases, an application calls for one or more of the favorable properties of copper. However, pure copper is quite soft—too soft for many tasks. Therefore, it is often hardened using one or more of the metallurgy processes discussed in Chapter 16: alloying, cold working, or precipitation hardening. These processes can also be used to bring about other desirable properties, such as increased ductility and malleability.

Alloying

Alloying elements are often added to the copper to increase its strength. Increased strength may mean that less metal is needed. Less metal translates to lower costs. Certain copper alloys have very high ductility and malleability.

Bronze and brass are two common copper alloys. Bronze is an alloy of copper and tin, arsenic, aluminum, or other metals. It was the first metal alloy discovered. Brass, an alloy of copper and zinc, resists seawater corrosion well. Brass was developed before 600 BC.

Copper and Copper Alloy Classifications		
UNS Designation	**Family**	**Principal Alloying Elements**
	Wrought Alloys	
C1xxxx	coppers, high-copper alloys	—
C2xxxx	brasses	Zn
C3xxxx	leaded brasses	Zn, Pb
C4xxxx	tin brasses, naval brasses	Zn, Sn
C5xxxx	phosphor bronzes	Sn, P
C6xxxx	aluminum, silicon, other bronzes	Al, Si
C7xxxx	copper-nickel alloys	Ni, Zn
	Cast Alloys	
C80xxx–C82xxx	coppers, high-copper alloys	—
C83xxx–C85xxx	brasses	Sn, Zn, Pb
C86xxx	manganese bronzes	Zn, Mn, Al, Fe, Pb
C87xxx	silicon brasses and bronzes	Si
C90xxx–C95xxx	tin bronzes	Sn
C96xxx	copper nickels	Ni, Fe
C97xxx	nickel silvers	Ni, Zn, Sn, Pb
C98xxx	leaded coppers	Pb
C99xxx	special alloys	various

Figure 18-6. *This table summarizes designations for copper alloys.*

Bronze

Bronze was the first metal purposely alloyed by the smith (the name for metallurgists in that day). Technically, "bronze" is a mixture of copper and another element, but people usually use the word "bronze" when referring to a copper-tin alloy. Bronze may contain up to 25% tin.

Early smiths discovered that mixing tin-bearing rocks and copper-bearing rocks together in the smelter produced a new material that had desirable properties. This new metal could be hammered into a hatchet more easily than pure copper, and it would hold an edge better than pure copper. The development of this alloy launched the "Bronze Age" of human development in about 3000 BC. See Figure 18-7.

The earliest bronzes were arsenical bronzes (Cu-4% As), but it was discovered that adding 5–10% tin to copper worked better. Tin bronze was also less toxic for the people casting it. Tin lowers the melting point of the copper and improves its strength. The addition of tin to the copper increases strength by formation of a fine, hard, second phase.

Bronze flows into a mold better than pure copper and is able to fill smaller spaces. The alloy also work hardens better than pure copper.

Other alloying metals, such as lead and zinc, can be added to bronze to enhance its properties. For example, lead is added to bronze to make the material easier to machine.

Today, tin bronze (copper and tin) and aluminum bronze (copper and aluminum) are used as both cast and wrought alloys. Aluminum bronze contains from 5–10% aluminum, plus small quantities of iron, and sometimes nickel or manganese. This material can be processed so it is as hard as steel.

Bronze alloys offer corrosion resistance equal to copper, are easier to cast than copper, and are stronger than copper. Most bearing blocks are made from bronzes because of these superior properties.

Brass

When 5–30% zinc is added to copper, an alloy known as *brass* is produced. The zinc increases the strength of the copper through solid solution hardening. Ductility and formability are also increased. For these reasons, brass alloys are used wherever very high formability is needed. Typical applications for brass include electrical fixtures, hardware, jewelry, and musical instruments.

The radiator headers that were illustrated in Figure 18-2 are formed into complex shapes and contain specially formed holes to hold the tubes. The ridges stiffen the large flat regions and hold the pressure of the hot water inside. Of course, this application also takes advantage of the property of corrosion resistance. Finally, ease of welding, soldering, and brazing ensures that the radiators are sound and water tight, Figures 18-8 and 18-9.

Other alloying elements can be added to brass to alter its properties as necessary. For example, small quantities of lead (1–3%) added to brass make it easier to machine. If superior corrosion resistance is needed, tin and nickel can be added.

Naval brass is an alloy originally developed for the British Navy. It is an alloy of 60%

Figure 18-7. *In early civilizations, bronze was used to make helmets and other strong materials. (Milwaukee Public Museum)*

Figure 18-8. *Brass metal is formed, welded, and soldered to make the header of a Beta-Weld radiator. (Modine Manufacturing Company.)*

Figure 18-9. *The vertical tubes in this close-up photo were flared to make a tight fit; then they were brazed to make the tube-header joint water-tight in the header of a radiator. (The International Copper Association, Ltd.)*

Figure 18-10. *These brass fittings for a small sail-boat are made of naval brass and chrome plated for appearance. (West Marine Corporation.)*

copper, 39.2% zinc, and 0.8% tin. Naval brass resists corrosion very well and can be cast easily to make cannon barrels for use on ships.

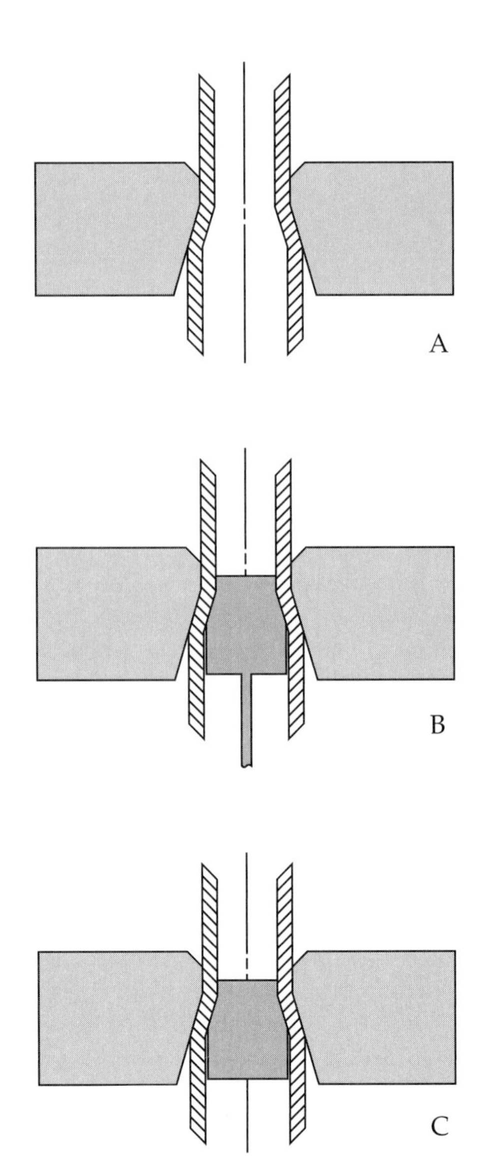

Figure 18-11. *Three different drawing die designs.*

Today, naval brass is called C46400. Figure 18-10 show brass fittings for a sailboat.

Other copper alloys

Several other alloying elements can be added to copper when specific properties are required. For example, some copper alloys contain small amounts of silver. The silver improves the strength of the alloy through solution hardening without hurting electrical conductivity. The addition of silver also helps prevent the alloy from undergoing *stress relaxation* over time. Copper-silver alloys are generally used for electric motors and large generator windings.

Cold Working

As discussed in Chapter 16, cold working is the repeated application of force to harden or strengthen metal. Cold working techniques can be used to change the characteristics of copper alloys.

As an example of cold working, consider the two types of copper tubing available at most hardware stores. Small-diameter tubing is found in large coils, and it can be cut easily with hand shears. This tubing can be bent and twisted to fit the path needed by the plumber.

The second kind of copper tubing is usually larger in diameter and is often called copper pipe or hard copper tubing. Copper pipe is available in long, straight sections. It does not bend easily. In fact, it is not intended to bend or dent. The wall thickness is less than that of the smaller tubing, yet it is clearly stronger than the tubing.

Hard copper tubing is a prime example of the strength added by cold working. This tubing is made by casting a billet, extruding a thick-walled tube, and then drawing it down to its final size. During the drawing operation, Figure 18-11, the metal is pulled through a die to reduce its diameter. A "plug" inside the die may be used to press outward and reduce the metal's thickness, Figure 18-11B. When a "floating plug" is used, seamless tubing of almost any length can be produced, Figure 18-11C.

Precipitation Hardening

When extreme strength and good conductivity are needed, C68800 or C17200 may be used. C68800 is a complex alloy of copper that uses zinc, aluminum, and cobalt to form the fine precipitates that produce high strength. It is usually supplied to the manufacturer in the as-quenched and aged condition due to the complexity of the heat treatment. It is often used in blade-type electrical contacts, where it must press firmly against the contact for a long time and stress relaxation is not permitted.

As discussed previously, C17200 contains mostly copper, with 1.9% beryllium and 0.20% cobalt. It can be purchased as-fabricated, which means it is shipped in the condition it had at the end of the mill fabrication process. It can also be purchased in the as-quenched condition or in the final high-strength condition. If the alloy is received in the maximum strength condition, no bending or forming can be done. The alloy will have very low ductility.

If the C17200 is received in the as-quenched condition, the manufacturer can stamp and form parts to shape. After forming, the parts will be aged to very high strengths that hold their shapes well. For some small edge card connectors in computers, this process is necessary for the computer to work well for a long time.

The table in Figure 18-12 lists some of the properties of common copper alloys.

Test Your Knowledge

Write your answers on a separate sheet of paper. Do not write in this book.

1. List five desirable properties of copper.
2. On average, what percent of copper ore is pure copper?
3. Summarize the process used to extract copper from copper ore.
4. Name the two most common copper alloys.
5. Which copper alloy is a mixture of copper and tin?
6. Aluminum bronze is an alloy of _____ and _____.
7. What effects does zinc have when added to copper?
8. List four applications for brass.
9. Adding _____ to copper alloys makes them easier to machine.
10. Adding tin and nickel to brass will improve its _____.

Selected Properties of Copper Alloys

Material	Description	Condition	Tensile strength (ksi)	Yield strength (ksi)	Elongation (% in 2")	Modulus of elasticity (thousands of ksi)	Specific Gravity	Melting point (°C)	Melting point (°F)
C10100	Oxygen free electronic	annealed full hard	32 66	10 53	55 4	17 17	8.94 8.94	1083 1083	1981 1981
C11000	Electrolytic tough pitch	annealed	32	10	55	17	8.89	1065–1083	1950–1981
C11620	silver bearing tough pitch copper	annealed full hard	32 66	10 53	55 4	17 17	8.92 8.25	1080 1080	1980 1980
C17200	beryllium copper	annealed precipitation heat treated	68 212	25 195	48 1	18.5 18.5	8.25 8.25	865–980 865–989	1590–1800 1590–1800
C22000	commercial bronze	annealed cold worked full hard	37 72	10 62	50 3	17 17	8.8 8.8	1020–1045 1020–1046	1870–1910 1870–1911
C26000	yellow brass	annealed cold worked full hard	44 130	11 65	66 3	16 16	8.53 8.53	915–955 915–956	1680–1750 1680–1751
C51000	phos-bronze	annealed cold worked full hard	47 140	19 80	64 2	16 16	8.86 8.86	975–1060 975–1060	1785–1945 1785–1945
C90500	tin bronze	as cast	45	22	25	15	8.86	854–1000	1570–1830

Figure 18-12. *Selected properties of copper alloys.*

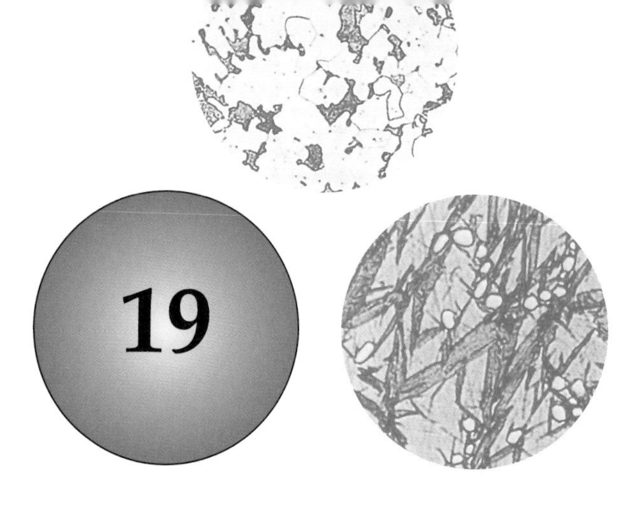

19 Magnesium, Zinc, Tin, and Specialty Metals

After studying this chapter, you will be able to:

- Identify the properties and commercial applications of magnesium, zinc, tin, and other nonferrous metals.
- Describe how the die casting process is used to produce parts from magnesium and zinc.
- Explain the use of tin alloys in wave soldering and other joining processes.
- List the properties and applications of nickel-based superalloys and titanium.
- Recognize how intermetallic compounds are formed.

In modern industry, metallurgists can choose from more than 60 metallic elements when designing a metal, depending on the nature of the application. Every metal has a unique set of properties that may be useful. Often, the desired properties for nonferrous metals can be attained through alloying, cold working, and precipitation hardening. These processes were covered in Chapter 16.

Three metallic elements, magnesium, zinc, and tin, are often used for nonferrous metallurgical applications. Specialty metals (such as nickel, titanium, and gold) are used in applications where the higher cost of processing is justified to produce special performance. These metals and their related properties are the focus of this chapter.

Magnesium

Magnesium is a very light metal used for small parts and many applications requiring low density and corrosion resistance. It has a high strength-to-weight ratio, making it comparable in strength to aluminum and some types of alloy steel. Magnesium has very low density. Its ductility is also less than that of other metals. Small amounts of aluminum, zinc, or zirconium are alloyed with magnesium to increase the strength of the metal alloy.

Magnesium is electrolytically refined from molten magnesium salts. Magnesium electrolysis is similar to aluminum electrolysis. During refinement, magnesium chloride or magnesium oxide is reduced to magnesium and chlorine or oxygen gas through the introduction of electrical current in a production cell. Since the magnesium is less dense than the molten electrolyte, it floats to the top of the cell for recovery.

Large amounts of electricity are needed for the commercial production of magnesium. There are major electrolytic refineries located in the far north of the Canadian province of Quebec, where electricity is abundant.

Using Magnesium for Commercial Applications

Magnesium sheet is made by casting an ingot in a furnace, then hot rolling the metal to a thickness close to the final gauge. A small amount of reduction through cold rolling may also be performed. Most magnesium alloys are not very ductile, so the amount of reduction possible through cold work is limited. For this reason, many magnesium parts are die cast to the final shape. *Die casting* is a process in which liquid metal is injected into a die or mold under high pressure to cast a part. Magnesium castings can fill narrow spaces and thin walls, reducing the weight of the finished part even further.

Magnesium is commonly used for specialty wheel rims for automobiles, Figure 19-1. The low density of the metal helps reduce the weight of the part and provides smooth handling. However, magnesium has low ductility. This is why you should never strike the edge of a magnesium wheel rim with a tool; the piece can be easily chipped and ruined.

Some machines are equipped with a lever arm or a similar component that must move quickly to accomplish tasks. If the arm is made from a heavy material, it will take more force to accelerate (and a larger brake to slow it down). Arms or gripper ends made of magnesium are light in weight and therefore require less force to move. The positioning of the component can also be controlled more precisely because of the lower weight. Some of the components used in laser printers and copy machines are made of magnesium.

Handheld devices, such as portable video cameras, must be as light as possible to minimize user fatigue. The outer case of the camera shown in Figure 19-2 is made of cast magnesium.

Die casting presents an added advantage for magnesium and other metals. A part with many details can be molded as one piece. The inside of a magnesium video camera case, for example, has a complex design. See Figure 19-3.

Strengthening Magnesium

Magnesium is not very ductile, so its strength cannot be improved by cold working. The strength of magnesium can be increased by the addition of alloying elements, such as aluminum, zinc, and calcium. Strong magnesium alloy parts can be cast, extruded, or hot rolled. Then, they can be machined to the final shape.

One of the most commonly used strengthening processes for magnesium alloys is precipitation hardening. Many small machine parts are hardened in this manner, Figure 19-4.

Two stages of heat treatment are used in precipitation hardening of magnesium alloys

Figure 19-1. *This Indy car racing wheel is made of cast magnesium. (ITMg, Inc.)*

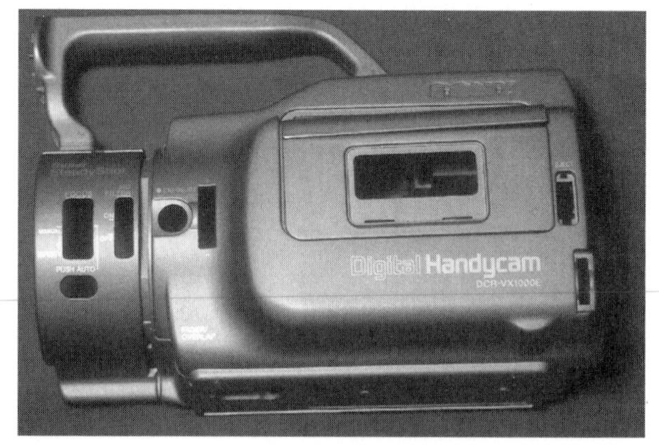

Figure 19-2. *This video camera enclosure is made of cast magnesium to minimize the weight of the device. (ITMg, Inc.)*

Figure 19-3. *The inside of a magnesium camera case reveals many fine details molded into a single part. (ITMg, Inc.)*

Figure 19-4. *This magnesium frame is used to hold a circuit board during assembly of the board components. After precipitation hardening, the frame remains light, but it is strong enough to hold the board in precise position. (ITMg, Inc.)*

such as M16600, which contains small amounts of zirconium and zinc. First, the metal parts are solutionized for a short time at 700°F (370°C) in a furnace filled with inert gas or carbon dioxide (to prevent oxidation). Then, the parts are quenched in air. If aging at room temperature does not add enough strength, the parts can be artificially aged at 300°F (150°C) for 24 hours. The final parts are much stronger than the original magnesium alloy parts.

Other alloying elements used to strengthen magnesium include aluminum, zinc, zirconium, and occasionally manganese and thorium. Some of the common properties of magnesium alloys are listed in Figure 19-5.

Using Magnesium Safely

Magnesium must be handled very carefully during processing and other operations for several reasons. At high temperatures, magnesium is a fire hazard in itself. Powdered magnesium burns very easily and can produce a very intense, hot fire.

Magnesium should be ground and cut only under the close supervision of a knowledgeable safety expert. Extra care must be taken to prevent sparks from the metal fines. This is true for any type of metal, but it is very important with magnesium.

Fires caused by metal are difficult to control. Water *cannot* be used to extinguish a metal fire. It makes the metal burn more intensely! Fire extinguishers that spray water, or a foam of carbon dioxide, are not designed for metal fires. Only a special Class D fire extinguisher can be used to extinguish burning metal.

Zinc

Zinc is an inexpensive, heavy metal most often used as an alloying element for materials requiring corrosion protection. It is commonly alloyed with aluminum, copper, and titanium for products that can be readily machined or soldered. The most widespread use of zinc, however, involves the surface coating of metals through galvanizing. *Galvanizing* is the application of zinc to coat a metal and protect it from corrosion.

Zinc is refined from its ore by roasting finely crushed zinc blende (zinc sulfide) or other forms of zinc ore in a furnace. The zinc is converted to vapor in an oxygen-free atmosphere and then condensed into liquid zinc in a container above the roasted ore. Zinc

Selected Properties of Magnesium Alloys											
UNS Designation	Type	Condition	Tensile strength (ksi)	Yield strength (ksi)	Elongation (% in 50 mm)	Impact strength (ft-lb.)	Shear strength (ksi)	Modulus of elasticity (thousands of ksi)	Density (g/cm³)	Melting point (°C)	Melting point (°F)
M11914	sand mold	precipitation hardened	40	21	6	1	21	6.5	1.827	470-595	875-1105
M10600	die casting	as cast	32	19	6			6.5	1.78	540-615	1005-1140
M13210	wrought	precipitation hardened	34	25	11	5.87	18	6.5	1.78	605-650	1120-1200
M13310	wrought	annealed	29	18	30.5	4		6.5	1.8	590-650	1090-1200
M13310	wrought	cold worked	37	29	9	3	20	6.5	1.8	590-650	1090-1200
M13310	wrought	precipitation hardened	37	29	9		20	6.5	1.8	590-650	1090-1200

Figure 19-5. *A list of common magnesium alloys and their various properties.*

can also be refined electrolytically in processes similar to aluminum and magnesium electrolysis.

Galvanizing

Galvanizing is a widely used commercial procedure that layers the surface of steel with a protective coating. In this process, a thin layer of zinc is applied to steel sheet, strip, or wire. The thin layer protects the base metal through *sacrificial corrosion.* In other words, air and moisture corrode the layer of zinc, rather than the base metal. Steel is galvanized by dipping a sheet or metal part into a molten bath of zinc, Figure 19-6.

When the liquid zinc cools, it freezes onto the steel. Large zinc crystals may form on the surface. The zinc keeps the steel free of rust until most of the coating has corroded away. If a final coat of paint is applied to galvanized steel, the combination of paint and zinc increases the overall corrosion resistance of the coating.

Figure 19-6. *Galvanized steel is made by dipping the steel into a bath of molten zinc. After galvanizing, the steel is transported to a cooling tower. (U.S. Steel Group/USX Corp.)*

Zinc Die Casting

Zinc die casting has widespread use in commercial applications for several reasons. Zinc melts at a relatively low temperature. It also flows very well when it is molten, so it can be cast into parts with very thin walls and complex shapes. See Figure 19-7. These properties allow longer use of the casting molds. Lower operating temperatures help dies retain the smooth surfaces and detail that the designer intended. An optics bracket made from cast zinc for use in a laser scanner is shown in Figure 19-8.

In zinc die casting, liquid metal is poured into a shot chamber, then injected into a hollow die under high pressure. See Figure 19-9. After a few seconds, the metal is frozen, the die opens, and the part is removed by hand (or the part may drop out of the die). Accurately sized parts can be made of complex shapes. Often, the parts can be used as is, without further surface finishing.

A die casting machine in operation is shown in Figure 19-10. Bars of zinc are melted in the holding furnace. The liquid metal is pressed into the die by the ram. When the die is opened at the end of the casting cycle, the parts fall into the cardboard box in front of the operator.

The force applied by the ram on the liquid zinc must be high enough to ensure that the casting is sound and free of *porosity* (the presence of internal pores caused by trapped air or gas). Sometimes, this high force pushes liquid metal out between the die halves, where it freezes in the cold die. This excess metal is called *flashing*. The next processing step is to remove the flashing with a heavy power brush or a shear punch. A zinc cast part with flashing around its edges is shown in Figure 19-11.

Most parts produced of zinc die casting are strengthened by alloying the metal with aluminum and magnesium. Two common zinc die casting alloys, Z33520 and Z35531 (known as Alloy 3 and Alloy 5), both contain aluminum and magnesium. The addition of

Figure 19-7. *Small fittings and housings made of zinc. (Stroh Casting Company)*

Figure 19-8. *This optics bracket is cast from zinc and designed for use in a laser scanner. The bracket must hold the laser and rotating mirrors very precisely to ensure proper placement of the laser beam. It must also resist the bending forces applied during manufacture and use. (Interzinc)*

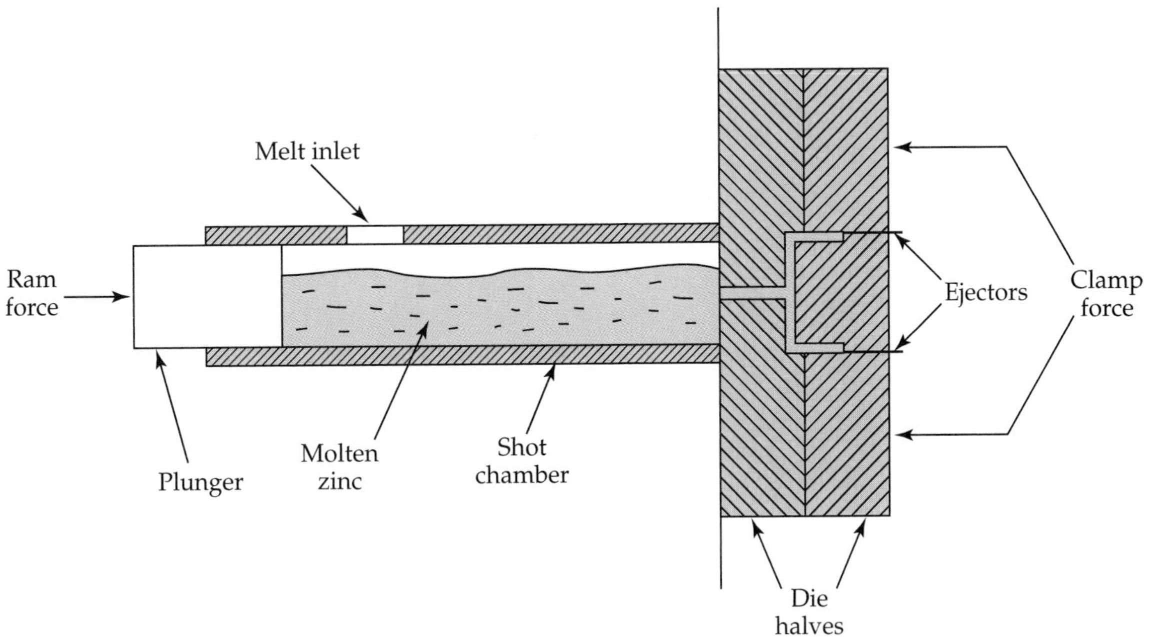

Figure 19-9. *The die casting process. Parts are ready for use after leaving the die.*

aluminum, and sometimes copper, reduces the melting temperature of zinc and makes it easier for the alloy to flow into complex shapes. Very small additions of magnesium (0.02% to 0.08%) improve the strength of the cast part.

At room temperature, parts made from these two zinc alloys may have higher impact strength than parts made from cast aluminum and magnesium alloys. They do have higher impact strength than similar iron castings at room temperature and at lower temperatures. The various properties of the two zinc alloys are listed in Figure 19-12.

In recent years, the strength of engineered plastics has improved significantly. Many parts that were once die cast are now molded of plastic. However, many detailed comparisons of material properties must be considered before a material selection or change is made. The needs of a specific application must also be considered carefully.

About 30% of the zinc used in the United States is recycled from scrap. This is lower than the percentage of recycled aluminum or copper due to the volume of zinc used for thin, corrosion-resistant coatings. Zinc applied as a protective coating cannot be recycled economically.

Tin

Tin is a soft metal that is most commonly used in the production of tin-based solders and for tin plate in food and beverage containers. Tin normally serves as an alloying element. It is not ordinarily used as an unalloyed structural material because it has very low strength at room temperature. Tin is alloyed with many different metals, including lead, silver, and copper. When tin is alloyed with copper, bronze is formed.

Tin is extracted from *cassiterite,* an ore of tin oxide. The ore is recovered by using a dredge or pump mining equipment. It is broken up into gravel and then refined into concentrate. The concentrate is smelted with a mix of metallurgical coal and limestone to form a charge. The charge is heated in a furnace, which reduces the oxide to impure tin.

Figure 19-10. *This die casting machine is used to make small parts from melted bars of zinc. Note the cleanliness of the work area. (Stroh Casting Company)*

The metal is then reheated in cast iron pots and purified by the introduction of steam or compressed air and a number of other elements. Liquation or electrolytic refining is often used to further purify the metal (until it is almost 99.99% pure). In this condition, the tin is cast into ingots for final processing.

Tin is most commonly electroplated onto steel to produce *tinplate* (steel sheet coated with tin). Plated steel is used to manufacture metal cans for foodstuffs and containers for other products, such as paint. The addition of tin protects materials from reacting with the steel of the can. Tinplate accounts for the greatest percentage of commercial tin production.

Tin is a very useful alloying element for metals requiring corrosion resistance and electrical conductivity. Tin alloys are fre-

Figure 19-11. *After die casting, flashing around the outside of the part shown can be removed with a brush or punch. (Stroh Casting Company)*

quently used to manufacture soldering products for use in the electronic and plumbing industries. These applications are discussed next.

Tin Solders

Tin alloys are used extensively in the production of solders for electronic circuits and other joining applications. *Solders* are alloys designed to melt at a lower temperature than the melting point of the base metal (the metal being joined). When mixed with lead, tin makes an excellent solder. An alloy of 60% tin and 40% lead, called an *electrical solder,* has a low melting point (361°F, 183°C) and bonds to copper wire easily. This alloy joins copper wires and leads to produce sound electrical connections. Any surface corrosion that occurs on a properly soldered joint will not disturb the electrical connection.

Tin-lead alloys were formerly used to join water pipes made of copper. Alloys composed of 50% tin and 50% lead were useful because they could bond to copper easily. These alloys cooled to a "mushy" condition during freezing, allowing plumbers to remove excess material with a wet rag. Thus, 50% tin and 50% lead alloys were called "wiping" solders.

Tin-lead solders are no longer used for pipes that carry drinking water because small amounts of lead can dissolve and contaminate the water. Today, lead-free tin solders are required for plumbing applications.

Wave Soldering

Wave soldering is a joining process that uses a low-temperature molten solder bath to connect electronic components in a printed circuit board. This process is shown in Figure 19-13. A circuit board consists of many copper-plated tracks that carry current from electronic chips to other components. Tin solders can be used to produce mechanical and electrical joints on the board. In wave soldering, the board is passed over the solder so that it just contacts the molten metal. As

Selected Properties of Zinc Alloys									
Alloy	Tensile strength (ksi)	Yield strength (ksi)	Elongation (% in 50 mm)	Impact strength (ft-lb.)	Shear strength (ksi)	Modulus of elasticity (thousands of ksi)	Density (g/cm³)	Melting point (°C)	Melting point (°F)
Z33520 (Alloy 3)	41	32	10		31	15.2	6.60	381-387	718-728
Z35531 (Alloy 5)	47.5	33	7	48	38	15.2	6.70	380-386	717-727

Figure 19-12. *A list of selected properties for two zinc alloys, Z33520 and Z35531.*

Figure 19-13. *A dual-wave soldering machine. The machine is used to make mechanical and electrical connections in printed circuit boards. As the circuit board passes over the two waves of solder, surface tension pulls the liquid metal up to the joining locations and holes on the board. (SEHO Seitz & Hohnerlein GMBH)*

solder is pumped from the bath, it flows over a *weir.* The flow of solder is similar to the movement of water over the spillway of a dam.

The solder "wets" the exposed copper leads, forming the electrical joints. The temperature of the solder must be kept as low as possible because the board components cannot withstand high temperature for very long. Also, the depth and time of immersion must be kept to a minimum.

Electronic chips joined by wave soldering are shown in Figure 19-14. The large chip has extended leads for making electrical connections on a circuit board. The leads are spaced 0.100" apart. This type of chip is called a *stuffed chip* because the leads can be inserted into holes in the board. During wave soldering, the solder will fill the holes to make the electrical joint.

Some chips are made with smaller leads that are bent into flat *lands* that rest on the board surface. The smaller chips in Figure 19-14 are called *surface-mount chips.* The leads for these chips are spaced apart 0.050" or 0.025" (or even less). They are soldered to sit on flat, tin-plated copper pads on the board. For this design to work well, the solder must physically hold the parts in place. See Figure 19-15.

In a dual-wave soldering machine, the different types of chips are soldered in different stages or "waves." Refer to Figure 19-13. The first wave of solder to the circuit board is called the *chip wave.* It is used to spread rippled solder to the many small surface-mount chips on the bottom of the board. In the second wave, called the *delta wave,* solder flows all the way up through the holes, making sound electrical and mechanical joints with the stuffed chips on the top of the board.

Higher temperatures during soldering can produce stresses on joints in the board. Sometimes, antimony is alloyed to the solder to increase its strength. It is impossible to use solutionizing or other heat treatment processes with tin alloys. The melting point

Figure 19-14. *Electronic chips are commonly joined to circuit boards by wave soldering. (Warner Consulting, Inc.)*

Figure 19-15. *Surface-mount chips soldered to a circuit board. (Warner Consulting, Inc.)*

of the alloy is so low that even at room temperature, the metal behaves the way other metals do at high temperature. The common metallurgical properties of tin are listed in Figure 19-16.

Nickel and Nickel Alloys

Nickel is a tough metal that has very strong resistance to corrosion and oxidation. It is widely used as an alloying element with metals such as copper, molybdenum, and iron. Nickel is most commonly alloyed with chromium to produce stainless steel.

Nickel is extracted from nickel-rich ores and then refined into pure nickel, ferronickel (an alloy of iron and nickel), and nickel oxide. These nickel products are then melted and combined with other metallic elements and alloys to produce nickel alloys. In a molten state, nickel alloys are cast into ingots. To achieve high corrosion resistance, ingots of nickel are typically cast in special vacuum arc furnaces. See Figure 19-17.

The resistance of nickel to oxidation at high temperatures makes the metal very suitable for jet engine parts and other applications involving hot, corrosive environments. Specialized nickel alloys strengthened to resist corrosion and high stresses at extreme operating temperatures (1200°F–2000°F, 650°C–1100°C) are known as *superalloys*.

Applications of Nickel Alloys

Nickel-based superalloys are commonly used in the manufacture of aircraft components. At temperatures above 1300°F (700°C), most ordinary steel oxidizes significantly in a few hours. Even stainless steel cannot survive

Selected Properties of Tin								
Metal	**Tensile strength (ksi)**	**Yield strength (ksi)**	**Elongation (% in 50 mm)**	**Shear strength (ksi)**	**Modulus of elasticity (thousands of ksi)**	**Density (g/cm³)**	**Melting point (°C)**	**Melting point (°F)**
Pure tin		1.8	64		6.03	7.3	231.9	449.4
Electrical solder (60Sn-40Pb)	7.61	8.2	60	5.38	4.35	8.92	183	361.4

Figure 19-16. *A list of the common properties of tin and a tin-based electrical solder.*

Figure 19-17. *This high-purity nickel ingot is being removed from a vacuum arc remelt furnace after casting. (Consarc Corporation)*

Figure 19-18. *Turbine blades and burn chamber shells in jet engines use nickel-based superalloys to survive the very high temperatures of burning jet fuel. The large titanium blades at left are compressor blades, while the smaller blades in the mechanism at far right are the driven blades. The driven blades are made from nickel alloys. (Pratt & Whitney)*

for long at these temperatures. But nickel can withstand such severe conditions. Among the applications for nickel-based superalloys are gas turbine rotor blades in jet engines, where exhaust gases flow at temperatures exceeding 1800°F (980°C). See Figure 19-18. The blades must resist corrosion to prevent excessive wear and any change to their shape. A typical turbine blade is designed to operate for well over 1000 hours of run time.

To achieve the properties necessary for operation, most turbine blades today are cast and then ground and polished to the final dimension. Nickel-based superalloys are very hard, and they have little ductility at room temperature. The use of cold-working processes to harden these alloys is impossible. Turbine blades are usually cast in a special fashion so that all of the grains in the metal are oriented in the long, stressed direction. Cooling methods are controlled to form long, rod-shaped grains in the metal, Figure 19-19. The final grain structure is very strong and allows the blades to resist elongation at high temperatures.

Two strong and commonly used nickel-based superalloys are Inconel® 600 and Inconel® 718. The properties of these alloys are listed in Figure 19-20.

Titanium

Titanium is a low-density, corrosion-resistant metal designed for specialized applications. Titanium has a relatively high melting temperature and a high strength-to-weight ratio, so it can be used for parts that must withstand stress at high temperatures. It is commonly alloyed with aluminum, tin, and vanadium to produce strong, lightweight materials.

The most common titanium alloy, R56400, is made up of mostly titanium, plus 6% aluminum and 4% vanadium. It is also known as Ti-6Al-4V. The alloy additions permit great increases in strength through precipitation

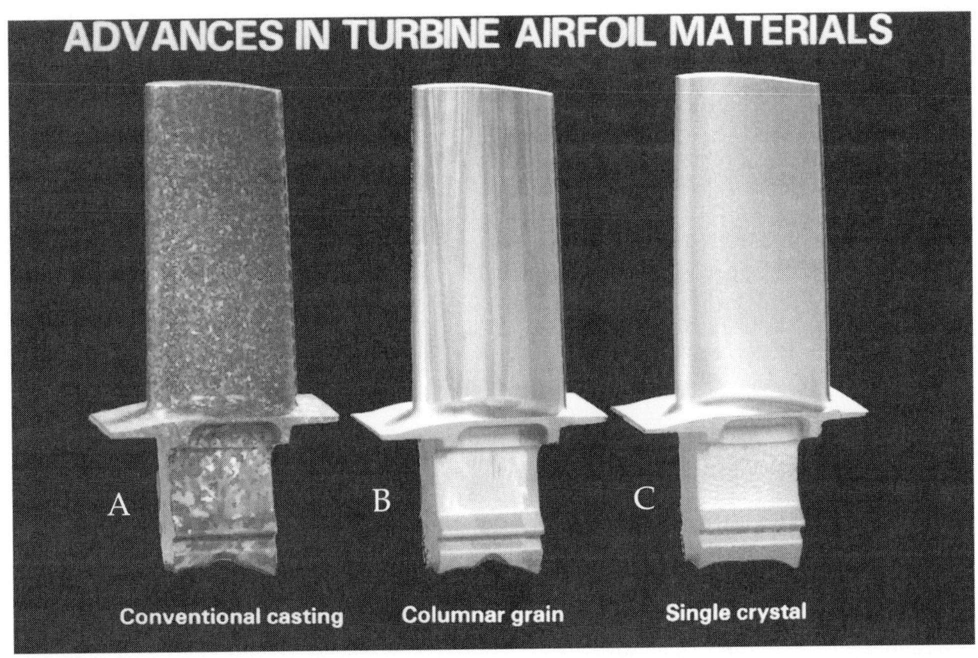

Figure 19-19. *These three nickel turbine blades illustrate the effects of different casting processes on the crystal structure of the metal. A—A conventional polycrystalline structure produced by conventional casting. B—A directionally cast crystal structure. C—A single crystal structure. (Pratt & Whitney)*

Selected Properties of Nickel-based Superalloys								
Alloy	**Condition**	**Tensile strength (ksi)**	**Yield strength (ksi)**	**Elongation (% in 50 mm)**	**Modulus of elasticity (thousands of ksi)**	**Density (g/cm^3)**	**Melting point (°C)**	**Melting point (°F)**
Alloy 600	Alloy strengthened	95	45	40	30	8.47	1355-1413	2470-2575
Alloy 718	Precipitation hardened	180	150	12	30.6	8.19	1260-1336	2300-2437

Figure 19-20. *A listing of common properties for Inconel® 600 and Inconel® 718, two nickel-based superalloys.*

hardening. Selected properties of this alloy are listed in Figure 19-21.

The use of titanium is limited in commercial applications because it is difficult to extract from *rutile* (a common mineral of titanium). Titanium processing must be closely controlled to prevent contamination of the metal. For these reasons, titanium is quite expensive as compared to steel or aluminum. Titanium is suitable only for specialized applications where its superior properties are critically necessary.

Titanium is processed from its ore in three stages. It is first reduced from titanium ore to sponge, a porous metal. The sponge is then melted into ingots in an electric arc vacuum furnace. Melting must occur in vacuum or inert gas conditions to prevent the presence of gases dissolved in the metal. In the last stage of processing, ingots are converted into

Selected Properties of Titanium Alloys						
Alloy	Condition	Tensile strength (ksi)	Yield strength (ksi)	Elongation (% in 50 mm)	Impact strength (ft-lb.)	Modulus of elasticity (thousands of ksi)
R56400 (Ti-6Al-4V)	Annealed	145	138	15	16	15.2
R56400 (Ti-6Al-4V)	Precipitation hardened	171	157	6	11	15.2

Figure 19-21. *A listing of properties for two forms of R56400, the most commonly used titanium alloy.*

primary products (such as billets) for production of final shapes. Each stage of processing must be closely regulated to ensure high purity of the final product.

Applications of Titanium

Titanium and its alloys are used for many aircraft components and other applications requiring light weight and strength at high temperatures and speeds. Compressor blades in jet engines are commonly made from titanium. Refer to Figure 19-18. These blades must withstand very high stresses at elevated temperatures. Large amounts of titanium are also used in the manufacture of wings for fighter planes, Figure 19-22. The skin of the wing shown reaches temperatures as high as 300°F (150°C).

Titanium can be formed easily when it is heated above 1400°F (760°C). Complex airframe sections for fighter planes are often made in this manner. However, titanium reacts readily with air when it is heated, so high-temperature forming procedures must occur in vacuum or inert (shielded) atmospheres. Titanium can also be welded if it is shielded with an inert gas cover or welded in a vacuum. Electron beam welding, which uses a beam of high-energy electrons to make joints in a vacuum chamber, works very well with titanium. Complex structures can be formed in this process, Figure 19-23. The titanium forward boom shown is strong but lightweight.

The high strength and low density of titanium make the metal suitable for many other applications requiring light frame construction. Titanium bicycle frames are the result of modern advances in engineering, Figure 19-24. This design allows the frame to maintain its rigidity at less weight.

Gold

Gold is a very precious, corrosion-resistant metal reserved for specialized use. Gold has been used to make jewelry since the beginning of civilization. Modern uses of gold include casting alloys in dental applications, brazing and soldering alloys, electronic components, and electroplating solutions.

Gold has high ductility and density. It does not tarnish normally, and it can be easily hammered into different shapes. Nickel or copper is commonly alloyed with gold to improve its strength and resistance to wear.

Gold is an expensive metal because of its rarity. Different mining methods are used to derive gold from its ores, depending on the depth of the location. Gold can be found in the metallic form of fine particles, vein deposits, and nuggets.

The weight percentage of gold in an alloy is given in *karats.* Pure gold is referred to as 24 karat (24K). A *karat* is equal to one twenty-fourth part of gold in an alloy; the percentage of gold can be determined by multiplying the weight fraction of a sample by 100%. Thus,

Figure 19-22. *A wing section of an F-22 fighter plane during assembly. Together, the wing and blue holding tool weigh about 14 tons. Over 40% of the assembled wing is made from titanium. (Boeing)*

14K gold is equal to 14/24, or 58% gold. An alloy containing 18K gold is 18/24, or 75% gold. Most jewelry is made from 14K or 18K gold for durability, Figure 19-25. Pure gold is very soft and wears when it is rubbed against one's skin.

Applications of Gold

Gold has a large number of uses in modern industry. It is an excellent metal for electroplating removable electronic connectors because it rarely tarnishes or corrodes. Electronic connectors made from copper are often

Figure 19-23. *This titanium forward boom designed for an F-22 fighter plane was made using the electron beam welding process. (Boeing)*

Figure 19-24. *The frame for this advanced bicycle design is made of titanium. (Seven Cycles)*

Figure 19-25. *A 14K gold ring. This ring won the 1997 Spectrum Award, issued by the American Gem Trade Association. The luster and finish of the gold are major contributors to the appearance. (Rasmussen Diamonds)*

formed into pins and clips. First, nickel is electroplated onto the pins and clips. Then, gold is electroplated onto the nickel plating, Figure 19-26. An alloying addition of 1%–2% cobalt to the gold increases the wear resistance of the plating.

Gold alloys are commonly used to make connections in solid state electronic devices. Because gold does not tarnish or oxidize easily, it is relatively easy to join without a flux or filler metal. Components for electronic chips can be joined by using very small spot welds and gold wires. Inside a chip is a small die made of silicon, with aluminum, boron, and other elements carefully added in

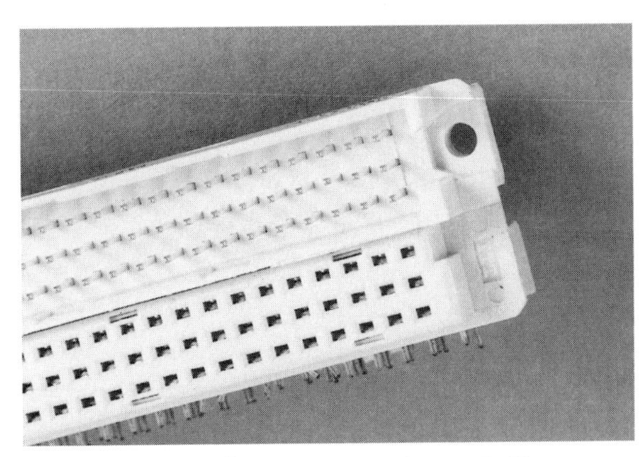

Figure 19-26. *The connector pins and clips on this electronic connector are electroplated with gold. (Warner Consulting, Inc.)*

selected regions. Different regions form different semiconductor devices, and some regions are electrical conductors. These conducting regions are connected to the lead frame, which contains copper leads for connection to a circuit board. Very fine gold wires are welded to join the silicon die to the frame, Figure 19-27. Each chip may need hundreds of these small connections. Thus, the spot welders must work very quickly (and precisely).

Gold has very good resistance to oxidation, so it is useful for brazing and soldering alloys. Alloys of gold and nickel and gold and platinum are typically used in brazing. When other metallic elements, such as silver and copper, are added, gold can be used as a soldering alloy. These alloys offer very good resistance to corrosion. Other common properties of gold alloys are listed in Figure 19-28.

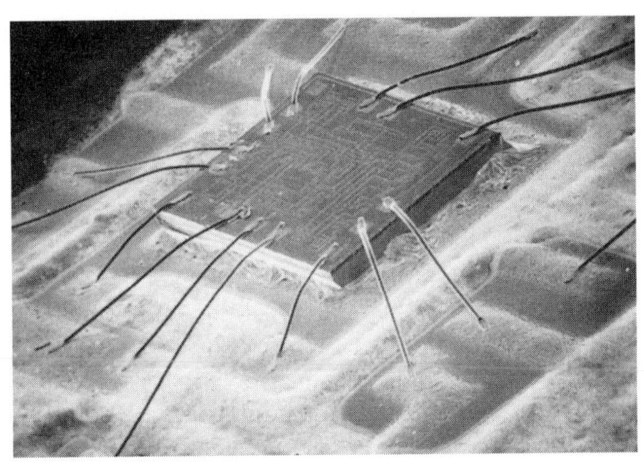

Figure 19-27. *Small gold wires, about 0.001" in diameter, connect the silicon die to the lead frame in this solid state electronic device. The die shown measures approximately 0.1" wide. (Warner Consulting, Inc.)*

Intermetallic Compounds

Intermetallic compounds are chemical compounds made up of two or more metals that exhibit different properties than the original metallic elements. Intermetallic compounds are specific *phases* (structures) within an alloy system and are classified separately from other nonferrous metals. These alloys act differently than other alloys because they have precise atomic ratios between the metallic elements and a combination of metallic and ionic bonds.

Intermetallic compounds have various uses in nonferrous metallurgy because their behavior is unique to the alloy system. For example, Al_2Cu, an intermetallic compound of aluminum and copper, is not ductile (like aluminum or copper). It is also not very conductive.

When titanium reacts with nitrogen, it forms titanium nitride (TiN). This is a very hard, wear-resistant compound that can be applied as a coating on cutting tools and selected abrasives. Cutting tools with a flat yellow or gold finish are coated with titanium nitride.

Titanium aluminide ($TiAl_3$) is a compound of titanium and aluminum. This alloy is much harder and more resistant to wear than both titanium and aluminum. For this reason, titanium aluminide is commonly used in the manufacture of automobile engine valves, Figure 19-29.

Test Your Knowledge

1. List three processes that can be used to produce desired properties in nonferrous metals.
2. Define *die casting*.

Selected Properties of Gold Alloys						
Alloy	**Condition**	**Yield strength (ksi)**	**Modulus of elasticity (thousands of ksi)**	**Density (g/cm^3)**	**Melting point (°C)**	**Melting point (°F)**
Yellow gold (60Au-20Ag-20Cu)	precipitation hardened			19.3	1064	1947
58Au-12.2Ni-23.5Cu-6Zn	annealed	35.4	11.5			
precipitation 58Au-12.2Ni-23.5Cu-6Zn	hardened	82.5	11.5			

Figure 19-28. *A listing of common gold alloys and their related properties.*

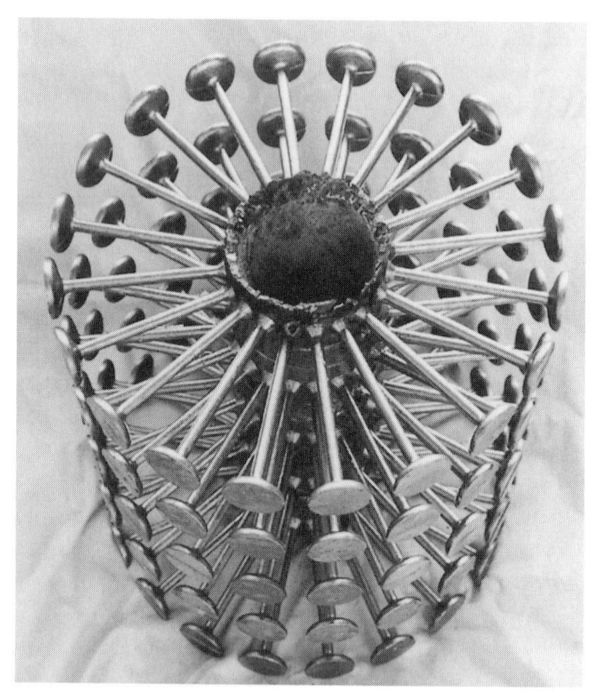

Figure 19-29. *A group of cast titanium aluminide engine valves, before they are cut from the casting "tree." (Consarc Corporation)*

3. Why are many parts made from magnesium alloys die cast to a final shape?
4. What are two properties of magnesium that make it suitable for the production of wheel rims?
5. What is *galvanizing?*
6. How is zinc refined from its ore?
7. Name two metals that are commonly alloyed with zinc.
8. What is *porosity?*
9. Why is tin primarily used as an alloying element?
10. What is *tinplate* and what is it commonly used to manufacture?
11. Describe the operation of a dual-wave soldering machine.
12. Which metal is alloyed with nickel to produce stainless steel?
13. What are *superalloys?*
14. Nickel-based superalloys are commonly used in the manufacture of gas turbine blades in jet engines. What properties of these alloys make them preferable to other metals for this application?
15. What properties of titanium make the metal very suitable for aerospace applications?
16. Why is the use of titanium limited in commercial applications?
17. List four modern applications of gold and its alloys.
18. Explain how gold alloys are used to make connections in solid state electronic devices.
19. What are *intermetallic compounds?*
20. What compound is typically applied as a protective coating for cutting tools and abrasives?
21. What properties of titanium aluminide make it a better choice than aluminum, copper, or nickel for the manufacture of automobile engine valves?

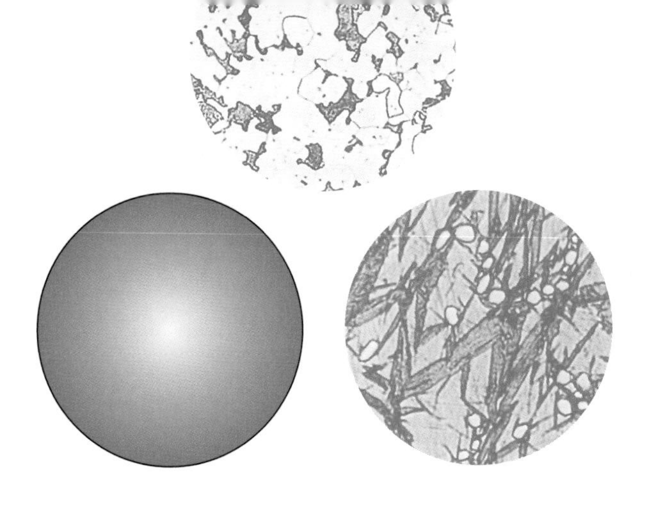

Glossary

A

agitation. Rapid vibration of metal parts during quenching in order to attain a high degree of hardness.

alloy. (1) A material dissolved in another metal in a solid solution. (2) Material that results when two or more elements combine in a solid solution.

alloy steel. A steel that contains more than the average amount of alloying elements, which improve the properties of the steel.

alloying. The act of creating an alloy by mixing two or more elements, usually by melting.

alloying elements. Material added to a metal.

annealing. (1) Slow, controlled cooling of metal within a furnace or oven. Furnace or oven temperature is reduced very slowly in order to attain a high degree of softness in the metal. (2) Full annealing.

anode. In an electrolytic bath, the electrode where electrons leave the solution to enter the electrode. *Contrast* **cathode.**

atom. Smallest part of an element that still has the characteristics of that element.

austempering. A special tempering technique in which dwell temperature is maintained slightly above the martensitic transformation temperature for a long period of time.

austenite. One of the basic steel structures wherein carbon is dissolved in iron. Austenite occurs at elevated temperatures.

austenitic. Steel which has the structural form of austenite.

B

bainite. A steel structure that is harder than pearlite, cementite, or ferrite and more ductile than martensite.

basic oxygen furnace. A basic steel-making furnace that uses an oxygen blast at supersonic speeds to intensify the heat. Today, the basic oxygen furnace has become the most widely used process in the manufacture of steel. Also referred to as *BOF.*

bauxite. The primary ore of aluminum.

Bessemer converter. One of the basic steel-making furnaces. This process uses a furnace in which molten pig iron is refined by a burning gas. At one time, the Bessemer converter was used extensively in the manufacture of steel; however, today it is used minimally.

BHN. Unit of hardness used in the Brinell hardness testing method. *BHN* stands for Brinell hardness number.

billet. A semi-completed metal form made from a bloom and smaller in cross-sectional size than a bloom.

blast furnace. A large furnace used to convert iron ore into pig iron.

blister copper. An impure form of copper produced in the copper refining process.

bloom. A semi-completed metal form with a relatively square cross section.

body-centered cubic. One of the common types of unit cells. This arrangement is typical of the ferritic form of iron.

body-centered tetragonal. One of the common types of unit cells. This arrangement is typical of the martensitic form of iron.

BOF. *See* **basic oxygen furnace.**

brass. A copper-rich alloy of copper and zinc. It may also contain small amounts of other elements for special properties.

brine. A solution of salt in water, used as a quenching medium in the cooling of metals.

Brinell hardness test. A common testing method using a ball penetrator, in which the diameter of the indentation is converted to units of *Brinell hardness number (BHN).* This method is extremely accurate for soft metals.

brittleness. Tendency to stretch or deform very little before fracture.

bronze. A copper-rich alloy of copper and tin, plus small amounts of other elements.

C

C curve. Transformation curve used on an isothermal transformation diagram.

carbon steel. Steel that contains comparatively less alloys than other steels.

carbonitriding. A gaseous surface-hardening process that introduces both carbon and nitrogen to steel.

carburizing. A surface-hardening process that impregnates the outer surface of low-carbon steel with considerable amounts of carbon before secondary heat treatment.

case hardening. *See* **surface hardening.**

cast iron. A material containing primarily iron, 2 to 6 percent carbon, and often small amounts of silicon and other elements.

cathode. The electrode of a battery or metal reduction cell where electrons leave the electrode and enter the rest of the system. For a metal reduction cell, the electrons enter the solution, which contains dissolved metal oxide. *Contrast* **anode.**

cementite. A very hard structural form of low-temperature steel that contains more than 0.8% carbon. Cementite occurs in steel that has not been previously heat treated or in steel that has been cooled slowly after being transformed into austenite.

checker chamber. Brick structure on either side of an open hearth furnace used to retain the heat of the exhaust gases.

cleavage failure. A brittle failure of metal in which atoms break apart and little stretch occurs.

close-packed hexagonal. A type of internal atomic arrangement that occurs in the formation of nonferrous metal.

coefficient of thermal expansion. Characteristic of a material that describes the amount of expansion a material undergoes due to heat.

coke. Purified coal used in the manufacture of iron and steel.

cold working. The process of forming, bending, or hammering a metal well below the melting point to improve strength and hardness. Also referred to as *work hardening.*

compound. A material composed of two or more chemically joined elements.

compressive strength. A material's ability to withstand a "pressing" or "squeezing together" type of stress.

conductor. A metal connection, usually a wire, that carries electrical current between two points.

continuous casting. A modern method of steel manufacture in which steel is continuously poured into shapes, blooms, slabs, or billets. Also referred to as *strand casting.*

corrosion. Deterioration of a metal by chemical reaction with its environment. Usually the end product is metal oxide or inorganic salts.

corrosion resistance. Resistance of a material against chemical attack by the environment.

creep. Slow plastic flow that occurs over time in a material when it is stressed.

cryolite. Sodium aluminum fluoride, which can dissolve aluminum oxide so the oxide can be electrolytically reduced to metal.

crystal. A lattice structure of atoms in solidified metal.

crystallization. Formation of crystals as a metal solidifies. The atoms assume definite positions in the crystal lattice.

cupola. A furnace commonly used in the manufacture of cast iron. It uses coal as its fuel.

cyaniding. A surface-hardening technique in which the surface is impregnated with both carbon and nitrogen. This is a liquid process.

D

deformation. The amount that a material increases or decreases in length when it is loaded.

dendrites. Branches of unit cells that grow during the formation of a crystal structure.

density. The mass of an object divided by its volume, expressed in terms of mass per unit volume.

die. A tool, usually containing a cavity, that gives the workpiece a desired shape. Usually a die is made of tool steel or other hard, strong metal.

die cast. Made by forcing liquid metal into a metal cavity, or die. Upon cooling, the metal is called a die cast part.

die casting. A process in which liquid metal is injected into a die or mold under high pressure to cast a part.

dielectrical strength. Electrical property of a material, a measure of its resistance to break down from a large voltage over a prolonged time period.

DPH. Unit of hardness used in the Vicker's hardness testing method. DPH stands for diamond-pyramid hardness. *See also* **Vicker's hardness test.**

drawing. A shaping method in which metal is pulled (drawn) cold through a die to produce a reduced final size.

drop casting. Method of casting metal in a shallow mold by lowering the bottom of the mold to allow more liquid metal into the mold. Thus a long ingot can be made using a small mold.

ductile cast iron. A type of cast iron that has high ductility and tensile strength. Also referred to as *nodular cast iron.*

ductility. Tendency to stretch or deform appreciably before fracture.

E

elastic range. The portion of a stress-strain curve between zero and the elastic limit stress point.

elasticity. The ability of a material to return to its original length after being deformed without any permanent deformation.

electric arc furnace. One of the basic steel-making furnaces. It is especially valuable in the production of high-alloy steel.

electric induction furnace. A furnace used in the manufacture of cast iron. It uses electricity as its source of power.

electrical conductivity. Ability of a material to permit electricity to flow through it. *Contrast* **electrical resistance.**

electrical resistance. Ability of a material to resist the flow of electricity. *Contrast* **electrical conductivity.**

electrolysis. A chemical change that occurs in an electrolytic cell when an electric current passes through an electrolyte.

electrolytic cell. An assembly, including a vessel, electrodes and an electrolyte, in which electrolysis is performed.

electrowinning. The recovery of metal from an ore by electrochemical methods, such as electrolysis.

element. A simple, pure substance made up of one kind of material.

endurance strength. Ability of a material to withstand a repeated stress loading.

equilibrium. State at which no further change occurs in a system.

equilibrium diagram. *See* **phase diagram.**

etching. The application of acid on a smooth, polished metal surface to make details of a microstructure more visible.

eutectoid point. On an iron-carbon phase diagram, the point where the upper transformation temperature line, the lower transformation temperature line, and the 0.8% carbon line intersect.

extruding. The process of making an extrusion.

extrusion. A finishing process in which metal is reduced in thickness as it is forced by a ram through a die orifice.

F

face-centered cubic. One of the common types of unit cells in which atoms are located on each corner and the center of each face of a cube. This arrangement is typical of the austenitic form of iron.

fatigue strength. Ability of a material to withstand repeated loading.

ferrite. A structural form of low-temperature steel that contains a very small percentage of carbon. Ferrite occurs in steel that has not been previously heat treated or in steel that has been cooled slowly after being transformed into austenite.

ferritic. Steel containing ferrite.

file hardness test. A fast, simple hardness test in which the edge of a file is scraped across a test material to see if it will scratch the surface. The extent of the scratch provides a relative measure of the material's hardness. Also referred to as *scratch hardness test.*

flame hardening. A surface-hardening process that heats selected areas of steel with a direct flame before quenching.

flashing. Excess metal around the edges of a cast part, formed when liquid metal is forced out from parts of the die.

flexibility. The ability of a material to bend, stretch, or distort without breaking.

flexure strength. Bending strength, generally causing tensile stress on one side of the material and compressive stress on the other.

full annealing. A stress-relief method in which metal is heated above its upper transformation temperature and then oven-cooled to room temperature. *Contrast* **process annealing.** Also referred to as *annealing.*

G

galvanizing. The application of zinc to coat a metal and protect it from corrosion.

gas carburizing. A gaseous surface-hardening process in which carbon is transferred to oven-heated steel before secondary heat treatment.

grain. Any portion of a solid that has external boundaries and a regular internal atomic lattice arrangement.

gray cast iron. The most widely used type of cast iron. Gray cast iron is less expensive than other types of cast iron and has comparatively less tensile strength.

H

hard copper. Type of rigid copper pipe or tube strengthened by cold working. *Contrast* **soft copper.**

hardness. A measure of resistance to deformation or penetration.

heat capacity. The amount of heat required to raise the temperature of a material one degree.

heat-treatable alloy. An alloy that can be strengthened by heat treatment.

heat treatment. The process of heating and cooling metal to prescribed temperature limits for the purpose of changing the properties or behavior of the metal.

hematite. One of the common types of iron ore.

high-carbon steel. Steel that contains 0.50%–2.0% carbon.

hot line. A production facility consisting of furnaces, rolling mills and run out tables, which together reduce an ingot in thickness.

hot metal. Semi-refined molten iron that is a product of the blast furnace. It is kept in the molten state and then transferred directly to a steel-making furnace.

hypereutectoid region. On an iron-carbon phase diagram, the region representing steel that contains more than 0.8% carbon.

hypoeutectoid region. On an iron-carbon phase diagram, the region representing steel that contains less than 0.8% carbon.

I

I-T diagram. *See* **isothermal transformation diagram.**

impact strength. Ability of a material to resist shock, dependent on both strength and ductility of the material.

induction hardening. A rapid surface-hardening process that heats steel by electromagnetic induction before quenching.

ingot. The initial cast rectangular or cylindrical shape of steel before it is rolled into a workable shape.

intermetallic compound. Chemical compound that exhibits different properties than the original metallic elements.

iron ore. A mineral that has a high iron content used as a basic ingredient in the manufacture of iron and steel.

iron-carbon phase diagram. A graph used to identify different structures of steel and percentages of carbon that occur in steel at various temperatures.

isothermal quenching and tempering. A special tempering technique in which the metal is tempered before quenching has completely taken place.

isothermal transformation diagram. A graph that identifies different austenitic transformation products occurring over a period of cooling time at isothermal conditions. Also referred to as *I-T diagram* and *time-temperature-transformation (T-T-T) diagram.*

K

karat. A unit of measure for gold equal to one twenty-fourth part of gold in an alloy.

knapping. The process of removing small chips from a stone (such as flint) to make a sharp edge.

Knoop hardness test. A common microhardness testing method that uses an elongated pyramid penetrator and provides Knopp hardness values for the tested material.

L

limestone. A mineral mined from the ground used to remove impurities in the manufacture of iron and steel.

liquid carburizing. A surface-hardening process that uses a liquid source of carbon, such as a molten salt bath, to carburize steel parts.

load-cell tester. A hardness tester that incorporates a closed-loop load-cell to eliminate friction in the Rockwell hardness testing machine.

localized heating. A heating method used in surface hardening to heat selected areas of steel.

low-alloy structural steel. Steel with a lower alloy content than other types of alloy steel, designed for structural work and applications requiring light weight.

low-carbon steel. A type of steel that contains 0.05%–0.35% carbon.

lower transformation temperature. The temperature at which the transformation of iron to austenite begins and the body-centered cubic structure starts to change to face-centered cubic.

M

machinability. The ease with which a material can be cut.

magnatite. One of the types of iron ore.

magnetic susceptibility. The ability of a material to be magnetized.

malleability. The ability of a material to permanently change after being hammered, forged, pressed, or rolled.

malleable cast iron. Cast iron that has superior ductility and greater impact strength as compared to other types of cast iron.

maraging steel. A special type of very strong, tough alloy steel containing a high quantity of nickel and a low percentage of carbon.

martempering. A special tempering process in which steel is held at a specific temperature during quenching, cooled to room temperature, and then reheated to a tempering temperature to produce tempered martensite.

martensite. A very hard, brittle structure of steel produced when steel is rapidly quenched after being transformed into austenite.

martensitic. Steel that has the structural form of martensite.

matte. Concentrated ore produced in the smelting of copper ore to copper metal.

medium-carbon steel. A type of steel that contains 0.35%–0.50% carbon.

melting point. Temperature at which a material turns from a solid to a liquid.

metallurgy. Science that explains the properties, behavior, and internal structure of metals.

microhardness tester. A type of hardness testing machine in which the size of the indenture is very small.

mixture. (1) A material composed of two or more elements or compounds mixed together, but not chemically joined. (2) An alloy produced when the addition of an alloying element to a metal results in the formation of a new compound.

modulus of elasticity. The ratio of stress to strain in a material. Also referred to as *Young's modulus.*

Mohs scale. An inaccurate, relative hardness testing scale in which the test specimen is scratched by ten different types of material.

molecule. Smallest part of a compound that still has the characteristics of that compound.

N

native copper. Form of copper found in nature.

naturally age. Aging of a heat-treatable alloy at room temperature after it has been solutionized and quenched.

naval brass. A brass alloy with good resistance to corrosion in seawater.

nitriding. A gaseous surface-hardening process that fills the outer surface of steel with nitrogen.

nodular cast iron. *See* **ductile cast iron.**

nonferrous metallurgy. The study of metals that do not use iron as their principal alloying element.

normalizing. The process of heating a metal above a critical temperature and allowing it to slowly cool under room temperature conditions to obtain a softer and less distorted material.

O

open hearth furnace. A basic steel making furnace consisting of a giant hearth exposed to a powerful gas flame. At one time, the open hearth furnace was the primary process in the manufacture of steel. It is still used today, but not as extensively as the basic oxygen furnace.

ore benefaction. The process of separating the metallic portion of ore from the nonmetal portion to increase the concentration of the desired metal.

oxides. Compounds of metals with oxygen.

P

pack carburizing. A surface-hardening process in which the parts to be heated are loaded into a metal box along with a solid carbonaceous material.

pearlite. A mixture of ferrite and cementite that contains approximately 0.8% carbon. Pearlite occurs in low-temperature steel that has not been previously heat treated or in steel that has been cooled slowly after being transformed into austenite.

penetration hardness. Hardness obtained by a testing method that uses very accurate techniques. A precision penetrator is used to indent the metal surface.

penetrator. Ball or pointer used in a hardness testing machine to penetrate or indent the sample.

percent elongation. A common measure of the ductility of a material: the percent a material will stretch before it breaks.

phase diagram. A graph that identifies alloy phases occurring at various temperatures and percentages of alloying elements. Also referred to as *equilibrium diagram.*

photomicrography. The process of creating a photograph from a microscope image.

pig iron. Semi-refined iron produced by the blast furnace.

plastic range. The portion of a stress-strain curve beyond the elastic limit point.

plasticity. The ability of a material to permanently deform without breaking.

Poisson's ratio. The ratio of lateral strain to axial strain.

porosity. The presence of internal pores in metal, caused by trapped air or gas.

pot line. An assembly of electrolytic cells, electrical conductors, and furnaces that produce pure metal from metal oxides or salts.

precipitates. Minute particles, consisting of metallic compounds, that form in an alloy during production and heat treatment.

precipitation hardening. A multiphase heat treatment process that strengthens alloys by causing phases to precipitate at various temperatures and cooling rates.

process annealing. A stress-relief method in which metal is heated to a temperature below the lower transformation temperature and then slowly cooled to room temperature. *Contrast* **full annealing.**

Q

quench and temper structural steel. Alloy steel that has more strength and better impact resistance than low-alloy structural steel.

quenching. A rapid cooling process in which a heated metal part is plunged into a liquid or other medium to harden the metal.

quenching medium. The material (normally liquid) into which metal is plunged during the quenching process. The quenching medium may be liquid.

R

refractory material. Material capable of enduring high temperature.

region of transformation. An area where steel products are transformed from austenite on an isothermal transformation diagram. The four regions of transformation are the coarse pearlite region, the fine pearlite region, the bainite region, and the martensitic region.

Rockwell hardness test. Most common hardness testing method, uses a minor load to prevent surface irregularities from affecting results. There are nine different Rockwell hardness tests corresponding to combinations of three penetrators (1/8" ball, 1/16" ball, or diamond-point penetrator) and three loads (60 kg, 100 kg, or 150 kg).

Rockwell superficial hardness test. A common hardness testing method that produces a relatively small indentation on the sample being tested.

rolling. The process of passing a metal slab, plate, strip, or sheet between two large cylinders under pressure, so that the thickness of the metal workpiece is reduced.

rolling mill. The part of a steel-making plant where a series of large, hard rollers compress steel ingots into different shapes.

runout. A horizontal plane, or table, consisting of many small cylindrical rolls that carry ingots and plate between rolling mills or other operations.

S

sacrificial corrosion. Corrosion that is purposely allowed to occur so that the corroding part protects the remainder of the metal.

scleroscope. A common hardness testing machine used in the Shore scleroscope hardness testing method.

scratch hardness test. *See* **file hardness test.**

semicontinuous casting. Casting by pouring liquid metal into a small mold, so that the frozen metal is removed through the bottom to form an ingot much longer than the mold. If the solid ingot is cut off while liquid metal is added, the process is called *continuous casting.*

shear failure. A failure of metal that occurs when groups of atoms slide past each other within a crystal structure.

shear strength. A material's ability to resist a "sliding past" type of stress.

Shore scleroscope hardness test. A hardness testing method in which a small hammer is dropped on the test sample and its bounce is measured.

shot chamber. Area that holds liquid metal before a ram forces it into the die.

slab. A partially completed steel form in which the width is appreciably larger than the thickness.

slag. A product of the iron and steel-making furnaces. It is basically considered to be a waste product, although today there are a few practical uses for it.

slip. A ductile failure in which fracture takes place along certain crystal lines called slip planes.

smelt. Reduction of a metal ore to metal by the addition of thermal energy, usually involving special atmospheres and temperatures well over the melting point of the metal.

soaking pits. Large ovens where ingots are heated for several hours before being transported to the rolling mill. *Soaking pits* are used to obtain more uniform properties in the metal.

soft copper. Annealed copper tube that can be bent and flexed a number of times before cracking or fracture. *Contrast* **hard copper.**

solid solution. A solution in which both solvent and solute are solid at room temperature.

solid solution alloy. A metal produced when a small amount of an alloying element is added and the crystal structure of the original metal is retained.

solid solution hardening. A method for increasing the strength and hardness of an alloy by the addition of a second element, while keeping the single phase structure. Also referred to as *solid solution strengthening.*

solid solution strengthening. *See* **solid solution hardening.**

solute. Substance dissolved in a solution. *Contrast* **solvent.**

solution. A special type of mixture in which one substance (solute) is thoroughly dissolved in the other (solvent).

solution extraction. The process of extracting the desirable metal portion from an ore by forming, and often precipitating a metal salt from, a liquid solution.

solution heat treatment. The process of thermally cycling a metal to dissolve small precipitates. Sometimes loosely used to indicate the entire solutionizing, quench, and precipitation cycle.

solutionizing. The process of heating a metal to a high temperature and allowing the precipitates to dissolve in the metal alloy.

solvent. Substance that dissolves another material in a solution. *Contrast* **solute.**

Sonodur hardness test. A hardness testing method in which the natural resonant frequency of the sample is determined and converted to a hardness value.

space lattice. The organized arrangement of atoms in a crystal.

specific heat. The ratio of the heat capacity of a material to the heat capacity of water.

specific weight. The ratio of the weight of a material to its volume.

spheroidizing. A stress-relief method used for high-carbon steel in which metal is heated to a temperature below the lower transformation temperature and is then slowly cooled to room temperature.

spring steel. A classification of alloy steel that has unusually high elasticity and strength.

stainless steel. A classification of special alloy steel that has high quantities of chromium or nickel and outstanding corrosion resistance.

steel. A material composed primarily of iron, less than 2% carbon, and (in alloy steel) small percentages of other alloying elements.

strain. The ratio of the deformation of a material to its original length.

strain hardening. A hardening and strengthening method that occurs as metal is deformed and reshaped by cold working.

strand casting. *See* **continuous casting.**

strength. Ability of a metal to resist forces or loads.

stress. The force per unit area imparted to a material. Stress also refers to the available strength or ability to withstand a force per unit area.

stress relaxation. The process of removing strain hardening. Usually this is done by heating the material.

stress relieving. Same as tempering, but for the primary purpose of relieving internal stresses.

stress-strain diagram. A graph of the stress versus the strain for a material.

stripping. Removing ingots from the ingot molds.

sulfides. Compounds of metals with sulfur.

superalloys. Specialized alloys strengthened to resist corrosion and high stresses at extreme operating temperatures.

supersaturated solution. An alloy that contains more of a second element than it can hold at equilibrium.

surface hardening. A heat-treating process that creates a thin, hardened, wear-resistant layer on the outer surface of a material while maintaining a soft and ductile inner core. Also referred to as *case hardening.*

T

T-T-T diagram. *See* **time-temperature-transformation diagram.**

taconite. A popular form of iron ore that is very plentiful but has a low iron content compared to other iron ores.

teeming. Pouring molten steel into ingot molds.

temperature-time line. Line on an isothermal transformation diagram that follows the path of the temperature of the steel with respect to time after quenching begins. Also referred to as *time line.*

tempering. The process of reheating a quenched steel to a temperature below its lower transformation temperature to increase ductility and relieve stress.

tensile strength. A material's ability to withstand stress in tension, or pulling apart.

thermal conductivity. The ability of a material to transmit heat.

time line. *See* **temperature-time line.**

time-temperature-transformation (T-T-T) diagram. *See* **isothermal transformation diagram.**

tinplate. Steel sheet coated with tin.

tool steel. A classification of high-alloy steel used for applications requiring high strength, hardness, and wear resistance at high temperatures.

torsional strength. A material's ability to resist shear stress in rotation.

toughness. Ability of a material to resist shock. Toughness depends on both strength and ductility of a material.

transfer temperature range. On an iron-carbon phase diagram, the area that comprises the two regions of structural change between the upper and lower transformation temperature lines.

tundish. Top part of a continuous casting machine, where molten metal is held.

tunnel oven. An enclosed oven through which parts ride on a moving belt and are heated in a controlled manner.

twinning. A shear failure of a ductile material in which two lines of atoms simultaneously slide past each other, forming mirror lines and twinning planes.

U

unit cell. The most fundamental arrangement of atoms in a space lattice.

upper transformation temperature. The temperature at which the transformation of iron to austenite is complete and the body-centered cubic structure has completely changed to face-centered cubic.

V

Vickers hardness test. A common microhardness testing method that uses a relatively small load and a diamond-point penetrator. Results of this test are expressed in units of diamond-pyramid hardness (DPH).

W

wave soldering. The process of soldering an entire circuit board at one time by passing it over a bath of liquid solder.

wear. Ability of a material to withstand wearing away by frictional scratching, scoring, galling, scuffing, seizing, pitting, or fretting.

weldability. The ability of a material to be welded.

white cast iron. A very hard, brittle type of cast iron used to produce malleable cast iron.

work hardening. *See* **cold working.**

workpiece. Metal being shaped to a desired condition.

wrought iron. A material composed almost entirely of iron, with very little or no carbon.

Y

Young's modulus. *See* **modulus of elasticity.**

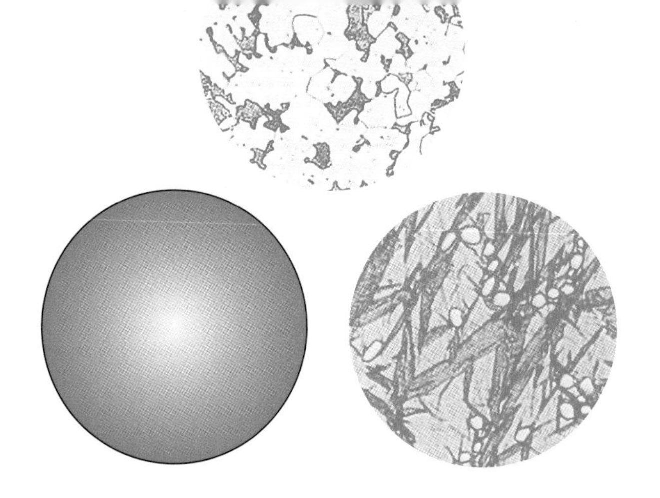

Index

D

E

O

P

Q

R

S

T

U

V

W

Y

Z